Somatic Embryogenesis in Ornamentals and Its Applications

A. Mujib
Editor

Somatic Embryogenesis in Ornamentals and Its Applications

Springer

Editor
A. Mujib
Cellular Differentiation
and Molecular Genetics Section,
Department of Botany
Hamdard University
New Delhi, India

ISBN 978-81-322-3814-0 ISBN 978-81-322-2683-3 (eBook)
DOI 10.1007/978-81-322-2683-3

Springer New Delhi Heidelberg New York Dordrecht London

Softcover reprint of the hardcover 1st edition 2016

Printed on acid-free paper

Springer (India) Pvt. Ltd. is part of Springer Science+Business Media (www.springer.com)

Preface

The incidence of somatic embryogenesis (SE) was described over fifty years back, and since then the knowledge on in vitro embryogenesis has been accumulated in a wide range of plants including monocot, dicot, and other groups of plants. This is a unique process by which vegetative or somatic plant cell transforms into an embryo, reconfirming cells' totipotency in culture. This in vitro embryogeny has immense fundamental and practical potential such as in understanding cellular differentiation of plants and in producing transgenic en masse. The entire process of SE has been, however, complex and is seemingly controlled by a variety of external and internal triggers. In this book, an attempt will be made to collect and compile the current development of embryogenic research particularly in ornamentals. Ornamentals are important groups of plants and constitute a good component in floriculture industry. These include bulbous, tuberous, and foliage plants, used mostly as indoor decorative, landscape, and potted plants. The plants are often propagated vegetatively, which is a slow process. Besides propagating raw materials at fast pace, SE may offer advantages in generating new traits/cultivars in ornamentals more quickly. The embryo precursor/mother cells in embryogenic cells are considered to be a good source in raising transgenic in several investigated plants. This book will therefore highlight the importance of embryogenic cultures/tissues in genetic transformation studies especially in ornamentals. The role of genotype, plant growth regulator, environment, and other controlling factors will be described in establishing embryogenic cultures and in developing protocol for obtaining plantlets. The morphology, maturation, and germination ability will also be discussed by investigating somatic embryos by scanning electron microscopy (SEM) and histological preparations. A few chapters will focus on comparative biochemical and physiological differences during the acquisition and development stages of SE. In recent times, several genes are found to be associated with specific embryogenic development; therefore, a few chapters will discuss molecular characterization, expression analysis of gene like somatic embryogenesis receptor kinase (*SERK*) during SE. Currently, proteomic study has been conducted widely at various stages of plant growth and development; here in this book, the importance of plant proteome as a source of somatic embryogenesis marker will also described very briefly. Recently, there has been a drastic decline in the number of cultivars and loss of genetic diversity of plant germplasm due to various socioeconomic reasons. The erosion of genetic diversity can be prevented by mass multiplication and simultaneous conservation of germplasm by utilizing in situ and in vitro methods. Preservation, i.e., storage of living tissues at low temperatures in laboratory conditions, has long been used to conserve plant germplasm for short-, medium-, or long-term basis. In this book, cryopreservation in the protection of genetic resources of ornamentals will be discussed using somatic embryo/embryogenic culture as the tissue of choice. The role of various cryoprotectants during preculture, pretreatment, and freezing will be discussed by monitoring cellular regrowth. The survival of cryopreserved SE tissues and plant regeneration will also be described in detail especially in ornamentals.

Although some of the information are available in websites, the need to accumulate fractured past years' knowledge and to organize embryogenesis research in ornamentals in structured book format is always necessary. This updated information on embryogenesis in ornamentals will be a very useful resource material to a wide range of people especially to researchers, graduate students, teachers, and many others professionals in various disciplines like biotechnology, botany, horticulture, pomology, agriculture, and other related fields.

New Delhi, India A. Mujib

Contents

1 Application of Cryogenic Technologies and Somatic Embryogenesis in the Storage and Protection of Valuable Genetic Resources of Ornamental Plants

Dariusz Kulus

Abstract

Cryopreservation is the most safe long-term conservation method of non-orthodox seed species. Somatic embryogenesis, on the other hand, is considered to be the most efficient (micro)propagation technique. By combining in vitro tissue culture techniques with cryoconservation, it is possible to develop highly diverse gene banks on a small surface at reduced costs from cutting down on labor and laboratory consumables. The application of embryonic tissue for storage in liquid nitrogen is very beneficial, especially with endangered species, since it does not require injuring the mother plant. The seeds are very often stored at sub-zero temperatures. Over time also zygotic embryos or their axes of about 100 species and somatic embryos of approximately 40 plant species have been cryopreserved with variable success. The share of ornamental plants, however, is low. The cryopreservation procedures are developed best for somatic embryos. For several species, an attempt to freeze embryogenic callus has been also made. There are even some reports referring to embryogenic potential or metabolic activity growth of proembryogenic masses in some ornamental species observed after freezing. Over time various cryopreservation techniques have been applied. As for seeds, direct immersion in liquid nitrogen or simple air drying is possible. With some species these techniques can be used with embryos. There are also reports on employing slow freezing for embryogenic tissues. Still, the so-called modern methods (e.g., vitrification, droplet vitrification, encapsulation–dehydration) are usually more efficient. The protocols, however, need to be adjusted not only to the individual species but also to single cultivars.

D. Kulus (✉)
Department of Ornamental Plants and Vegetable Crops – Laboratory of Biotechnology, UTP University of Technology and Life Sciences in Bydgoszcz, Bernardyńska 6, PL-85-029 Bydgoszcz, Poland
e-mail: dkulus@gmail.com

A. Mujib (ed.), *Somatic Embryogenesis in Ornamentals and Its Applications*,
DOI 10.1007/978-81-322-2683-3_1

Keywords

Cryopreservation • Embryogenic tissue • Liquid nitrogen • Ornamental plants • Somatic embryogenesis

1.1 Introduction of the Meaning of Ornamentals, Micropropagation, and Long-Term Storage Methods of Plant Genetic Resources

With the societies getting richer and richer, the importance of and demand for higher-order goods are increasing, e.g., plants. Due to their aesthetic value, ornamental plant species have been important for both human life and the economy since the ancient times (Ozudogru et al. 2010). They are around us on every stage of our life, from birth until death (Sengar et al. 2010). In 2007 only 1.67 % of the residents surveyed in Warsaw (Poland) did not buy even a single cut flower (Jabłońska and Perzyńska 2009). Not surprisingly the demand for ornamentals is constantly growing. The value of the world flower market is estimated at €70–90 billion and it is still increasing (Sharma and Agrawal 2012). Traditional reproduction methods, however, no longer meet the growing demands of the market. Nowadays in vitro tissue culture-based micropropagation is considered to be the most efficient plant reproduction method. The discovery of the possibility of stimulating the development of somatic cells into embryoid structures turned out to be groundbreaking for the propagation and protection of plants. Today somatic embryogenesis is believed to be the most efficient propagation technique allowing for the mass reproduction of plants in a relatively short time at a competitive price (Kulus 2014). Due to the possible occurrence of somaclonal variation during indirect regeneration via callus or multicellular origin of the somatic embryos, the technology may also facilitate increasing the gene pool with healthy plants, which is important with endangered species (Lema-Rumińska and Kulus 2012). The use of somatic embryogenesis for the reproduction of superior ornamental plant species generates a need for a long-term embryogenic culture storage.

Besides the improvement in quantity, the development of modern horticulture is also associated with the introduction of new cultivars. Due to the development of such in vitro-based biotechnological tools as mutation breeding, haploid production, protoplast fusion, embryo rescue, somaclonal variation, and genetic transformation, the number of the cultivars available is constantly increasing (Teixeira da Silva and Kulus 2014). For example, there are about 35,000 rose, 30,000 chrysanthemum, and 15,000 tulip cultivars available, which are only the world's top three ornamental plants (Chen et al. 2009; Bendahmane et al. 2013), while work is being performed with another 160 genera of which 87 % are ornamentals (Rout et al. 2006). As market trends keep on changing, the development of new cultivars is required. Diverse flower types, petal colors, and plant architecture/habit are important to the horticultural industry (Zalewska et al. 2007). New cultivars are replacing the current assortment. However, due to the change in consumer preferences, the cultivars not fashionable today may once again be attractive for potential buyers. Furthermore, very often they constitute a great breeding material source, especially wild varieties, when it comes to stress-resistance genes. Very often ornamental plants are also an important source of medicinal-valuable secondary metabolites useful in cancer and AIDS treatment, e.g., cacti and chrysanthemums. However, considerable effort and time are often required to produce high-quality embryogenic tissues, and when produced, the elite cell lines must be maintained by frequent subcultures, with the risk of loss of regeneration potential and somaclonal variation occurrence (Dixit et al. 2003). Similarly, often mature somatic embryos differentiate into seedling without any lag phase (Tessereau et al. 1994), which might be the main limiting factor for commercial application (including storage and delivery) of somatic embryogenesis and one of its main applications, engineered/synthetic seeds technology; hence, the protection and storage of valuable genetic resources are gaining importance. Nevertheless it is difficult for breeders and horticulturalists to provide enough space and finance for the traditional cultivation of such numerous cultivars (associated with the expansion of glasshouses/arable land, introducing new equipment and care treatments), which is laborious and threatened with biotic and abiotic stresses (Sekizawa et al. 2011). Traditional genetic conservation in the field or greenhouse requires intensive care of pot cultures or carefully separated field plots (Reed 2006). Haploids (important in breeding) and transgenic cultivars, gaining popularity among ornamental plants (*Rosa*, *Dianthus*, *Gladiolus*), require isolation to be protected from crossbreeding (Rajasekharan et al. 1994; Joung et al. 2006). Additionally many ornamental species (e.g., representing families Orchidaceae, Cactaceae, and Gentianaceae) due to being over-collected (for horticultural purposes, or illegal contraband), development of urbanization, and habitat destruction are on the brink of extinction (Lema-Rumińska and Kulus 2014). The progress of agriculture; the use of herbicides, pesticides, and the allocation of new areas to tillage, urbanization, and infrastructure progress; the construction of roads, dams, and mines; as well as introducing new exotic expansive species and excessive use in folk medicine pose a serious threat to many ornamental plant species (Lema-Rumińska and Kulus 2014). Another serious issue is the developing genetic erosion, a decrease in biodiversity. Some ornamentals (e.g., *Mammillaria gaumeri*) are especially susceptible to environmental variation and land-use changes due to a narrow geographic range, high habitat specificity, and low population numbers (Ferrer-Cervantes et al. 2012). For example, *Rhipsalis pentaptera*, although commonly cultivated, is probably extinct in the wild due to human activity (Lema-Rumińska and Kulus 2014). There are also only few populations of magnificent *Astrophytum asterias* left (Lema-Rumińska and Kulus 2012).

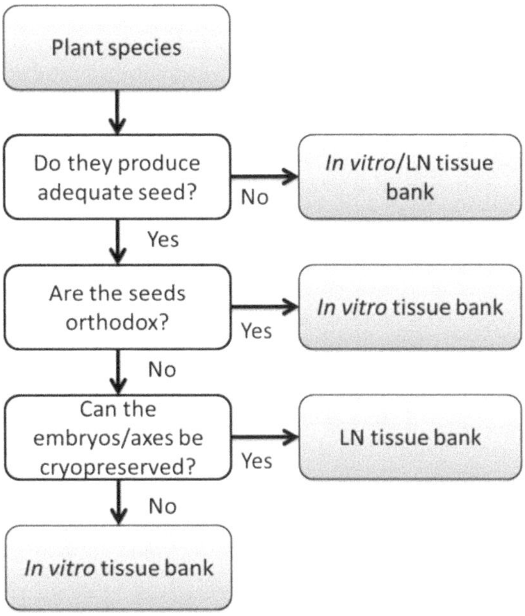

Fig. 1.1 Identification of species which need in vitro propagation and/or liquid nitrogen (LN) storage for long-term ex situ preservation of their biodiversity (Mikuła et al. 2013)

A fast and easy access to high-quality gene banks of a great material variety is the key for ornamental plant producers, and so an efficient method for long-time conservation of the plant material may be extremely valuable for breeding and horticultural production. Today cryopreservation, developing rapidly for the last 25 years, is believed to be the most promising and valuable long-term storage method. The identification of species which need in vitro propagation and/or liquid nitrogen (LN) storage for long-term ex situ preservation of their biodiversity is shown in Fig. 1.1.

1.2 Cryopreservation of Embryo(ge)nic Tissue

Cryobiology (Greek *kryos*, cold; *bios*, life; *logos*, science) is a branch of biotechnology which examines the effect of low temperatures (hypothermic, below physiological temperature, and cryogenic, −150 °C below) on the functioning of living organisms. Cryopreservation techniques are based on biological material storage (usually with high cell division rates including meristems, shoot tips, seeds, zygotic, and somatic embryos but also callus cultures and cell suspensions) at ultralow temperature of liquid nitrogen (LN, −196 °C) or seldom its vapor phase (−150 °C). Rarely somatic embryos isolated or clustered (of 3–4; 2–11 mg) are frozen (−70 °C)/thawed (20 °C) in N_2 with cooling/warming rates of ±10 °C min^{-1} (Álvarez et al. 2007). At the temperature below −150 °C, the biochemical metabolic and cell division activities get arrested, allowing for long-term storage. By these means the technique circumvents the loss of totipotency in plant systems (Quainoo 2009). The main advantage of this method is the reduction of in vitro culture costs, required space, contamination, and somaclonal variation risk. Freezing can also improve the flowering efficiency of some species (e.g., *Chrysanthemum hypargyrum* and *C. oreastrum*) which normally require vernalization to enter the generative growth phase, and therefore so far their role was minor on the ornamental plant market (Zhao et al. 2009). By combining in vitro tissue culture techniques with cryoconservation, it is possible to develop highly diverse gene banks of both vegetatively and generatively propagated species on a small surface at reduced costs. At the beginning of the twentieth century, there was no or little attention paid by both scientific and commercial laboratories to the possibility of applying tissue cultures for long-term plant genetic resources storage (Ríordáin 1992). Today cryopreservation is a very popular issue raised by plant biotechnologists, being a potentially useful tool for conserving endemic indigenous plant species.

A properly protected embryo(ge)nic explant can be stored in liquid nitrogen for virtually indefinite time without affecting its properties (provided continuous replenishment of the rapidly evaporating liquid nitrogen). Therefore, cryopreservation can be utilized in storage, conservation, and, finally, ornamental species breeding. As a result, more and more laboratories have been interested in that technology and incorporate it successfully (Reed 2008). The wide interest in cryopreservation has resulted in the establishment of multi-annual international research programs (such as CRYOPLANET, CRYMCEPT – Cryopreservation Methods for Conserving European Plant Germplasm Collection), which aim at finding scientific grounds for the plant cell behavior under stress conditions, implementation of cryopreservation techniques, evaluation of technical problems, and sharing experience (Mikuła and Rybczyński 2006). Even though the first concepts about the impact of low temperatures on the functioning of living organisms emerged already in the seventeenth century, proposed by a British physicist and chemist Robert Boyle, the true development of plant cryobiology began in the first half of the twentieth century (Fuller 2003). The origins of cryopreservation date back to the 1930s when the first reports on successful cryopreservation of moss tissues were published (Luyet and Gehenio 1938). Next, in the 1950s, attention was drawn to the possibility of woody plant tissue storage in liquid nitrogen (Sakai 1956). The first evidence of the cryopreservation of ornamental species came from Fukai (1989) and considered *Dianthus hybrida*. Next there were some experiments conducted with *Chrysanthemum* × *grandiflorum* by Fukai (1990) and other ornamentals (Kulus and Zalewska 2014a). The following decades brought a promising progress in terms of plant cryobiology. At the beginning of the twenty-first century, about some 200 species could be mentioned which, due to the development of cryopreservation techniques, storage costs were reduced (Gamez-Pastrana et al. 2011). Cryopreservation has been successfully used for many agricultural/horticultural (Forsline et al. 1998; Zhao et al. 2005; Xue et al. 2007), medicinal (Ai et al. 2012; Sharaf et al. 2012; Rabba'a et al. 2012), and threatened (Ashmore et al. 2011; Johnson et al. 2012; Ma et al. 2012) species. With its short history, the significance of ornamental plants (unlike micropropagation) is smaller, although growing every year. In the past few years, considerable progress has been made in terms of freezing of somatic embryos.

1.3 Basic Conditions and Material Selection for Cryopreservation or Why Is It Beneficial to Freeze Embryonic Tissue?

Cryopreservation involves a number of successive steps, promoting effective protection against the adverse effects of dehydration stress, low temperature, and rehydration. The changes observed in nature during the acquisition of tolerance to stress occur also under in vitro conditions. The factors, which stimulate stress tolerance, can be applied to reduce cryo-injuries and thus to minimize the risk of unwanted changes (Mikula 2008).

The final effect of cryopreservation depends on a number of factors, such as the type and size of the explant (embryo or cell aggregate); the age of the embryogenic strain from which the embryos have originated (indicated also by the number of subcultures); the cell suspension density; the condition of the donor plant and its climatic origin, pretreatment, and rate of cooling and thawing; and the regeneration conditions (Rout et al. 2006). Also seasons may influence the final cryopreservation efficiency. It has been observed by Seo et al. (2007) with *Paeonia lactiflora* that a higher survival is observed after freezing the material in late autumn/winter (November–February) when the plants are naturally preparing themselves for winter dormancy by accumulating abscisic acid (ABA), amino acids, sugars, etc. Thawing and recovery, on the other hand, should be carried out in the spring or summer, when the regeneration potential of the plant is the greatest (Zalewska et al. 2011).

The selection of the explant for cryostorage depends on the species and its tolerance to stress, as well as its regeneration potential and stability under in vitro conditions. Over the years it has been demonstrated that the usefulness of the plant material to cryopreservation is determined not only by its tolerance to low temperatures but mostly to dehydration (Sarasan et al. 2006). It is freezable water removal (to the level of 20–35 %), which is the most basic requirement that has to be fulfilled to achieve embryo survival and embryogenic tissue induction when applying LN. That way the formation of lethal ice crystals, which, by increasing their volume (in comparison with the liquid phase), results in cell bursting, is avoided. A small amount of water in the tissue is acceptable since it forms nonlethal ice crystals or the so-called nonfreezable/vitrified water. Therefore, it is essential to establish the moisture threshold below which all freezable water is removed and explants can survive storage in LN. This is a complex process which can be affected by multiple factors such as temperature, relative humidity, airflow rate, and tissue quantity (Kong and Aderkas 2011). The optimal water content (WC) is species dependent and can be defined as the value that allows freezing with the least injury to cellular components as measured by survival after cryopreservation (Fernandez et al. 2008). As for *Gentiana cruciata* and *G. tibetica* PEMs, it accounted for 28 % (85 and 50 % viability, respectively) (Mikuła and Rybczyński 2007). Water removal can be possible through direct dehydration which has been mentioned to be suitable for zygotic embryos or, by using chemical dehydration, recommended for soft tissues like PEMs and somatic embryos (Suranthran et al. 2011). Still variation in moisture between individual embryos should be considered. Sometimes various parts of the same plant have different cold resistance (Pearce 2009). For example, there is different sensitivity to the desiccation of root and shoot tips of embryos, which is a serious problem for developing effective cryopreservation procedures (Chmielarz et al. 2011). On the other hand, Fang et al. (2004) suggest that cryopreservation competence of embryos relies only partially on the acquisition of dehydration tolerance as in their experiment, and only a small share of desiccation-tolerant embryos was recovered after LN storage. This suggests that factors other than ice crystals may have been responsible for embryo damage during cryopreservation.

Over time, different parts of plants were used for cryopreservation, e.g., apical or axillary shoot tips, seeds, somatic and zygotic embryos or their axes, spores, pollen, gametophytes, rhizomes, or even protoplasts (Kulus and Zalewska 2014a).

Among them, shoot tips are most frequently used (Kulus et al. 2013; Zalewska and Kulus 2013). However, their acquisition is low efficient and time consuming; a microshoot produces only one apical bud (the species of elongative growth: chrysanthemum, crocus, rose), while the use of axillary buds is often less efficient when it comes to cryopreservation, as shown with *Chrysanthemum × grandiflorum* (Fukai 1990), *Colocasia esculenta* (Takagi et al. 1997), and *Rosa × hybrida* (Pawłowska and Bach 2011). Furthermore, high-precision and technical skills are necessary to isolate shoot tips or meristems, since these structures are tightly covered with leaves. Any injury of the tissue at that point may be fatal. Moreover, isolation must be performed rapidly, in order to protect the explants from excessive desiccation. Also shoot tips of some species (e.g., chrysanthemum) are covered with hairs, which impede the penetration of cryoprotectants or covering with alginate. Therefore, it is recommended to find a source of easier-to-obtain explants. The production of somatic embryos, on the other hand, is a much more efficient method. As for *Cyclamen persicum* Mill., it was possible to produce 40,000 progeny plants over 10 weeks just from 1 liter of the medium (Hvoslef-Eide et al. 2005), while with *Catharanthus roseus*, Mujib et al. (2014) produced nearly 100 somatic embryos from just 50 mg of callus mass. Due to a bipolar structure and no physical connection with the mother tissue, somatic embryos are also easier to isolate. Another advantage of applying somatic embryos is the fact that very often they have a single-cell origin, and, as such, they can be used for separating chimera components (especially sectorial and mericlinal types) and saving new genotypes/cultivars occurring as a result of spontaneous or induced mutations, which otherwise would be lost. Furthermore, as shown by Lema-Rumińska (2011), somatic embryos are the most stable structures (in terms of ploidy variation) under in vitro culture.

The storage in liquid nitrogen of cells from embryogenic callus lines (PEMs) is an application of relevant scientific interest. Embryogenic suspension is the best source of getting an unlimited number of totipotent cells/protoplasts ready to be used for any somatic cell hybridization or genetic manipulation (Tomiczak et al. 2009). The great variability of cell lines is a common feature of embryogenic cultures (Vágner et al. 2007). These morphogenetically active cells could play an important role in the studies of numerous chemical and physical stresses connected with low-temperature treatment (Mikuła and Rybczyński 2007). The primary advantages of using callus are tiny and synchronized materials compared with larger and differentiated explants (Shin et al. 2012). Proliferating embryogenic callus clumps are known to regenerate easily under in vitro conditions and to have the potential for genetic transformation and can be used in cell manipulation techniques (Ming-Hua and Sen-Rong 2010). Cryopreservation avoids the loss of embryogenic potential, which is frequently observed when cultures are aging after prolonged maintenance under in vitro conditions (Shin et al. 2012). The establishment of cryobanks of embryogenic cultures would allow for a safe and long-term storage of transgenic lines, during the time required for the verification of transgenic plant characteristics, as well as for the preservation of stock cell cultures used for secondary metabolite production, in order to avoid their decline due to repeated subculturing (Lambardi et al. 2005). On the other hand, even though the use of suspension cultures has the advantage of storing larger numbers of vials, however, it requires long periods to grow suspension culture and high labor for weekly culture maintenance (Ma et al. 2012).

The material selected for cryostorage should be young, demonstrating a high regeneration potential. The PEMs should be at the linear growth phase: 5–10 days after subculture. Under such conditions, the cells demonstrate characteristics optimal for freezing, including a high nucleolus–cytoplasmic ratio, dense cytoplasm, small vacuoles, large amounts of storage products, and the synchronization of cell growth and development. However, the length of lag phase strongly depends on the genotype (Salaj et al. 2007). The size of cryoprotected tissue depends on the type of suspension; some are very friable with small cell clusters, and others with embryogenic competence often have embryogenic structures at various levels of development (Mikuła 2006). The aggregate size should be >500 μm, especially if longer pretreatment with DMSO is applied. As for *Gentiana tibetica*, smaller aggregates (<300 μm) do not survive freezing (Mikuła 2006). With *G. cruciata*, on the other hand, smaller aggregates are better (Mikula et al. 2008). Also the origin of cell suspension may be important, although Mikuła et al. (2005) did not observe any significant differences in cell suspension viability derived from cotyledon and hypocotyl callus of *G. cruciata* subjected to cryopreservation applying two techniques (preculture and vitrification).

Somatic embryos acquire a higher desiccation tolerance late in their development (Wu et al. 2007). Generally, water content decreases as the embryo develops (Ma et al. 2012). If a tissue is properly protected during the pre-freezing step, even large embryos can survive freezing (Tessereau et al. 1994). Therefore, embryos at early cotyledonary stage are the best for cryostorage as they are less sensitive to water loss than premature embryos (Fang et al. 2004). Smaller and more immature embryos may not have sufficient vigor to survive and recover from the liquid nitrogen treatment, although they can grow well in the absence of freezing (Hor et al. 1993). Still successful cryopreservation of globular embryos is also possible. It was even observed that small embryos (0.25–0.5 mm in length) can rapidly repair any damage, unlike the large ones. This may be due to the increased cellular differentiation in larger embryos leading to different cell types with various water contents making the controlled osmotic dehydration for all these tissues much more difficult to achieve (Tessereau et al. 1994). Somatic embryos before the cotyledonary stage are also characterized with higher embryogenicity (Kong and Aderkas 2011). The cryopreservation of somatic embryos has become an important tool in reforestation, reintroduction, and breeding programs, especially when conifers are concerned, many of which are important ornamental park and garden species (Marum et al. 2007). Also transgene stability has been proven in cryopreserved somatic embryos (Álvarez et al. 2007). Significant cryopreservation clone banks have been established for somatic embryogenesis clonal propagation programs in Canada, France, New Zealand, Sweden, and the USA (Sharma 2005).

As for species producing "non-orthodox" seeds, zygotic embryos (entire when cotyledons are not flashy) or their axes are suggested to be the primary choice for ex situ conservation (Mikuła et al. 2013). After partial desiccation, they can be successfully cryopreserved, overcoming the problem of long-term storage. These means they facilitate reducing the costs associated with tissue culture initiation and maintenance and the introduction of the plant material to LN banks. Many plant species (especially recalcitrant) produce large seeds (e.g., palms), impractical to conserve and likely to be killed during freezing. Excised embryos survive better and can be conserved in small (2.0 ml) cryovials (Hor et al. 1993). Embryo oxidation observed sometimes at this point may be overcome by imbibing seeds in water or on agar before excision and the addition of 2 % active charcoal to in vitro germination medium (Davies et al. 2009). Excision of zygotic embryos prior to desiccation is important. This is due to the fact that dehydration of the embryo is slow within the confined, moist environment of the perisperm, which can additionally encourage microbial infection and inhibit aerobic respiration. Further, the isolation of embryos after drying seeds is difficult, and many can get injured during this process (Hor et al. 1993).

As for species with fleshy cotyledons, it is more beneficial to isolate axes only. Leaving the segments of cotyledon attached to the axis increases the mass of the tissue and so decreases both drying and cooling rates, contrary to the requirements for successful cryopreservation (Perán et al. 2006). The isolation of axes could avoid the problems of cotyledon damage and subsequent contamination of whole seeds during germination; however, it may be complicated due to decreased shoot production (Normah et al. 1994). Also axis isolation leads to an immediate burst of O_2^- production (Roach et al. 2009). Furthermore, the acquisition of embryonic axes may be sometimes difficult. Generally in recalcitrant seeds, embryonic axes account for 0.25 % of the dry weight of the whole seed, as compared to an average of 1.4 % for orthodox seeds (Chin 1989; cited after Normah et al. 1994).

Often meristematic heads are the only part of the embryos which remain alive after freezing. Suspensor cells are damaged by cryostorage, yet they recover after thawing (Vágner et al. 2007). Therefore, in order to reduce the amount of excessive tissue, which does not survive freezing, only a meristematic part of the embryo can be isolated. That way the risk of polyphenol secretion by the necrotic tissue, which can kill the viable cells, is avoided. This also allows for a faster and more efficient dehydration. In the experiments reported by Chmielarz et al. (2011), the small size of *Quercus robur* plumules (shoot apical meristems of embryos) facilitated achieving a higher drying rate during desiccation over silica gel and in nitrogen gas, in comparison with the whole embryonic axes, and ensured a better survival.

Based on those reports, it can be assumed that embryogenic tissue (especially PEMs and somatic embryos) is a suitable material for cryopreservation provided that effective conversion systems are developed, which is not always easy. Somatic embryos of *Paeonia lactiflora* showed similar survival and regrowth rates after cryopreservation as zygotic embryos and dormant buds (Kim et al. 2004, 2006; Seo et al. 2007). Over time zygotic embryos or their axes of about 100 species and somatic embryos of approximately 40 species of plants from different climates have been cryopreserved with variable survival and/or regrowth rates. The cryopreservation procedures are developed best for the latter ones (Engelmann 2011).

1.4 Importance of Preculture

Some explant types (e.g., orthodox seeds or pollen) are characterized by a naturally low water content (which is associated with their biological function). These explants do not require any pretreatment prior to freezing and can be directly plunged into LN (Popov et al. 2004; Jitsopakul et al. 2012). However, embryonic axes of *Camellia sinensis* and *Melia azedarach* did not survive such treatment (Kaviani 2010). Most of the plant tissues (including cell suspensions, callus, embryos) contain high levels of intracellular water, especially under in vitro conditions, since the air humidity in the culture vessel reaches even 100 %, and, therefore, these explants must be artificially dried. Some species (e.g., of the *Kalanchoe* and other succulent genera) are (highly) desiccation tolerant; they are the easiest ones for cryopreservation. As for these species, water loss leads to saturation of the fatty acids in the cell membranes, which results in stiffening and decreased permeability (Walters et al. 2002; cited after Mikula 2008). Others, dehydration-vulnerable species, do not undergo the plasmalemma restructuration during drying; instead, the membranes agglutinate and, next, break during rehydration (after thawing). Therefore, the explants of these species require preparation for both dehydration and cold stress, which can be achieved by applying a proper preculture.

In natural conditions, in the summer, resistance to cold is similar in various species. During autumn, however, various species acquire different resistance to cold by thermal and photoperiodic factor. Tropical species do not develop cold-tolerance mechanism and are thus highly sensitive to low temperatures. Temperate species undergo a number of physiological and biochemical changes, caused by the activity of AFP (antifreeze proteins), LEA (late-embryogenesis abundant), ABA, proline, etc. These compounds alter the dynamics of water crystallization, inhibit the growth of ice crystals, and thus increase the resistance of plants to frost (Mikuła and Rybczyński 2006). Under in vitro conditions, stress resistance in developing embryos must be artificially induced by exogenous signals by applying adequate culture conditions (medium composition, temperature, and light intensity) (Zalewska and Kulus 2013).

Preculture (hardening) is one of the most important factors affecting the efficiency of a cryopreservation protocol (Mikula et al. 2005; Antony et al. 2011a, b; Zalewska and Kulus

2014). As for somatic embryos of some species, it is the most essential step (Dumet et al. 1993). The precise mechanism of tolerance induction is not yet clear (Popova et al. 2010); however, it stimulates the accumulation of endogenous sugars (starch), polypeptides, ABA, proline, and bounded water, which increase the cell tolerance to osmotic excursion and dehydration as well as the glass-forming tendency within the cytoplasm (Suzuki et al. 2006). The most common preculture for cryopreservation of embryogenic cells includes the use of sugars (sucrose, maltose, glucose) or sugar alcohols (mannitol, sorbitol) (Salaj et al. 2011). According to many authors, a 2-week preculture in/on a medium with an increased level of sucrose provides the best survival (Kulus and Zalewska 2014a). High sugar concentration and prolonged treatment in the cytoplasm help to establish a vitrified state during cryopreservation and to stabilize membranes by the interaction of sucrose OH groups with membrane phospholipids and enable cells to tolerate dehydration that can cause freezing damage. The higher the sucrose concentration in the preculture medium, the greater the extent of sucrose accumulation in the embryos (Fang et al. 2009). The sucrose concentration increase from 0.09 to 0.27 M succeeded in the survival growth of *Gentiana cruciata* PEMs (from 12 to 61 %, the results recorded right after thawing) (Mikuła et al. 2005). Similar results were observed in PEMs of certain chrysanthemum cultivars after the addition of 0.5–0.7 M sucrose (for 1–2 days) to the culture medium (Halmagyi et al. 2004). Also Winkelmann et al. (2004) demonstrated a positive effect of higher sucrose concentrations (0.6 M for 48 h) in preculture medium on *Cyclamen persicum* PEM regeneration capacity. Partial dehydration lowers the freezing temperature, elevates the glass transition temperature, and increases the intracellular viscosity. Furthermore, preculture allows for reducing the heterogeneity of cells (different in size, vacuolization, and cytoplasm content) facilitating a more balanced dehydration (Mikuła 2006). The ultrastructural changes of *Gentiana* embryogenic aggregates resulting from increasing the sucrose concentration in the medium showed the replacement of large vacuoles by the numerous small ones, condensation of cytoplasm, accumulation of starch, and the fragmentation of the endoplasmic reticulum (Mikuła et al. 2005; Mikuła and Rybczyński 2007). Most of those changes, however, are reversible 2 days after thawing. Pregrown embryos display a lower enthalpy variation during freezing and rewarming than the non-pregrown ones (Dumet et al. 1993). Various authors showed that many excellent cryoprotectants such as glycerol, proline, or glucose had no stabilizing effect during extreme dehydration of embryos (Dumet et al. 1993). The only solutes that did have such properties were disaccharides (trehalose, sucrose, maltose). However, despite its well-known cryoprotective action, sucrose can have deleterious effects on delicate embryo viability if applied inappropriately (>0.5 M) as a result of progressing plasmolysis or osmotic shock as observed with some species (Chmielarz et al. 2011). As for these plants, the application of nonreducing sugars may facilitate tolerance to desiccation by protecting membranes and proteins. Also stepwise preculture and recovery plating, as well as the use of semisolid instead of liquid medium, may be essential for preserving embryo viability (Fang et al. 2004).

The addition of ABA (10–50 μM, especially in combination with cytokinins), polyethylene glycol (PEG), or/and proline (0.25 M) may also be essential (especially for tropical species). Abscisic acid suppresses embryo multiplication and precocious germination, promoting its maturation instead (Kong and Aderkas 2011). ABA-treated embryos have small and densely cytoplasmic cells, whereas ABA-untreated embryos have vacuolated and expanded cells (Srinivas et al. 2006). The addition of ABA is especially beneficial since, besides conferring desiccation tolerance and promoting accumulation of storage lipids and proteins (including dehydrins), it induces a synchronized maturation that results in uniform high frequency germination of somatic embryos (Srinivas et al. 2006). ABA makes somatic embryos lose their chlorophyll, suspend growth, exhibit low rates of respiration, and maintain elevated sucrose content (Hoekstra et al. 2001). The presence of ABA also results in a lower desiccation rate of the embryos (Lu et al. 2009). Cryopreserved somatic embryos not precultured with ABA show signs of denaturation, which might be related to low sugar contents (Hoekstra et al. 2001).

The problem of excessive hydration can be also solved by lowering the amount of PGRs, or increasing the gelling agent concentration (e.g., agar 10 g·dm^{-3}) which reduces the amount of available water (Garcia-Osuna et al. 2011). Since calcium ions may inhibit the meristematic and embryogenic activity of the explant, whereas medium with little or no calcium increases the proportion of active cells with a high nucleocytoplasmic ratio and high soluble protein content, the reduction of $CaCl_2$ content in the preculture medium may also be beneficial (Lardet et al. 2007).

Beside medium manipulation, during preculture one can lower the temperature (down to 0–5 °C) and reduce the light intensity in the growth room for at least 7 days (Uemura et al. 2009). Usually cryopreservation is successfully induced when embryos are matured under 5 °C for adequate time, typically 4–8 weeks (Ming-Hua and Sen-Rong 2010; Kong and Aderkas 2011). Shibli (2000) achieved a better survival and regrowth (60 and 54 %, respectively) of *Iris nigricans* somatic embryos cryopreserved by encapsulation–dehydration after 3-day preculture at 22 °C, then at 30 °C for 1 day.

1.5 Cryoprotectants

Over time a few cryopreservation techniques have been developed (Kulus and Zalewska 2014a, Fig. 1.2, Table 1.1). Nearly all of them require the use of protective chemical substances, cryoprotectants. Cryoprotectants are interacting and altering water distribution inside/outside cells and dehydrating them (Uemura et al. 2009). The substances are increasing stability or intactness of the plasmalemma, lowering

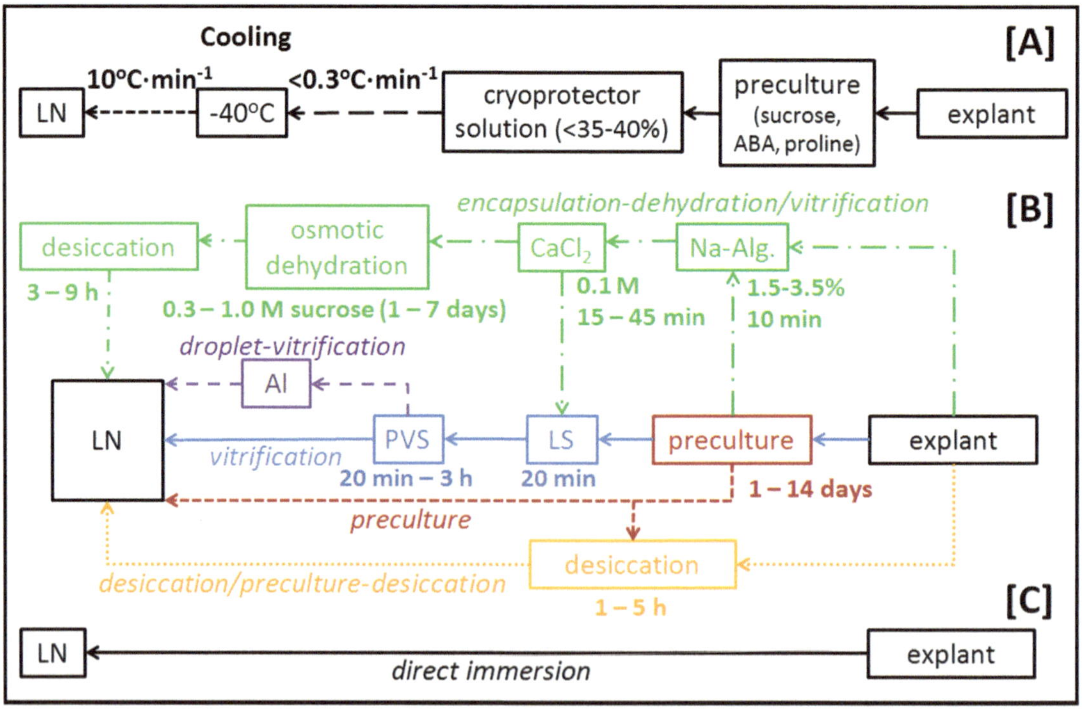

Fig. 1.2 Review of cryopreservation techniques: [**A**] traditional/program freezing, [**B**] vitrification-based – *LS* loading solution, *PVS* plant vitrification solution, *Al* aluminum foil strips, *Na-Alg*. Sodium alginate, *CaCl₂* calcium chloride, *ABA* abscisic acid, *LN* liquid nitrogen (Kulus and Zalewska 2014a)

Table 1.1 Cryopreservation of embryogenic tissue of different ornamental plant species by various techniques

Plant species	Material	Technique/pretreatment/treatment	Recovery (%)	Reference
Slow cooling				
Amaryllis belladonna	Zygotic embryos	Desiccation under air stream to water content <0.4 g $H_2O \cdot g^{-1}$ dry mass	70	Sershen et al. (2011)
		Preculture in 5 % glycerol (1 h), then in 10 % glycerol (1 h)		
		Direct immersion in LN		
Chrysanthemum cinerariaefolium	Callus	Preculture with 0.55 M sucrose (30 days)	92	Hitmi et al. (1997)
		Incubation in 5 % DMSO at 0 °C (1 h)		
		Cooling rate 1 $°C \cdot min^{-1}$ from 0 to −20 °C		
Cyclamen persicum	PEMs	Preculture with 0.6 M sucrose	75	Winkelmann et al. (2004)
		Incubation in 0.6 M sucrose + 10 % DMSO (1 h)		
		Incubation at −20 °C (2 h)		
Gentiana cruciata	PEMs	Preculture with 0.18 M sucrose (4 weeks)	2.5–2.7	Mikuła et al. (2005)
		Incubation in 0.2 M sorbitol (24 h) + 0.4 M sorbitol (24 h)		
		Incubation in 0.1 M DMSO at 0 °C (1 h)		
		Cooling rate 1 $°C \cdot min^{-1}$ from 0 to −40 °C, then to −150 °C at 10 °C/min		
Vitrification				
Crocus sp.	Calli	Preculture with 0.1 M sucrose (72 h)	50-67	Baghdadi et al. (2011)
		Loading with 2.0 M glycerol + 0.4 M sucrose at 25 °C (20 min)		
		Dehydration with 20–40–60–100 % PVS2 at 25 °C for 20 min (5 min for each concentration)		
Gentiana cruciata	PEMs	Preculture with 0.18 M sucrose (4 weeks)	86–91	Mikuła et al. (2005)
		Loading with 2.0 M glycerol + 0.4 M sucrose at 25 °C (20 min)		
		Dehydration with PVS2 at 0 °C (2 h)		
Encapsulation dehydration				
Catharanthus roseus	Suspension cells	Encapsulation in 1.5 % Ca alginate	48	Bachiri et al. (1995)
		Dehydration with a stepwise increased sucrose concentration over 2 days to final concentration of 0.3, 0.5, 0.7, and 1 M		
		Dehydration with 1.0 M sucrose (7 days)		
		Desiccation over silica gel for 4 h		
Camellia sinensis	Embryonic axes	Encapsulation in 3 % Ca alginate + 0.75 M sucrose (2 h)	100	Kaviani (2010)
		Desiccation with sterile airflow (1 h)		
Cosmos atrosanguineus	Suspension cultures	Encapsulation in Ca alginate	35	Wilkinson et al. (2003)
		Dehydration with liquid medium + 0.3 M (72 h) + 0.75 M (96 h) sucrose		
		Desiccation in sterile airflow (7 h)		
Galanthus elwesii	Somatic embryos	Encapsulation in 3 % Ca alginate	0	Pawłowska (2008)
		Gradual dehydration with 0.3, 0.5, 0.75, and 1.0 M sucrose (2 days each concentration)		
		Desiccation in sterile airflow (5 h)		

(continued)

Table 1.1 (continued)

Plant species	Material	Technique/pretreatment/treatment	Recovery (%)	Reference
Galanthus nivalis	Somatic embryos	Encapsulation in 3 % Ca alginate	0	Pawłowska (2008)
		Gradual dehydration with 0.3, 0.5, 0.75, and 1.0 M sucrose (2 days each concentration)		
		Desiccation in sterile airflow (5 h)		
Gentiana cruciata	PEMs	Encapsulation in 3 % Ca alginate	82	Mikuła et al. (2005, 2011a)
		Gradual dehydration with 0.3, 0.5, 0.75 (48 h), and 1.0 M (24 h) sucrose		
		Desiccation in sterile airflow for 5 h		
Gentiana tibetica	PEMs	Preculture MS + 0.18 M sucrose	63	Mikula et al. (2008)
		Encapsulation 3 % Ca alginate		
		Gradual dehydration with 0.3, 0.5, 0.75 (48 h), and 1.0 M (24 h) sucrose		
		Desiccation in sterile airflow for 4 h		
Melia azedarach	Embryonic axes	3 % Ca alginate + 0.75 M sucrose + 200 μM SA (1 h)	71	Kaviani (2007, 2010)
		Desiccation in sterile airflow (1 h)		
Paeonia lactiflora	Zygotic embryos	Preculture with 0.3 mg·dm^{-3} GA_3 (1 day)	85	Kim et al. (2004)
		Encapsulation in 3 % Ca alginate		
		Loading with 2.0 M glycerol + 0.5 M sucrose (1 h)		
		Desiccation in sterile airflow (5 h)		

the freezing point and increasing the viscosity of the cytosol, and, at the same time, protecting the cells from injuries throughout freezing. Cryoprotectants are divided into two groups: non-penetrating (sucrose and other carbohydrates or sugar alcohols) and penetrating the cells that have a rapid entrance rate and so require short incubation periods (e.g., DMSO, dimethyl sulfoxide used often in the range of 5–10 %; EG, ethylene glycol and glycerol used usually in the range of 10–20 %). A better efficiency of the latter one was proven with recalcitrant zygotic embryos of *Amaryllis belladonna* (Sershen et al. 2011). This is probably due to the fact that embryos are complex structures with a heterogeneous cellular composition and require penetration of the cryoprotectants to the deepest tissues to ensure conservation of their structural integrity. Also a combination of these two groups is applied in the so-called modern techniques. These are known as PVS (plant vitrification solutions). Effective cryopreservation requires the control of the procedure used for dehydration and cryoprotectant permeation and prevention of injury due to chemical toxicity or excessive osmotic stress during dehydration (Kaviani et al. 2012). Therefore, with every species and method used, it is crucial to establish an optimal concentration and the exposition time of the cryoprotectants applied since they determine the extent of cell dehydration and the amount of cryoprotectants permeated into the cells. Delaying permeation of cryoprotectants yet allowing them for adequate dehydration is critical. Therefore, the temperature in which plant germplasms imbibe into PVS is also important (Kaviani et al. 2012). Incubation on ice (0 °C) during pretreatment is usually beneficial, especially if longer dehydration is required. Such a treatment produced high recovery, even in tropical plants (Kulus and Zalewska 2014a). A direct exposure of germplasm to PVS2 reduces viability. The stepwise increase in PVS concentration reduces this toxic effect, as observed with embryogenic callus of wild crocus, *Crocus hyemalis* and *C. moabiticus* (Baghdadi et al. 2011).

1.6 Slow-Cooling Techniques

The first cryopreservation techniques developed (traditional or equilibrium freezing) in the 1940s were based on preliminary water removal by incubation in cryoprotectants followed by controlled slow-freezing (<0.3 °C·min^{-1}) of the material to a certain terminal temperature (usually −40 °C), stimulating freeze dehydration (cryo-desiccation, which removes most of the water), and later on rapid freezing (10 °C·min^{-1}) or direct immersion in LN. The essence of this technique is based on the fact that the water crystallization temperature depends on its purity. The higher the concentration of salt (dissolved phase), the lower the freezing point of the solvent. The cytoplasm is an aqueous solution of sugars, proteins, organic acids, etc. Thus it is characterized by a lower freezing point than the cleaner intercellular water. During slow cooling, at a temperature of approx. −9 °C, the intercellular water freezes first lowering at the same time the osmotic potential of the cytosol. The cell wall protects the cell membrane from a destructive influence of the extracellular ice. During further cooling, the density of the cytoplasm is increasing since the water is removed from the cell by osmosis and participates in a further development of extracellular ice. The cell content is not freezing, protected by cryoprotectants (the so-called unfreezable water). Slow cooling is considered appropriate for plant cells with high water content (Zeliang et al. 2010). Experiments performed by Heine-Dobbernack et al. (2007) with more than 700 undifferentiated cell lines from about 600 different plant species proved that this technique is most successful with undifferentiated materials (including PEMs). With *Gentiana cruciata* PEMs, however, the slow-cooling method turned out to be ineffective (Mikuła et al. 2005). Also with *G. tibetica* after controlled-rate cooling, the majority of embryogenic cells were lethally damaged, with only 3 % viability observed (Mikuła and Rybczyński 2007). Moreover, callus is more prone to somaclonal variation. As for more differentiated explants (zygotic and somatic embryos or their axes), only plants tolerant to low temperatures (e.g., members of the *Aster*, *Campanula*, *Chrysanthemum*, *Delphinium*, *Dianthus* genera) can be cryoconserved with this technique.

Despite some advantages, such as relatively high efficiency, two-step freezing shows several substantial disadvantages. First of all, it cannot be used with differentiated embryos of cold-susceptible (tropical) species, which are popular ornamental plants. Second of all, it is expensive for it requires a special freezing apparatus, although alternatives, which can replace the programmable freezer, have been developed. A simplified two-step freezing procedure of whole hydrated embryos has been elaborated by Lecouteux et al. (1991; cited after Tessereau et al. 1994). In that approach embryos in a cryoprotective solution are prefrozen in a domestic freezer before immersion in LN.

1.7 Rapid Cooling Techniques

Nowadays, the vitrification-based techniques (also known as one-step freezing) are much more popular. Those techniques involve most of the water being removed during osmotic/chemical dehydration. This approach, unlike the traditional one, is based on rapid cooling by direct transfer of the material from room temperature (optionally 0 °C) to liquid nitrogen. The sudden change in temperature together with a very high cellular viscosity causes a great slowdown of molecule motion and transformation of a liquid (remaining water and cryoprotectants) into an amorphous glass state (a few degrees below 0 °C when plunged into LN, the so-called glass transition temperature – Tg) without the development of crystalline which does not have time to form. Sugars are mainly involved in the formation of biological glass. Water in the glass state retains the physical properties of both liquid and solid. The concentration of cryoprotective solutions used in modern techniques is higher than in the traditional procedures. Vitrification techniques are more preferable than two-step freezing since they allow for the storage of more differentiated materials of both cold- and desiccation-susceptible/tolerant plants, guarantee higher freezing rates, and are less expensive than the traditional ones. For example, the highest *post*-thaw viabilities (roots and shoots produced) across amaryllid species were recorded for embryonic axes subjected to rapid rather than slow cooling (Sershen et al. 2007). Vitrification-based methods were also more efficient with *Gentiana tibetica* suspension culture (Mikuła 2006). There are several modern techniques.

Sometimes simple drying of the material under a stream of sterile air or over silica gel in an exsiccator for 1–5 h (possibly preceded by osmotic dehydration), before rapid immersion in LN, is the only requirement for developing a suitable cryopreservation protocol. The temperature of ice crystallization decreases progressively in line with the desiccation period of embryos (Dumet et al. 1993). This technique is mostly used with seeds but can also be applied for embryos as proven with *Paeonia lactiflora* somatic embryos (regrowth over 66 % after 1-h desiccation, Kim et al. 2006). Very often (especially with tropical species) the moisture contents of isolated embryos or their axes are higher than of seeds (Hor et al. 1993). Therefore, different desiccation periods may be required for these structures. Freezing of somatic embryos in liquid nitrogen is often optimal for the water content of 0.7 g $H_2O{\cdot}g^{-1}$ dry weight (dw). One should keep in mind though that somatic embryos survive dehydration upon slow drying. Fast drying, on the other hand, leads to poor survival of the embryos due to greater denaturation, no clearly defined glassy matrix, and much weaker strength of its hydrogen bonding (Wolkers et al. 1999). Slow drying leads to a partial conversion of sucrose into oligosaccharides and the expression of dehydrin transcripts. Slow-dried somatic embryos have stable membranes, retain their native protein secondary structure, and have a densely packed cytoplasmic glassy matrix. On the other hand, rapidly dried embryos experience some loss of phospholipids and an increase in free fatty acids (Hoekstra et al. 2001). The desiccation technique was effective for the storage of lily (a total duration of two hours) and *Encholirium pedicellatum* seeds (drying over silica gel for 4–6 weeks) (Tarre et al. 2007; Kaviani et al. 2009). With many species, it was even observed that desiccation had a positive effect on the conversion of somatic embryos to plantlets (Srinivas et al. 2006). However, cryopreservation of *Camellia sinensis* and *Melia azedarach* by applying a 1-h dehydration with liquid MS medium + 0.75 M sucrose, followed by 1-h air desiccation, was unsuccessful (Kaviani 2010). Hazubska-Przybył et al. (2010) made an attempt of improving the efficiency of this method by increasing the pace of freezing an embryogenic tissue of *Picea omorika*, by combining the 7-day preculture (in sucrose gradient) with a 2-h air desiccation and placing the material in vials filled with LN. This, however, resulted in higher losses due to contamination, as compared with the explants frozen in cryotubes without LN. Following desiccation in a laminar flow cabinet to the WC of 20–24 % (fresh weight basis), 90–100 % of zygotic embryo axes of several woody species survived cryostorage and over 60 % subsequently developed into whole plants (Chmielarz et al. 2011). Excised embryos desiccated in the laminar flow cabinet lost their moisture very rapidly from 40 to 10 % within 1.5 h. This may be a strong stress for some plants. As for more vulnerable species, to protect the explants from too rapid drying, seeds and embryos can be desiccated for a week in an air-conditioned room, which allows to reduce their moisture from 50 to 10–15 % (Hor et al. 1993). The most important advantage of the desiccation method is its short duration. However, due to its simplicity, the use of desiccation technique is strongly limited to orthodox seeds and the embryos of few desiccation-tolerant species. A typical vitrification-based protocol is a much more complex process involving several steps.

The first step of a typical vitrification procedure (after optional preculture) is loading the explants with a diluted/milder mixture of colligative cryoprotectants (usually 0.4 M sucrose + 2.0 M glycerol for 20 min at room temperature), in order to dehydrate and increase the material resistance to more concentrated and toxic compounds. Next, the material can be dehydrated with more efficient cryoprotectants, PVS at room temperature or at 0 °C – on ice. The most popular plant vitrification solution, PVS2, consists of 30 % (w/v) glycerol, 15 % (w/v) ethylene glycol (EG), 15 % (w/v), DMSO, and 0.4 M sucrose, although others, e.g., PVS3, 50 % glycerol (w/v) + 1.46 M sucrose, or PVS4, 35 % glycerol (w/v) + 20 % (w/v) EG + 0.6 M sucrose, with no toxic DMSO, are also known (Kulus and Zalewska 2014a). The PVS treatment time varies from 5 to 15 min or longer, depending on the species. Vitrification solution, by which cells are osmotically dehydrated at a nonfreezing temperature, provides a transition of intracellular water to an amorphous glass state without ice crystallization. Profiling of crystallization and recrystallization events using differential scanning calorimetry (DSC) confirmed that freezing injury is minimized in embryogenic callus samples after loading and cryoprotection with PVS (Shin et al. 2012). This technique secured 76 % embryogenic cell aggregates of *Camelia sinensis* (YaJun et al. 2009). Vitrification guaranteed the highest survival rate (over 80 %) of *Gentiana cruciata* and *G. tibetica* PEMs (Mikuła et al. 2005; Mikuła 2006), although the survival was nil with *Camellia japonica* somatic embryos (Janeiro et al. 1996), while for embryonic axes of *Lilium ledebourii*, it reached only 10 % (Kaviani et al. 2008). The main difficulty of vitrification is the high cytotoxicity of the concentrated cryoprotectant solutions employed. This problem may be overcome by utilizing alternative vitrification solutions, derived from original PVS2, PVS3, and PVS4, with modified cryoprotectant concentrations or by applying a PVS gradient.

The typical vitrification approach is provided with several modifications, e.g., the so-called droplet-vitrification technique. In this procedure, explants (incubated previously in LS and PVS) are placed in a drop of PVS on aluminum foil strips prior to direct immersion in LN. The technique is based on high thermal conductivity of aluminum, which increases the freezing and thawing pace. These two parameters are crucial for effective cryopreservation since the fastest the freezing/thawing pace, the lower the risk of ice crystallization. The droplet-vitrification technique is more complex, as compared with the vitrification procedure; however, it allows to minimize the necessity of manipulating the explants and usually produces higher post-cryopreservation regrowth of embryogenic tissue (Shin et al. 2012). Still this approach is least popular with somatic embryos and PEMs of ornamental plant species.

Another possibility is closing the explants in a protective alginate-based bead. Alginate is a copolymer refined from Atlantic seaweeds (*Laminaria hyperborea*, *Ascophyllum nodosum*, and *Macrocystis pyrifera*) thrown on the shore. It is an unbranched polysaccharide built of (1–4)-linked β-D-mannuronate and its C-5 epimer α-L-guluronate residues, distributed widely in the cell walls of brown algae, where through binding with water it forms a viscous gum (Beneke et al. 2009). Its parameters depend on its viscosity and the ratio of D-mannuronic to L-guluronic acid (Kakita and Kamishima 2009), the compound gels in the presence of calcium cations. Alginate is nontoxic (it is used as a food additive) and cheaper in comparison to cryoprotectants. Therefore, it can be used to protect explants efficiently. In addition the provision of an artificial endosperm to the encapsulated somatic embryo has been shown to increase vigor (Srinivas et al. 2006). Also the addition of salicylic acid (200 μm) in complement with sucrose may play a role in the process of resistance to desiccation as well as stimulate the growth of cryopreserved embryonic axes/embryos as observed with *Melia azedarach* (Kaviani 2007). Encapsulation dehydration is a vitrification-based procedure and comprises the gradual osmotic (0.3–1.0 M sucrose) and evaporative/physical dehydration (3–9 h) under sterile conditions (silica gel or laminar flow chamber) of encapsulated (in 3 % sodium alginate) plant cells prior to LN exposure. Dehydration must be long enough to ensure sufficient cell dehydration, without cytotoxic effects. Studies show that alginate beads dried to 20 % moisture vitrify on exposure to LN and form stable glasses that do not form ice crystals on rewarming (Kaviani et al. 2012). Embryos in closer contact with the underlying silica gel lose more water and faster than those dried in a laminar air flow chamber, which are further away from the drying agent (Fang and

Wetten 2011). This approach accompanied by high desiccation tolerance achieved in somatic embryos may facilitate the commercial production of synthetic seeds available directly after cryostorage, which can be sown under in vitro or ex vitro conditions, provided the bead is supplemented with MS salts (Murashige and Skoog 1962) and fungicides (Kulus and Zalewska 2014b). However, since intensive desiccation can be harmful for some explants, other less-typical procedures are developed, however, with a various degree of success. The desiccation of plumules (i.e., shoot apical meristems of embryos) of *Quercus robur* over silica gel resulted in significantly higher survival after cryopreservation (58 %) in comparison with desiccation in nitrogen gas (29 %), with regrowth (shoots with leaves) 5–18 % (Chmielarz et al. 2011). The encapsulation–dehydration technique is the most popular among ornamental plants and has been utilized for the cryopreservation of somatic embryos of a range of tropical species (Fang et al. 2004). The use of somatic embryos is especially beneficial since, as observed by Kulus and Zalewska (2014b), encapsulation greatly inhibits the formation of roots when shoot tips are applied. Embryos are bipolar structures with a well-formed root pole, and so this phenomenon should not be observed. Encapsulation and desiccation often induce a better survival of embryos than vitrification (Fernandez et al. 2008). As for *Iris nigricans*, this technique secured 69 % of somatic embryos and a 54 % conversion rate was obtained (Shibli 2000). Also with zygotic embryos of *Paeonia lactiflora*, the technique ensured 85 % survival (Kim et al. 2004). The encapsulation procedure gave faster recovery of the *Gentiana tibetica* culture suspension (4 weeks) than did vitrification or slow cooling (4 months) and ensured culture homogeneity and embryogenic competence (Mikuła 2006; Mikuła and Rybczyński 2007). This was also observed by Mikuła et al. (2005) with *Gentiana cruciata* PEMs, which is an important advantage of the technology.

An interesting alteration, considering the usage of electrostatic droplet generator (electrostatic extrusion directly from a plastic syringe or a needle of various diameter), allowing for the automation of the procedure, was introduced by Al-Hajry et al. (1999) for *Saintpaulia ionantha* callus. Applying this approach, the positive electrode wire is connected to the needle, and the ground wire is attached to the 0.1 M $CaCl_2$ collecting solution. As the alginate solution is forced out of the end of the needle by the syringe pump or the air pressure from the air tank, the droplets are pulled off by the action of electrostatic forces (Al-Hajry et al. 1999; Demirci and Montesano 2007).

There is also a possibility of combing the procedures. In the encapsulation–vitrification method, the most popular combined technique, encapsulated explants are incubated in LS and/or PVS and then immersed in LN. This technique combines the advantages of a fast vitrification procedure, both chemical and physical protection, and the facility of manipulating small explants, hence, its growing popularity with usable plant species. However, as for embryonic axes of *Lilium ledebourii*, this approach was unsuccessful, while with seeds a 10 % survival was observed (Kaviani et al. 2010).

1.8 Thawing and Regeneration

The thawing procedure is as important as the freezing stage. In the literature, little attention has been paid to this step. In the publications from the 1980s and 1990s, slow thawing at room temperature was relatively often used. Such procedure was mostly correlated with the slow-cooling technique. Nowadays it is known that thawing should proceed rapidly at a high temperature as proven with *Freesia refracta* vitrified callus (Wen et al. 1999). That way the recrystallization of ice and devitrification are prevented. Usually, regardless of the method used, cryovials or straws are placed for 2–3 min in a water bath at 35–42 °C (usually 38 °C, although this also should be optimized). Only the encapsulation–dehydration technique offers a possibility of thawing the beads at room temperature (Sharma 2005). After thawing, cryoprotectants should be removed (unloaded) properly to minimize the toxic effect of DMSO by washing in liquid MS medium (possibly with 1.2 M sucrose for 20 min). It is also important to remember that the bigger the amount of biological material in the cryovial, the slower the thawing pace, which is not recommended (Mazur and Seki 2011). A higher level of survival, additional 15 %, was achieved when the smaller than 500 μm of *Gentiana cruciata* embryogenic cell aggregates were taken (Mikuła and Rybczyński 2007). Therefore, it is recommended to insert a smaller number of embryos (10–15) in the vial.

After thawing and washing off the cryoprotectants and/or possible bead removal, samples are inoculated on recovery media. Long-term storage of plant germplasm requires an efficient and reliable plant regeneration system following cryopreservation. Regeneration medium has to be supplemented with plant growth regulators (PGRs; usually auxins and cytokinins and/or gibberellins) necessary to stimulate germination (in the case of seeds), to produce embryos (as for cryopreserved PEMs), or to stimulate the conversion of frozen embryos to plantlets. However, even though cytokinins are recorded as essential components of in vitro media for shoot tips after cryopreservation by a number of authors (Kulus et al. 2013; Zalewska and Kulus 2013, 2014), there is little information about the concentration optimization of those, or other, PGRs for embryonic tissues retrieved from LN (Perán et al. 2006). The level of sucrose may be important since controlled rehydration is also required. The presence of sucrose may serve as a signal for the synthesis of storage proteins, resulting in improved quality and germinability of the somatic embryos, in addition to serving as a source of carbon, energy, and osmoticum (Singh and Chand 2010). On the other hand, with some species, mature somatic embryos may require a progressive decrease of sucrose content in the (liquid) medium for further development into a plantlet (Tessereau et al. 1994). As shown by Berjak et al. (1999), retrieving embryo axes from cryostorage in a solution containing Ca^{2+} and Mg^{2+} resulted in a fast rehydration, and a 70 % initiated organized shoot development on the 6th day after thawing. The composition of these media should be developed prior to storing the biological material in LN. It is also necessary that these protocols of embryo regeneration and conversion are highly efficient.

As for embryogenic cell suspension, it is necessary to transfer them after thawing to an agar-solidified medium for at least 3 weeks. A premature transfer of tissue from agar to liquid medium resulted in the total loss of viability in *Gentiana tibetica*, especially of smaller aggregates unless they were encapsulated (Mikuła 2006). This was confirmed by Mikuła and Rybczyński (2007) who studied the regrowth of cryopreserved *Gentiana* spp. cell aggregates by culturing on agar, liquid, and agar/liquid culture. The combined type of culture significantly affected the biomass increase and stimulated the recovery of suspension culture. Cryopreservation did not influence the dynamics of regrowth and morphological appearance of *Gentiana* PEMs during the first 3 weeks of culture. On average 4–5 somatic embryos per capsule were regenerated after 3 weeks of culture. Many of them reached maturity, converted, and developed into normal green plantlets (Mikula et al. 2008).

When developing a new cryopreservation protocol, it is necessary to consider several criteria: the survival and regeneration rates, the duration of the protocol, its economy, and the difficulties encountered with the manipulation. It is assumed that a cryopreservation protocol is successful if the survival (and regeneration) rate reaches at least 50 %. However, this is not always possible with dehydration-susceptible species. Therefore, if the thawed material can be rapidly, clonally multiplicated, then a lower survival rate is acceptable.

A lack of germination after cryostorage is a common problem. The low survival and/or conversion of the somatic embryos following cryopreservation may be attributed to the desiccation process, sublethal injuries, and stress associated with LN exposure (Quainoo 2009). Each step of the cryopreservation procedure (except encapsulation) may contribute to the loss of embryo integrity. The damage occurring at the outer portion of the embryo initiates due to excessive dehydration, whereas those at the inner portion are results of ice crystal damage in insufficiently dehydrated cells (Fang and Wetten 2011). A lack of germination after cryoexposure of zygotic embryos is often due to the damage to the cotyledons, depriving the embryonic axes of the nutrients necessary to develop into plants and resulting in axis death. As for somatic embryos, the apical meristem of the embryonic axes is poorly developed, it lacks any protective covering layer, and during the partial drying treatment, it may dry to a lower water concentration than other parts of the axis and, consequently, suffer more desiccation damage allowing only root growth (Perán et al. 2006). *Post*-thaw regrowth of injured embryos is possible through the proliferation of surviving cells; however, such embryos often fail to convert into plantlets (Fang and Wetten 2011). Cryostorage may also induce explant dormancy (Normah et al. 1994). As for *Iris nigricans*, only 10 % of the cryopreserved embryos showed secondary embryogenesis (Shibli 2000). The regrowth of the cryopreserved suspension cultures is influenced by the culture period before cryopreservation (Lardet et al. 2007). With cell suspension, some of the unsuccessful cultures may have become non-embryonic over the longer time they were maintained prior to cryopreservation (Ma et al. 2012). Still, Mikuła (2006) successfully cryopreserved 10-year-old PEMs of *Gentiana tibetica*.

It is sometimes observed that cryopreservation may slow the growth rate of the frozen embryos (especially the mature ones) due to the denaturation of chromosomal proteins and a change in the number of "housekeeping" genes. This phenomenon, however, is usually no longer observed after acclimatization and longer cultivation in the field (Tessereau et al. 1994; Sisunandar et al. 2010). Reversible changes after freezing were also observed in the embryogenic tissue of *Gentiana tibetica* pretreated with vitrification solution: dilation of cell membranes, mitochondria with electron-lucent vessels, aggregation of numerous vesicles, and degradation of starch in amyloplasts, and anomalies were no longer visible after 48 h of *post*-thawing culture (Mikuła and Rybczyński 2007). Zeliang et al. (2010) have reported that initial phenotypic changes observed after the recovery from cryopreservation were not detected after longer growth ex vitro or culturing, although there are reports of atypical morphology, e.g., fasciated cotyledons, stunted hypocotyls, blackened radicles, etc.

On the other hand, sometimes conversion of frozen embryos is better than that of the untreated ones (Tessereau et al. 1994). This can be due to the fact that freezing selects the embryos with a better capability for normal development. There are also some reports referring to embryogenic potential growth of encapsulated PEMs of *Cyclamen persicum* (Winkelmann et al. 2004), *Gentiana cruciata*, and *G. kurroo* Royle (Mikuła et al. 2011a, b) and a greater pyrethrum production capacity in *Chrysanthemum cinerariaefolium* cell lines (Hitmi et al. 1997), after cryostorage. This can be explained by the fact that cryopreservation represents a strong selection pressure which can eventually change the embryogenic capacity of the culture by eliminating nonembryogenic cells from the cultures (Vágner et al. 2007). Another possible explanation might be higher synchronization of embryo development after cryopreservation (Popova et al. 2010). The regeneration potential and metabolic activity of the explants can be controlled in the future, by manipulating the protocol parameters, since Mikuła et al. (2011b) observed that osmotic dehydration alone increased the somatic embryogenesis efficiency of *Gentiana kurroo* about 9–20 times (depending on the material source, even though cryotreatment did not change the genome size or DNA sequence). Further steps of cryprocedure, air desiccation, and LN submersion did not affect it significantly.

Genetic stability of the regenerated material is another issue. There are five types of factors that cause abnormal development of plants. These include unusual climatic conditions, hereditary load, parasites, and chemical and physical agents. Accessions should be kept genetically unchanged for a long time until the clones are evaluated in field test. The arrest of metabolic functions at ultralow temperature theoretically decreases the possibility of genetic changes of the material stored (Vágner et al. 2007). In practice, however, the effect of cryo-injury is often unknown

(Zeliang et al. 2010). Every step of a cryopreservation procedure including preculture, cryoprotection (pretreatment), freezing, thawing, and recovery may be a source of somaclonal variation. The true-to-type status of the plant material regenerated after cryopreservation can be assessed in morphological, biochemical, ultrastructural/cytological, and chromosomal and molecular levels (Popova et al. 2010). Among these, PCR-based markers (e.g., RAPD, ISSR, AFLP fingerprinting) have been routinely used for monitoring genetic stability of germplasm after cryopreservation. There is increasing documentation on the stability analysis of the plants recovered from cryopreserved embryogenic tissues (Dixit et al. 2003). Usually the time of storage in liquid nitrogen does not influence the stability of the properly protected explants. *Gentiana cruciata* PEMs maintained constant viability and regeneration potential during the 2-year storage (Mikuła et al. 2011a). Flow cytometry analyses showed that the cryopreservation did not induce alteration in nuclear DNA content of *Gentiana* spp. proembryogenic cells and plantlets regenerated from them (Mikuła and Rybczyński 2007). Ultrastructural irregularities were minimal in rapidly cooled glycerol cryoprotected embryos of *Amaryllis belladonna* (Sershen et al. 2011). However, sometimes some perturbations occur. DMSO, the most popular cryoprotectant, has been described in the literature as a potential mutagenic compound (Marum et al. 2007). The mutagenic nature of DMSO is due to its effect on the metabolic activity of the cells, causing the change in permeability and fluidity of the cell membrane, enzyme activity, and oxygen consumption. Furthermore, it interacts directly with the nucleic acid by interfering with the metal cations which are components of eukaryotic chromatin, leading to a dispersion of chromosome structure and reduction of thermal stability (Aronen et al. 1999). The data presented by Zeliang et al. (2010) suggest that cryopreservation can induce heritable genetic changes in the plant material.

Attention should also be paid to problem of epigenetic stability (mainly cytosine methylation level) and its heredity as it is the most common alternation observed after cryopreservation, however, not always. Such changes may influence the phenotype and growth of the frozen plants. The variation is theoretically nonheritable; nevertheless, heritable changes in DNA methylation have been increasingly referred to in recent years (Vázquez 2007).

1.9 Conclusion

The past 30 years have witnessed a series of systematic biotechnological advances made in horticultural production. These encompass optimization and establishment of in vitro culture techniques including somatic embryogenesis, synthetic seed production, and cryopreservation. These techniques are now considered to be the future of horticulture. At the same time many plant genotypes are threatened with loss. Developing conservation practices for species producing recalcitrant seeds is important for scientific and practical reasons. Most plant species need to be stored at three levels: ecosystem level (in situ), genotype level (ex situ), and gene level (molecular/DNA libraries) (Kaviani 2011). The great potential of somatic embryogenesis and cryopreservation offers an attractive possibility of creating large-scale gene banks of important ornamental plant species. The methods prolong storage life and minimize storage space requirements. Cryopreservation of organized structures (i.e., somatic and zygotic embryos) has progressed significantly. In the last decade, desiccation-, vitrification-, and encapsulation-based techniques have proved effective for cryopreserving a broad range of embryogenic materials of plant species of both temperate and tropical origin. However, a general protocol for cryopreservation of embryogenic cultures has not yet been developed since the steps involved in the process require standardization for individual species or even cell lines. Still all cryopreservation techniques have a dehydration step in common. Water removal is necessary to reduce or avoid lethal intracellular ice formation during the freeze–thaw cycle. Major problems associated with somatic embryo cryopreservation include the following: (1) the establishment and maintenance of embryogenic cell suspensions in some species (especially woody) are difficult, or possible only from juvenile material with unknown genetic properties; (2) the different cell lines show different cryotolerance; (3) the regeneration of plants from non-cryopreserved somatic embryos of some species is generally low or asynchronous; (4) the lack of starch and protein reserves as well as dormancy and lower stress tolerance, as compared with zygotic embryos (Fang et al. 2004); and (5) the problems with conversion since the shoot apical meristem of somatic embryo is poorly developed (Bandyopadhyay and Hamill 2000). Progress in the large-scale production of somatic embryos relies on improvement at diverse phases of the process, resulting also in their stress tolerance (Tessereau et al. 1994). More study is needed to increase viability of embryogenic tissue of ornamentals, before and after exposure to LN. Cell integrity is currently being assessed at different stages of cryopreservation as a basis for improvement in the conversion process (Fang et al. 2004). A better understanding of cryoprotectant actions, embryo development, and recovery pathway may further contribute to the development of the cryopreservation of numerous ornamental plant species with both orthodox and recalcitrant seeds. Also more attention should be paid to the possibility of utilizing the droplet-vitrification and encapsulation–vitrification techniques as they have not been yet applied with embryonic structures of ornamental plants even though they are very successful with other plant groups. Still, cryopreservation is a reliable procedure for conserving the viability, recovery, and embryogenic competency of embryogenic tissue, and it can be routinely applied for long-term storage of ornamental plants.

References

Ai PF, Lu LP, Song JJ (2012) Cryopreservation of *in vitro*-grown shoot-tips of *Rabdosia rubescens* by encapsulation-dehydration and evaluation of their genetic stability. Plant Cell Tissue Organ Cult 108:381–387

Al-Hajry HA, Al-Maskry SA, Al-Kharousi LM, El-Mardi O, Shayya WH, Goosen MFA (1999) Electrostatic encapsulation and growth of plant cell cultures in alginate. Biotechnol Prog 15:768–774

Álvarez R, Revilla MA, Ordás RJ (2007) Transgene stability in cryopreserved cork oak somatic embryos. COST European Cooperation in Science and Technology, Oviedo, pp 46–47

Antony JJJ, Keng C, Rathinam X, Marimuthu S, Subramaniam S (2011a) Effect of preculture and PVS2 incubation conditions followed by histological analysis in the cryopreserved PLBs of *Dendrobium* Bobby Messina orchid. Aust J Crop Sci 5(12):1557–1564

Antony JJJ, Sinniah UR, Keng CL, Pobathy R, Khoddamzadeh AA, Subramaniam S (2011b) Selected potential encapsulation-dehydration parameters on *Dendrobium* 'Bobby Messina' protocorm-like bodies using TTC analysis. Aust J Crop Sci 5(13):1817–1822

Aronen TS, Krajnakova J, Haggman H, Ryynanen L (1999) Genetic fidelity of cryopreserved embryonic cultures of open-pollinated *Abies cephalonica*. Plant Sci 142:163–172

Ashmore SE, Hamilton KN, Offord CA (2011) Conservation technologies for safeguarding and restoring threatened flora: case studies from Eastern Australia. In Vitro Cell Dev Biol Plant 47:99–109

Bachiri Y, Gazeau C, Hansz J, Morisset C, Dereuddre J (1995) Successful cryopreservation of suspension cells by encapsulation-dehydration. Plant Cell Tissue Organ Cult 43:241–248

Baghdadi SH, Makhadmeh I, Syouf M, Arabiat A, Shibli RA, Shatnawai MA (2011) Cryopreservation by vitrification of embryogenic callus of wild crocus (*Crocus hyemalis* and *Crocus moabiticus*). Acta Hortic 908:239–246

Bandyopadhyay S, Hamill JD (2000) Ultrastructural studies of somatic embryos of *Eucalyptus nitens* and comparisons with zygotic embryos found in mature seeds. Ann Bot 86:237–244

Bendahmane M, Dubois A, Raymond O, Le Bris M (2013) Genetics and genomics of flower initiation and development in roses. J Exp Bot 64(4):847–857

Beneke CE, Viljoen AM, Hamman JH (2009) Polymeric plant-derived excipients in drug delivery. Molecules 14:2602–2620

Berjak P, Walker M, Watt MP, Mycock DJ (1999) Experimental parameters underlying failure or success in plant germplasm cryopreservation: a case study on zygotic axes of *Quercus robur* L. CryoLetters 20:251–262

Chen F-D, Li F-T, Chen S-M, Guan Z-Y, Fang W-M (2009) Meiosis and pollen germinability in small-flowered anemone type chrysanthemum cultivars. Plant Syst Evol 280:143–151

Chmielarz P, Michalak M, Pałucka M, Wasileńczyk U (2011) Successful cryopreservation of *Quercus robur* plumules. Plant Cell Rep 30:1405–1414

Davies R, Nadarajan J, Pritchard HW (2009) Cryopreservation of difficult-to-handle palm seeds. CryoLetters 30(5):386–387

Demirci U, Montesano G (2007) Cell encapsulating droplet vitrification. Lab Chip 7:1428–1433

Dixit S, Mandal BB, Ahuja S, Srivastava PS (2003) Genetic stability assessment of plants regenerated from cryopreserved embryogenic tissues of *Dioscorea bulbifera* L. using RAPD, biochemical and morphological analysis. CryoLetters 24:77–84

Dumet D, Engelmann F, Chabrillange N, Duval Y, Dereuddre J (1993) Importance of sucrose for the acquisition of tolerance to desiccation and cryopreservation of oil palm somatic embryos. CryoLetters 14:243–250

Engelmann F (2011) Cryopreservation of embryos: an overview. Methods Mol Biol 710:155–184

Fang J-Y, Wetten A (2011) Importance of structural integrity of somatic embryos for long-term cryopreservation of cocoa (*Theobroma cacao* L.) germplasm. Afr J Agric Res 6(17):3954–3961

Fang J-Y, Wetten A, Hadley P (2004) Cryopreservation of cocoa (*Theobroma cacao* L.) somatic embryos for long-term germplasm storage. Plant Sci 166:669–675

Fang JY, Sacande M, Pritchard H, Wetten A (2009) Influence of freezable non-freezable water and sucrose on the viability of *Theobroma cacao* somatic embryos following desiccation and freezing. Plant Cell Rep 28(6):883–889

Fernandez P, Rodriguez E, Pinto G, Roldan-Ruiz I, De Loose M, Santos C (2008) Cryopreservation of *Quercus suber* somatic embryos by encapsulation-dehydration and evaluation of genetic stability. Tree Physiol 28:1841–1850

Ferrer-Cervantes ME, Méndez-González ME, Quintana-Ascencio PF, Dorantes A, Dzib G, Durán R (2012) Population dynamics of the cactus *Mammillaria gaumeri*: an integral projection model approach. Popul Ecol 54(2):321–334

Forsline PL, Towill LE, Waddell JW, Lamboy WF, McFerson JR (1998) Recovery and longevity of cryopreserved dormant apple buds. J Am Soc Hortic Sci 123(3):365–370

Fukai S (1989) Plant regeneration from shoot tips of *Dianthus hybrid* cryopreserved in liquid nitrogen up to 2 years. Plant Tissue Cult Lett 6:177–178

Fukai S (1990) Cryopreservation of chrysanthemum shoot tips. Sci Hortic 45:167–174

Fuller B (2003) A dip into history: science and publication in cryobiology 150 years ago. CryoLetters 24:133–134

Gamez-Pastrana R, Gonzalez-Arnao MT, Martinez-Ocampo Y, Engelmann F (2011) Thermal events in calcium alginate beads during encapsulation dehydration and encapsulation-vitrification protocols. Acta Hortic 908:47–54

Garcia-Osuna HT, Benavides Mendoza A, Escobedo Bocardo L, Villarreal Quintanilla JA, Cornejo Oviedo E (2011) Hyperhydricity control of *in vitro* shoots of *Turbinicarpus valdezianus* (Moller) GL&F. Phyton 80:175–179

Halmagyi A, Fischer-Kluver G, Mix-Wagner G, Schumacher HM (2004) Cryopreservation of *Chrysanthemum morifolium* (*Dendranthema grandiflora* Ramat.) using different approaches. Plant Cell Rep 22:371–375

Hazubska-Przybył T, Chmielarz P, Michalak M, Bojarczuk K (2010) Cryopreservation of embryogenic tissues of *Picea omorika* (Serbia spruce). Plant Cell Tissue Organ Cult 102:35–44

Heine-Dobbernack E, El Banna A, Kiesecker H, Schumacher HM (2007) Dedifferentiated plant cell lines for the evaluation of physiological parameters of cryotolerance. COST European Cooperation in Science and Technology, Oviedo, pp 20–21

Hitmi A, Sallanon H, Barthomeuf C (1997) Cryopreservation of *Chrysanthemum cinerariaefolium* Vis. cells and its impact factor on their pyrethrin biosynthesis ability. Plant Cell Rep 17:60–64

Hoekstra FA, Golovina EA, Tetteroo FA, Wolkers WF (2001) Induction of desiccation tolerance in plant somatic embryos: how exclusive is the protective role of sugars? Cryobiology 43(2):140–150

Hor YL, Stanwood PC, Chin HF (1993) Cryopreservation of *Coffea liberica* seeds and embryos following desiccation and freezing treatments. Pertanika J Trop Agric Sci 16(2):75–80

Hvoslef-Eide AK, Olsen OAS, Lyngved R, Munster C, Heyerdahl PH (2005) Bioreactor design for propagation of somatic embryos. Plant Cell Tissue Organ Cult 81(3):265–276

Jabłońska L, Perzyńska K (2009) The level of demand for ornamental plants in Warsaw in 2007 and its determinants. Zeszyty Naukowe Instytutu Sadownictwa i Kwiaciarstwa 17:119–132

Janeiro LV, Vieitez AM, Ballester A (1996) Cryopreservation of somatic embryos and embryogenic axes of *Camellia japonica* L. Plant Cell Rep 15:699–703

Jitsopakul N, Thammasiri K, Yukawa C, Ishikawa K (2012) Effect of cryopreservation on seed germination and protocorm development of *Vanda tricolor*. Sci Asia 38:244–249

Johnson T, Cruse-Sanders J, Pullman GS (2012) Micropropagation and seed cryopreservation of the critically endangered species Tennessee yellow-eye grass, *Xyris tennesseensis* Kral. In Vitro Cell Dev Biol Plant 48:369–376

Joung HY, Cantor M, Kamo K (2006) Cryopreservation of *Gladiolus* cultivars. In: Hummer KE (ed) XXVII international horticultural congress – IHC2006: II international symposium on plant genetic resources of horticultural crops. ISHS Acta Hortic Seoul, pp 225–231

Kakita H, Kamishima H (2009) Some properties of alginate gels derived from algal sodium alginate. Dev Appl Phycol 2:93–99

Kaviani B (2007) Effects of salicylic acid and encapsulation on enhancing the resistance of embryonic axes of persian lilac (*Melia azedarach* L.) against cryopreservation. Int J Agric Biol 9(4):625–627

Kaviani B (2010) Cryopreservation by encapsulation-dehydration for long-term storage of some important germplasm: seed of lily [*Lilium ledebourii* (Baker) Bioss.], embryonic axe of persian lilac (*Melia azedarach* L.), and tea (*Camellia sinensis* L.). Plant Omics J 3:177–182

Kaviani B (2011) Conservation of plant genetic resources by cryopreservation. Aust J Crop Sci 5(6):778–800

Kaviani B, Safari-Motlagh MR, Padasht-Dehkaei MN, Darabi AH, Rafizadeh A (2008) Cryopreservation of Lily [*Lilium ledebourii* (Baker) Bioss.] germplasm by encapsulation-dehydration. Int J Bot 4(4):491–493

Kaviani B, Abadi DH, Torkashvand AM, Hoor SS (2009) Cryopreservation of seeds of lily (*Lilium ledebourii* / Baker/ Bioss): use of sucrose and dehydration. Afr J Biotechnol 8(16):3809–3810

Kaviani B, Dahkaei M, Hashemabadi D, Darabi A (2010) Cryopreservation of *Lilium ledebourii* (Baker) Bioss. by encapsulation-vitrification and *in vivo* media for planting of germplasm. Am Eur J Agric Environ Sci 8(5):556–560

Kaviani B, Darabi Hossein A, Roudposhti VR (2012) *In vitro* conservation of genetic resources of tea (*Camellia sinensis* L. cv. 100) using storage of germplasms (embryonic axes and shoot tips) in cryopreservation conditions. Ann Biol Res 3(7):3541–3546

Kim HM, Shin JH, Sohn JK (2004) Cryopreservation of zygotic embryos of herbaceous peony (*Paeonia lactiflora* Pall.) by encapsulation-dehydration. Korean J Crop Sci 49(4):354–357

Kim HM, Shin JH, Sohn JK (2006) Cryopreservation of somatic embryos of the herbaceous peony (*Paeonia lactiflora* Pall.) by air drying. Cryobiology 53(1): 69–74

Kong L, Aderkas P (2011) A novel method of cryopreservation without a cryoprotectant for immature somatic embryos of conifer. Plant Cell Tissue Organ Cult 106:115–125

Kulus D (2014) Biotechnological methods of ornamental plants reproduction. In: Mucha-Szajek E (ed) Jakość życia w badaniach młodych naukowców. Maiuscula, Poznań, pp 87–104

Kulus D, Zalewska M (2014a) Cryopreservation as a tool used in long-term storage of ornamental species – a review. Sci Hortic 168:88–107

Kulus D, Zalewska M (2014b) *In vitro* plant recovery from alginate encapsulated *Chrysanthemum × grandiflorum* /Ramat./Kitam. shoot tips. Prop Ornam Plants 14(1):3–12

Kulus D, Mikuła A, Zalewska M (2013) Cryopreservation: an efficient tool to combat genetic erosion phenomenon in agriculture and horticulture. In: Abstracts of IVth international scientific symposium for PhD students and students of agricultural colleges – innovative researches for the future of agriculture and rural areas development, UTP University of Technology and Life Sciences, Bydgoszcz, 19–21 September 2013

Lambardi M, Benelli C, De Carlo A (2005) Cryopreservation as a tool for the long-term conservation of woody plant germplasm: development of the technology at the CNR/IVALSA Institute of Florence. In: Abstracts of the role of biotechnology, Turin, 5–7 March 2005

Lardet L, Martin F, Dessailly F, Carron MP, Montoro P (2007) Effect of exogenous calcium on post-thaw growth recovery and subsequent plant regeneration of cryopreserved embryogenic calli of *Hevea brasiliensis* (Mull. Arg.). Plant Cell Rep 26:559–569

Lema-Rumińska J (2011) Flow cytometric analysis of somatic embryos, shoots, and calli of the cactus *Copiapoa tenuissima* Ritt. *forma monstruosa*. Plant Cell Tissue Organ Cult 106(3):531–535

Lema-Rumińska J, Kulus D (2012) Induction of somatic embryogenesis in *Astrophytum asterias* (Zucc.) Lem. in the aspect of light conditions and auxin 2,4-D concentrations. Acta Sci Polonorum Hortortorum Cultus 11(4):77–87

Lema-Rumińska J, Kulus D (2014) Micropropagation of cacti – a review. Haseltonia 19:46–63

Lu ZW, Popova EV, Wu CH, Lee EJ, Hahn EJ, Paek KY (2009) Cryopreservation of *Ginkgo biloba* cell cultures: effect of pretreatment with sucrose and ABA. CryoLetters 30(3):232–243

Luyet BJ, Gehenio PM (1938) The survival of moss vitrified in liquid air and its relation to water content. Biodynamica 2:1–7

Ma X, Bucalo K, Determann RO, Cruse-Sanders JM, Pullman GS (2012) Somatic embryogenesis, plant regeneration, and cryopreservation for *Torreya taxifolia*, a highly endangered coniferous species. In Vitro Cell Dev Biol Plant 48:324–334

Marum L, Rocheta M, Oliveira MM, Miguel C (2007) Genetic fidelity of *Pinus pinaster* somatic embryogenic cultures following cryopreservation. COST European Cooperation in Science and Technology, Oviedo, pp 48–49

Mazur P, Seki S (2011) Survival of mouse oocytes after being cooled in a vitrification solution to −196°C at 95°C to 70,000°C/min and warmed at 610° to 118,000°C/min: a new paradigm for cryopreservation vitrification. Cryobiology 62:1–7

Mikula A (2008) Cryopreservation as a method to facilitate preservation of life capacity of plant cells and tissues. Biotechnologia 2(81):41–57

Mikuła A (2006) Comparison of three techniques for cryopreservation and reestablishment of long-term *Gentiana tibetica* suspension culture. CryoLetters 27(5):269–282

Mikuła A, Rybczyński JJ (2006) Cryopreservation – as a tool for long-term storage of cells, tissues and organs from *in vitro* culture derived. Biotechnologia 4(75):145–163

Mikuła A, Rybczyński JJ (2007) Fundamental aspects of cell suspension cryopreservation of *Gentiana* spp.

COST European Cooperation in Science Technology, Oviedo, pp 30–31

Mikula A, Tykarska T, Kuraś M (2005) Ultrastructure of *Gentiana tibetica* proembryogenic cells before and after cooling treatments. CryoLetters 26(6):367–378

Mikula A, Olas M, Śliwińska E, Rybczyński JJ (2008) Cryopreservation by encapsulation of *Gentiana* spp. cell suspensions maintains regrowth, embryogenic competence and DNA content. CryoLetters 29(5):409–418

Mikuła A, Fiuk A, Rybczyński JJ (2005) Induction, maintenance and preservation of embryogenic competence of *Gentiana cruciata* L. cultures. Acta Biol Crac Bot 47:227–236

Mikuła A, Tomiczak K, Rybczyński JJ (2011a) Cryopreservation enhances embryogenic capacity of *Gentiana cruciata* (L.) suspension culture and maintains (epi)genetic uniformity of regenerants. Plant Cell Rep 30(4):565–574

Mikuła A, Tomiczak K, Wójcik A, Rybczyński JJ (2011b) Encapsulation-dehydration method elevates embryogenic abilities of *Gentiana kurroo* cell suspension and carrying on genetic stability of its regenerants after cryopreservation. Acta Hortic 908:143–152

Mikuła A, Makowski D, Tomiczak K, Rybczyński JJ (2013) Kultury *in vitro* i krioprezerwacja w zachowaniu różnorodności roślin – standardy dla banku genów. Polish J Agron 14:3–17

Ming-Hua Y, Sen-Rong H (2010) A simple cryopreservation protocol of *Dioscorea bulbifera* L. embryogenic calli by encapsulation-vitrification. Plant Cell Tissue Organ Cult 101:349–358

Mujib A, Ali M, Isah T, Dipti (2014) Somatic embryo mediated mass production of *Catharanthus roseus* in culture vessel (bioreactor) – a comparative study. Saudi J Biol Sci 21(5):442–449

Murashige T, Skoog F (1962) A revised medium for rapid growth and bio assays with tobacco tissue cultures. Physiol Plant 15:473–497

Normah MN, Reed BM, Yu X (1994) Seed storage and cryoexposure behaviour in hazelnut (*Corylus avellana* L. cv. Barcelona). CryoLetters 15:315–322

Ozudogru EA, Previati A, Lambardi M (2010) *In vitro* conservation and cryopreservation of ornamental plants. In: Jain SM (ed) Protocols for *in vitro* propagation of ornamental plants, Methods in molecular biology 589. Humana Press, Hertfordshire, pp 303–305

Pawłowska B (2008) Employment of encapsulation-dehydration method for liquid nitrogen cryopreservation of ornamental plant explants propagated *in vitro*. Folia Hortic 20(1):61–71

Pawłowska B, Bach A (2011) Cryopreservation by encapsulation-dehydration of *in vitro* grown shoot buds of *Rosa* 'New Dawn'. In: Panis B, Lynch P (eds) I international symposium on cryopreservation in horticultural species. Acta Hortic, Leuven, pp 303–308

Pearce RS (2009) Complexity and a regulatory role for sugars in the acclimation to cold and freezing of plants. CryoLetters 30(3):388

Perán R, Berjak P, Pammenter NW, Kioko JI (2006) Cryopreservation, encapsulation and promotion of shoot production of embryonic axes of a recalcitrant species *Ekebergia capensis*, Sparrm. CryoLetters 27(1):5–16

Popov AS, Popova EV, Nikishina TV, Kolomeytseva GL (2004) The development of juvenile plants of the hybrid orchid *Bratonia* after seed cryopreservation. CryoLetters 25(3):205–212

Popova E, Kim H-H, Paek K-Y (2010) Cryopreservation of coriander (*Coriandrum sativum* L.) somatic embryos using sucrose preculture and air desiccation. Scientia Hortic 124:522–528

Quainoo AK (2009) Effect of liquid nitrogen storage time on the survival and regeneration of somatic embryos of cocoa. Afr Crop Sci J 17(1):17–24

Rabba'a MM, Shibli AS, Shatnawi MA (2012) Cryopreservation of *Teucrium polium* L. shoot-tips by vitrification and encapsulation-dehydration. Plant Cell Tissue Organ Cult 110:371–382

Rajasekharan PE, Rao TM, Janakiram T, Ganeshan S (1994) Freeze preservation of gladiolus pollen. Euphytica 80:105–109

Reed BM (2006) Cryopreservation of bermudagrass germplasm by encapsulation dehydration. Crop Sci 46(1):6–11

Reed BM (2008) Cryopreservation – practical considerations. In: Reed BM (ed) Plant cryopreservation: a practical guide. Springer, Corvallis, pp 3–15

Ríordáin FÓ (1992) The European plant tissue culture industry – 1990. Agronomie 12:743–746

Roach T, Beckett RP, Minibayeva F, Kranner I (2009) Reactive oxygen species production from the initial stage of cryopreserving *Castanea sativa* embryonic axes. CryoLetters 30(5):383–384

Rout GR, Mohapatra A, Jain SM (2006) Tissue culture of ornamental pot plant: a critical review on present scenario and future prospects. Biotechnol Adv 24:531–560

Sakai A (1956) Survival of plant tissue at super-low temperature. Low Temp Sci Ser B 14:17–23

Salaj T, Panis B, Swennen R, Salaj J (2007) Cryopreservation of embryogenic tissues of *Pinus nigra* Arn. by a slow freezing method. CryoLetters 28(2):69–76

Salaj T, Matusikova I, Fraterova L, Pirselova B, Salaj J (2011) Regrowth of embryogenic tissues of *Pinus nigra* following cryopreservation. Plant Cell Tissue Organ Cult 106:55–61

Sarasan V, Cripps R, Ramsay MM, Atherton C, McMichen M, Prendergast G, Rowntree K (2006) Conservation *in vitro* of threatened plants – progress in the past decades. In Vitro Cell Dev Biol Plant 42:206–214

Sekizawa K, Yamamoto S, Rafique T, Fukui K, Niino T (2011) Cryopreservation of *in vitro*-grown shoot tips of carnation (*Dianthus caryophyllus* L.) by vitrification method using aluminium cryo-plates. Plant Biotechnol 28:401–405

Sengar RS, Chaudhary R, Tyagi SK (2010) Present status and scope of floriculture developed through different biotechnological tools. Res J Agric Sci 1(4):306–314

Seo MJ, Shin JH, Sohn JK (2007) Cryopreservation of dormant herbaceous peony (*Paeonia lactiflora* Pall.) shoot-tips by desiccation. CryoLetters 28(3):207–213

Sershen N, Pammenter NW, Berjak P, Wesley-Smith J (2007) Cryopreservation of embryonic axes of selected amaryllid species. CryoLetters 28(5):387–399

Sershen N, Berjak P, Pammenter NW, Wesley-Smith J (2011) The effects of various parameters during processing for cryopreservation on the ultrastructure and viability of recalcitrant zygotic embryos of *Amaryllis belladonna*. Protoplasma 249(1):155–169

Sharaf SA, Shibli RA, Kasrawi MA, Baghdadi SH (2012) Cryopreservation of wild Shih (*Artemisia herba-alba* Asso.) shoot-tips by encapsulation-dehydration and encapsulation-vitrification. Plant Cell Tissue Organ Cult 108(3):437–444

Sharma SD (2005) Cryopreservation of somatic embryos: an overview. Ind J Biotechnol 4:47–55

Sharma SD, Agrawal V (2012) Tissue culture aspects of ornamental plants. J Biotechnol 1(1):40–48

Shibli RA (2000) Cryopreservation of black iris (*Iris nigricans*) somatic embryos by encapsulation-dehydration. CryoLetters 21(1):39–46

Shin DJ, Kong H, Popova EV, Moon H-K, Park SY, Park SU (2012) Cryopreservation of *Kalopanax septemlobus* embryogenic callus using vitrification and droplet-vitrification. CryoLetters 33(5):402–410

Singh AK, Chand S (2010) Plant regeneration from alginate-encapsulated somatic embryos of *Dalbergia sissoo* Roxb. Ind J Biotechnol 9:319–324

Sisunandar RA, Turquay P, Samosir Y, Adkins SW (2010) Cryopreservation of coconut (*Cocos nucifera* L.) zygotic embryos does not induce morphological, cytological or molecular changes in recovered seedlings. Planta 232:435–447

Srinivas L, Ganapathi TR, Suprasanna P, Bapat LA (2006) Desiccation and ABA treatment improves conversion of somatic embryos to plantlets in banana (*Musa* spp.) cv. Rasthali (AAB). Ind J Biotechnol 5:521–526

Suranthran P, Gantait S, Sinniah UR, Subramaniam S, Alwee SSRS, Roowi SH (2011) Effect of loading and vitrification solutions on survival of cryopreserved oil palm polyembryoids. Plant Growth Reg 66:101–109

Suzuki M, Ishikawa M, Okuda H, Noda K, Kishimoto T, Nakamura T, Ogiwara I, Shimura I, Akihama T (2006) Physiological changes in *Gentian* axillary buds during two-step preculturing with sucrose that conferred high levels of tolerance to desiccation and cryopreservation. Ann Bot 97:1073–1081

Takagi H, Tien Thinh NT, Islam OM, Senboku T, Sakai A (1997) Cryopreservation of *in vitro*-grown shoot tips of taro (*Colocasia esculenta* (L.) Schott) by vitrification. 1. Investigation of basic conditions of the vitrification procedure. Plant Cell Rep 16:594–599

Tarre E, Pires BBM, Guimaraes APM, Carneiro LA, Forzza RC, Mansur E (2007) Germinability after desiccation, storage and cryopreservation of seeds from endemic *Encholirium* Mart. ex Schult. & Schult. f. and *Dyckia* Schult. & Schult. f. species (Bromeliaceae). Acta Bot Bras 21(4):777–783

Teixeira da Silva J, Kulus D (2014) Chrysanthemum biotechnology: discoveries from the recent literature. Folia Hortic 26(2):67–77

Tessereau H, Florin B, Meshine MC, Thierry C, Pétiard V (1994) Cryopreservation of somatic embryos: a tool for germplasm storage and commercial delivery of selected plants. Ann Bot 74:547–555

Tomiczak K, Rybczyński JJ, Wójciók A, Mikuła A (2009) Morphogenic potential of *Gentiana kurroo* (Royle) cell suspension after cryotreatment. CryoLetters 30(5):396

Uemura M, Minami A, Kawamura Y (2009) Effect of low temperature and cryoprotectants on plant plasma membrane. In: Panis B, Lynch P (eds) 1st international symposium: cryopreservation in horticultural species. ISHS Acta Hortic, Leuven, p 15

Vágner M, Spacková J, Eliasová K (2007) Cryopreservation of embryogenic cultures of Norway spruce. COST European Cooperation in Science and Technology, Oviedo, pp 34–35

Vázquez AM (2007) Cryopreservation and genetic instability. COST European Cooperation in Science and Technology, Oviedo, pp 42–44

Wen Y, Chunyan L, Xiuling B, Hengchun L, Li W (1999) Study on callus cryopreservation of *Freesia refracta* Klatt. J Northeast Norm Univ (Nat Sci Ed) 4:70–72

Wilkinson T, Wetten A, Prychid C, Fay MF (2003) Suitability of cryopreservation for the long-term storage of rare and endangered plant species: a case history of *Cosmos atrosanguineus*. Ann Bot 91:65–74

Winkelmann T, Mubmann V, Serek M (2004) Cryopreservation of embryogenic suspension cultures of *Cyclamen persicum* Mill. Plant Cell Rep 23:1–8

Wolkers WF, Tetteroo FA, Alberda M, Hoekstra FA (1999) Changed properties of the cytoplasmic matrix associated with desiccation tolerance of dried carrot somatic embryos. An *in situ* fourier transform infrared spectroscopic study. Plant Physiol 120(1):153–164

Wu Y-J, Huang X-L, Chen Q-Z, Li X-J, Engelmann F (2007) Induction and cryopreservation of embryogenic cultures from nucelli and immature cotyledon cuts of mango (*Mangifera indica* L. var Zihua). Plant Cell Rep 26:161–168

Xue SH, Luo XJ, Wu ZH, Zhang HL, Wang X (2007) Cold storage and cryopreservation of hairy root cultures of medicinal plant *Eruca sativa* Mill., *Astragalus membranaceus* and *Gentiana macrophylla*. Plant Cell Tissue Organ Cult 92:251–260

YaJun L, LiPing G, Tao X, KeJun G (2009) Study on cryopreservation of suspension culture cells by vitrification in *Camellia sinensis*. J Tea Sci 29(2):120–126

Zalewska M, Kulus D (2013) Cryopreservation of *in vitro*-grown shoot tips of chrysanthemum by encapsulation-dehydration. Folia Hortic 25(2):133–140

Zalewska M, Kulus D (2014) Improvement of *Chrysanthemum × grandiflorum* /Ramat./ Kitam. encapsulation-dehydration cryopreservation protocol. Acta Scientarum Polonorum Hortorum Cultus 13(2):97–108

Zalewska M, Lema-Rumińska J, Miler N (2007) *In vitro* propagation using adventitious buds technique as a source of new variability in chrysanthemum. Scientia Hortic 113:70–73

Zalewska M, Lema-Rumińska J, Miler N, Gruszka M, Dąbal W (2011) Induction of adventitious shoot regeneration in chrysanthemum as affected by the season. In Vitro Cell Dev Biol Plant 47:375–378

Zeliang PK, Pattanayak A, Iangrai B, Khongwir EA, Sarma BK (2010) Fertile plant regeneration from cryopreserved calli of *Oryza rufipogon* Griff. and assessment of variation in the progeny of regenerated plants. Plant Cell Rep 29:1423–1433

Zhao MA, Xhu YZ, Dhital SP, Khu DM, Song YS, Wang MY, Lim HT (2005) An efficient cryopreservation procedure for potato (*Solanum tuberosum* L.) utilizing the new ice blocking agent, Supercool X1000. Plant Cell Rep 24:477–481

Zhao EH, Liu ZH, Hu X, Yin JL, Li W, Rao GY, Zhang XH, Huang CL, Anderson N, Zhang QX, Chen JY (2009) Chrysanthemum genetic resources and related genera of *Chrysanthemum* collected in China. Genet Resour Crop Evol 56:937–946

Somatic Embryogenesis in Japanese Black Pine (*Pinus thunbergii* Parl.)

2

Tsuyoshi E. Maruyama and Yoshihisa Hosoi

Abstract

Somatic embryogenesis in Japanese black pine (*Pinus thunbergii* Parl.) was initiated from megagametophytes containing immature zygotic embryos. Embryogenic cultures were maintained and proliferated in a medium supplemented with 3 μ*M* 2,4-dichlorophenoxyacetic acid, 1 μ*M* 6-benzylaminopurine, 30 g l^{-1} sucrose, and 1.5 g l^{-1} L-glutamine. The somatic embryo maturation experiments were performed in darkness at 25 °C. Embryogenic tissues were cultured on maturation media containing 50 g l^{-1} maltose, 2 g l^{-1} activated charcoal, 100 μ*M* abscisic acid, and 100 g l^{-1} polyethylene glycol. Desiccation of somatic embryos at high relative humidity resulted not only in a marked increment in germination frequency but also subsequently improved plant conversion rate. In addition, this treatment resulted in a considerable improvement of synchronization of the germination period, compared to those of untreated control. Somatic plants were acclimatized and their growth has been monitored in the field.

Keywords

Conifers • Embryogenic cultures • Plant regeneration • Post-maturation treatments • Somatic embryos • Somatic plants

2.1 Introduction

Japanese black pine (*Pinus thunbergii* Parl.), locally named "kuromatsu," is one of the most important native forest tree species in Japan. This pine native to Honshu, Kyushu, and Shikoku Islands is widely used for reforestation, as windbreaks against sand and salt in coastal areas, as ornamental tree in parks and house gardens, and as one of the classic "bonsai" subjects. However, nowadays, Japanese black pine populations have further declined as a result of pine wilt disease, caused by the pinewood nematode, *Bursaphelenchus xylophilus* (Steiner et Buhrer) Nickle (Kiyohara and Tokushige 1971), which is transmitted in Japan by two cerambycid beetles,

T.E. Maruyama (✉) • Y. Hosoi
Department of Molecular and Cell Biology, Forestry and Forest Products Research Institute, Matsunosato 1, Tsukuba 305-8687, Japan
e-mail: tsumaruy@ffpri.affrc.go.jp

A. Mujib (ed.), *Somatic Embryogenesis in Ornamentals and Its Applications*,
DOI 10.1007/978-81-322-2683-3_2

Monochamus alternatus Hope (Mamiya and Enda 1972) and *M. saltuarius* (Gebler) (Sato et al. 1987). The pine wilt disease is one of the most serious epidemic tree diseases in Japan and has been a critical factor in the mass mortality not only in *P. thunbergii*, but also in other Japanese pine forests such as *P. densiflora* Sieb. et Zucc. (akamatsu), *P. luchuensis* Mayr (ryukyumatsu), and *P. armandii* Franch. var. *amamiana* (Koidz.) Hatusima (yakutanegoyou) (Mamiya 1983; Kishi 1995; Kanetani et al. 2001; Maruyama and Hosoi 2012). Therefore, the development of resistant clones to pine wilt disease is expected.

Genetic engineering offers a significant tool for improving forest trees in a relatively short time, with the advantage over traditional breeding that genes which are foreign to a population can be added to superior genotypes without the loss of any qualities of the parental lines. Although genetic transformation of conifers has been reported (Ellis et al. 1993; Shin et al. 1994; Charest et al. 1996; Klimaszewska et al. 1997; Walter et al. 1998), currently there are still many limitations in the genetic engineering of forest trees. One of the major limitations to forest tree transformation is the difficulty of regenerating whole plants from target cells. Thus, the development of an efficient and stable plant regeneration system is essential for genetic engineering of this species.

Somatic embryogenesis is one of the most promising techniques for mass propagation of selected trees. It allows, the ex situ conservation of genetic resources by cryopreservation techniques, and for the purposes of genetic transformation. However, for many species, the maturation of large numbers of somatic embryos and their plant conversion efficiency has been one of the limiting factors for widespread commercial use.

Since plant regeneration via somatic embryogenesis in *P. thunbergii* was first reported by Ishii and Maruyama (1999), subsequent other studies (Ishii et al. 2001; Hosoi and Ishii 2001; Taniguchi 2001; Maruyama et al. 2005a) were reported, but the achieved plant conversion frequencies of somatic embryos were limited by the low frequencies of root emergence. Later, an improved somatic embryo germination protocol for Japanese pines based on the desiccation of somatic embryos after the maturation step was reported (Maruyama and Hosoi 2012). In that study, post-maturation treatment markedly increased germination frequencies and considerably improved synchronization of the germination period. Similarly, in order to improve the somatic embryo germination protocols, post-maturation treatments based on the desiccation of somatic embryos have also been reported to successfully improve germination frequencies in conifers (Hay and Charest 1999; Klimaszewska and Cyr 2002; Stasolla et al. 2001).

This chapter describes a plant regeneration system for propagation of Japanese black pine through somatic embryogenesis, focusing on the results reported by Maruyama et al. (2005a) and Maruyama and Hosoi (2012).

2.2 Embryogenic Culture Initiation

Collected immature cones were disinfected by 3–5 min immersion in 99.5 % ethanol and then were dried in a laminar flow cabinet before dissection. Excised seeds were disinfected with 2–3 % (w/v available chlorine) sodium hypochlorite solution for 15–30 min and then rinsed five times with sterile distilled water. After the seed coats had been removed, the megagametophytes containing immature zygotic embryos were used as explants for the initiation of embryogenic cultures. Excised megagametophytes were cultured in 4 compartment plates (90 × 15 mm) (Kord-Valmark Labware, Ontario, Canada) containing somatic embryogenesis initiation medium. The initiation medium was EM medium (Maruyama et al. 2000), modified as follows: basal salts, vitamins, and myoinositol were reduced to half the standard concentrations; the concentrations of KCl was reduced to 40 mg l^{-1}; and 500 mg l^{-1} casein hydrolysate, 1 g l^{-1} L-glutamine, 10 g l^{-1} sucrose, 10 μM 2,4-dichlorophenoxyacetic acid (2,4-D), and 5 μM 6-benzylaminopurine (BA) were added. The medium was solidified with 3 g l^{-1} gellan gum (Gelrite®; Wako Pure Chemical, Osaka, Japan). The pH of the medium was adjusted to 5.8 before sterilization, and the cultures were kept in darkness at 25 °C. Medium without plant growth regulators (PGR) but containing 2 g l^{-1} activated charcoal (AC) was also tested for the induction of embryogenic cells. The presence of somatic embryos at distinct early stages characterized by an embryo part (smaller and denser cells) with a suspensor (long vacuolated cells) was determined weekly under an inverted microscope up to 12 weeks.

As shown in Fig. 2.1, the somatic embryogenesis initiation frequency from eight open-pollinated seed explants of Japanese black pine collected in July (Ibaraki Prefecture) varied from 0 to 6.3 %. Although the initiation of somatic embryogenesis was possible in the absence of exogenous PGR, the number of families with embryogenic response was higher when the medium was supplemented with 2,4-D and BA (Fig. 2.1). The extrusion of embryogenic tissue from the micropylar end of explants occurred mostly after about 3–6 weeks of culture, and the proliferation of embryonal masses was evident after about 6–8 weeks of culture (Fig. 2.2a). Achieved low initiation rate was consistent to the results reported for other pine species such as *Pinus banksiana* (0.37 %) (Park et al. 1999), *Pinus patula* (0.04 %) (Jones and van Staden 1999), and *Pinus rigida* × *P. taeda* (0.44 %) (Kim and Moon 2007). In contrast, Taniguchi (2001) reported initiation frequencies up to 17.3 % in Japanese black pine using modified DCR medium. Similarly, higher initiation rates were also reported for *P. taeda* (17.9 %) (Pullman et al. 2003), *P. strobus* (20–53 %) (Klimaszewska et al. 2001), *P. sylvestris* (up to 22 %), and *P. pinaster* (up to 40 %) (Lelu et al. 1999).

Low initiation frequency of somatic embryogenesis is one of the key problems to resolve for practical applications. Although the seed genotypes and culture procedures may impact on the somatic embryogenesis initiation (Park et al. 2006; Miguel et al. 2004; Klimaszewska et al. 2001; Lelu et al. 1999; Garin et al. 1998), the appropriate developmental stage of zygotic embryos seemed the most critical factor (Maruyama et al. 2000, 2002; Klimaszewska et al. 2007). The optimal developmental stage for many pine species has been reported in terms of seed collection date or time after fertilization (Becwar et al. 1990; Jain et al. 1995; Lelu et al. 1999). However, due to the difficulty in determining the precise time of fertilization in open-pollinated cones and the fact that the variation in the zygotic embryo development depends on weather and location, the explant collection criteria for somatic embryogenesis initiation cannot be easily generalized. In addition, variation in the developmental stage of embryos may be observed among trees and even the same tree when individual cones are compared. Observation of the developmental stage of individual embryos is thus likely to be the most appropriate method to determine the optimal time for embryo selection.

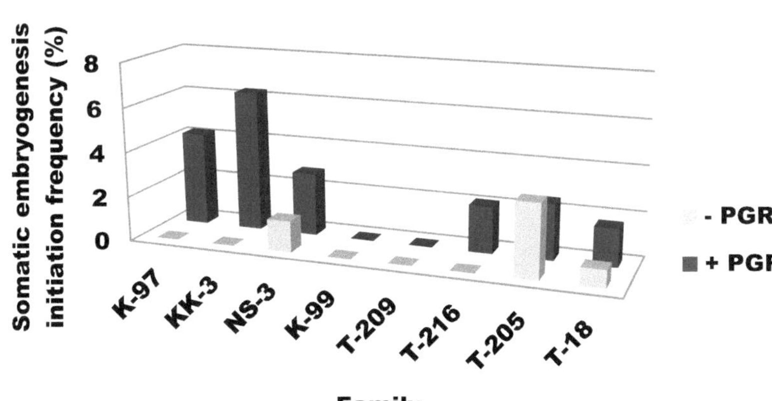

Fig. 2.1 Embryogenic response of eight open-pollinated seed families from megagametophytes containing zygotic embryos of Japanese black pine cultured on somatic embryogenesis initiation medium supplemented with (+ PGR) or without (− PGR) plant growth regulators

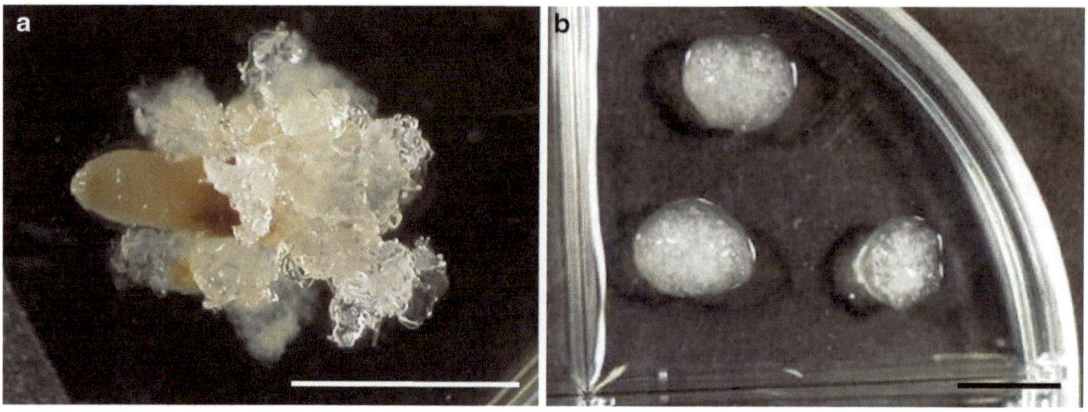

Fig. 2.2 Somatic embryogenesis initiation (**a**) and proliferation of embryogenic tissues (**b**) in Japanese black pine. *Bars* 1 cm

2.3 Maintenance and Proliferation of Embryogenic Cultures

Initiated embryogenic cultures were transferred onto the fresh proliferation medium (initiation medium supplemented with 30 g l^{-1} sucrose, 3 μ*M* 2,4-D, 1 μ*M* BA, and 1.5 g l^{-1} glutamine, and without casein hydrolysate). Once embryogenic cultures increased in mass, they were maintained and proliferated by subculturing 12 pieces of embryogenic tissue per plate at 2–3-week intervals (Fig. 2.2b). Although initiation of somatic embryogenesis was also possible without any additional auxin and cytokinin supplements required (Fig. 2.1), the presence of exogenous PGR was found essential for the continuous maintenance and proliferation of embryogenic tissues. The proliferation medium supported the growth of the initiated embryogenic cell lines, retaining their original translucent and mucilaginous appearance. In contrast, embryogenic tissues maintained on a medium with no PGR displayed spontaneous embryo development and declined in proliferation over the time (data not shown). This tendency was consistent with the results reported for other Japanese conifers (Maruyama et al. 2000, 2002, 2005b, c, 2007; Hosoi and Maruyama 2012).

In general, solid medium was used for maintenance and liquid medium for more rapid proliferation of the cultures. The cells were transferred to conical flasks containing liquid medium (proliferation medium without gellan gum) and cultured on a rotary shaker at approximately 70 rpm in darkness at 25 °C. The fresh weight of tissues in liquid proliferation medium increased 5–12-fold after about 2 weeks culture period (data not shown).

2.4 Maturation of Somatic Embryos

About 500 mg FW of embryogenic tissues suspended in about 2–3 ml of medium without plant growth regulators was plated over 90 × 15 mm diameter plates containing 30–40 ml of maturation medium. Maturation medium containing basal salts and vitamins from the original EM medium, 50 g l^{-1} maltose, 100 g l^{-1} polyethylene glycol 4000 (PEG), 2 g l^{-1} AC, 100 μ*M* abscisic acid (ABA), and EMM amino acids (Smith 1996) was used. Petri dishes were sealed with Novix-II film (Iwaki Glass Co., Ltd., Chiba, Japan) and kept in darkness at 25 °C for 8–12 weeks.

About 2 weeks after transfer onto maturation medium, embryogenic cells developed gradually to form an individual and compact mass going to mature stage. Cotyledonary embryos were first observed about 4 weeks after transfer of embryogenic cells and were clear at 6–8 weeks of culture (Fig. 2.3a, b). The results in somatic embryo production after 8 weeks of culture from eight cell lines are shown in Fig. 2.4. Somatic embryos were produced in all cell lines tested, but the abil-

Fig. 2.3 Somatic embryo maturation of Japanese black pine after 6 weeks of culture on maturation medium (**a**). Close-up of cotyledonary somatic embryo (**b**). *Bars* 1 mm

Fig. 2.4 Somatic embryo production from eight cell lines of Japanese black pine

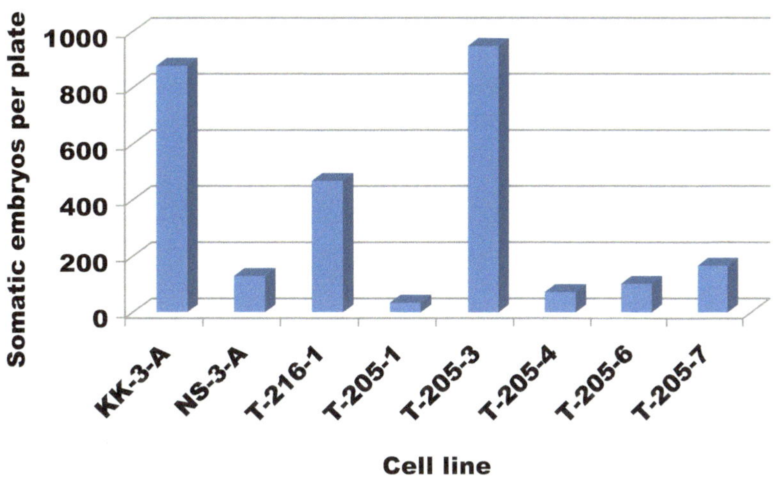

ity to produce mature embryos was notably different among the genotypes. The average number of somatic embryos per plate ranged from 35 to 949. Variation in the potential for mature embryo production and subsequently for plant regeneration is commonly reported among genotypes and families in conifers (Lelu et al. 1999; Igasaki et al. 2003; Maruyama et al. 2007).

2.5 Somatic Embryo Germination and Plantlet Conversion

Mature cotyledonary embryos were picked up from maturation medium and transferred to germination medium (1/2 EM medium without PGR and supplemented with 2 g l^{-1} AC). Cultures were kept at 25 °C under a photon flux density of about 65 μmol m^{-2} s^{-1} provided by cool, white fluorescent lamps (100 V, 40 W; Toshiba Company, Tokyo, Japan) for 16 h daily. The numbers of somatic embryos germinated (emergence of root) and converted into plantlets (emergence of both root and epicotyl) were recorded after 4 and 8 weeks, respectively.

The start of radicle emergence was observed 1–2 weeks after transfer, and after 4 weeks of culture, the germination frequencies varied from 10 to 60 % (Fig. 2.5). Subsequently, epicotyls developed, and after 8 weeks of culture, the frequencies of plantlet conversion from somatic embryos ranged from 3 to 51 % (Fig. 2.5).

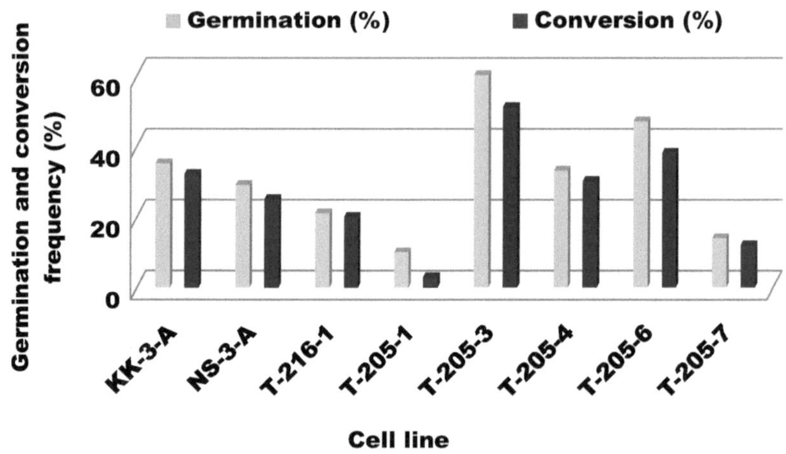

Fig. 2.5 Somatic embryo germination and plantlet conversion from eight cell lines of Japanese black pine

2.6 Improvement of Germination and Plantlet Conversion by Post-maturation Treatments (PMT) of Somatic Embryos

2.6.1 Effect of Desiccation on Water Content and the Germination Frequency of Somatic Embryos

To determine the effect of different periods of desiccation on the water content of somatic embryos and their subsequent germination rates, mature somatic embryos from one genotype of Japanese black pine were subjected to either fast or slow desiccation. For fast desiccation, embryos placed over 70 mm diameter filter paper disks (Advantec No. 2, Toyo Roshi Kaisha Ltd., Tokyo, Japan) into a 90 mm diameter opened plates containing 30–40 ml of maturation medium supplemented with no PEG and solidified with 30 g l^{-1} gellan gum were kept at 25 °C in a laminar flow cabinet for 0–4 h. For slow desiccation (desiccation at high relative humidity (HRH)), embryos placed over 30 mm diameter filter paper disks into 2 (central) wells of a 6-well plates (Iwaki, AGC Techno Glass Co., Ltd., Chiba, Japan) in which the remaining 4 (side) wells were filled with 5–6 ml of sterile water were sealed tightly with Novix-II film and placed in darkness at 25 °C for 0 (control) and 3 weeks (Fig. 2.6). Under these conditions, the generated relative humidity registered with a thermo-hygrometer recorder (RS-10, ESPEC MIC Corporation, Aichi, Japan) was approximately 98 %. The water content of somatic embryos was determined after various time periods of desiccation in a laminar flow cabinet and under desiccation at HRH conditions. After the desiccation, the fresh weight was determined before drying the embryos in an oven at 110 °C for 24 h and reweighing to measure the dry weight. The embryo water content was calculated on a wet weight basis.

The fast desiccation of somatic embryos in a laminar flow cabinet resulted in a drastic decrease of germination frequency after 0.5 h and 100 % mortality after 2 h (Table 2.1). This result suggests that rapid water loss caused by very fast drying was lethal to the somatic embryos of the Japanese black pine. In contrast, the slow desiccation of somatic embryos at HRH resulted in a mild loss of water content, which declined by only 10 % after 3 weeks of drying (Table 2.2). This treatment resulted in a considerable improvement of germination frequency and synchronization of the germination period. The germination percentage was improved more than threefold (21–71 %), and the germination period was synchronized over a half period (28–14 days) compared with the control.

2.6.2 Effect of Different PMT on the Germination Frequency of Somatic Embryos

To determine the effect of different PMT on germination frequency, mature somatic embryos from three genotypes of Japanese black pine

Fig. 2.6 Slow desiccation of Japanese black pine somatic embryos. *Bar* 1 cm

Table 2.1 Water content and germination frequency of Japanese black pine somatic embryos after different time periods of desiccation in a laminar flow cabinet

Desiccation period (hours)	Water content (%)	Germination frequency (%)
0	74.6±5.0	15
0.5	49.9±4.4	2
1	42.9±6.3	1
2	33.8±4.8	0
4	22.5±2.5	0

Water content values represent the mean±SE, $n=3$ replicates of 25 somatic embryos. Germination frequencies were calculated after 6 weeks of culture from 100 somatic embryos for each desiccation period

highly recalcitrant to germination were exposed to following treatments: (a) Control: embryos without any PMT were placed directly onto the germination medium and kept at 25 °C under a photon flux density of about 65 μmol m^{-2} s^{-1} for 16 h. (b) Slow desiccation at 25 °C: HRH desiccation at 25 °C (as described above). (c) Slow desiccation at 4 °C: HRH desiccation at 4 °C (as described above, but keeping the embryos in darkness at 4 °C for 3 weeks). (d) Desiccation on gelled medium: embryos placed over 70 mm diameter filter paper disks into a 90 mm diameter Novix-II film-sealed plates containing 30–40 ml of maturation medium supplemented with no PEG and solidified with 30 g l^{-1} gellan gum were desiccated in darkness at 25 °C for 12 weeks. (e) Desiccation in an empty plate: embryos placed over 70 mm diameter filter paper disks into a 90 mm diameter Novix-II film-sealed empty plates were desiccated in darkness at 25 °C for 3 weeks. (f) Cold-wet stratification: embryos placed over 70 mm diameter filter paper disks into a 90 mm diameter Novix-II film-sealed plates containing 30–40 ml of germination medium were kept in darkness at 4 °C for 12 weeks. As shown in Table 2.3, when somatic embryos without any PMT (control) were placed directly on the germination medium, the root emergence of embryos occurred at a very low frequency (an average of 3 %). Subsequently, the embryos became hyperhydric with little or no epicotyls and hypocotyl elongation (Fig. 2.7a). The best results were obtained with HRH desiccation at 25 °C, which saw markedly improved germination frequencies in the three genotypes tested (average of 85 %). Following the transfer of the treated embryos onto the germination medium, the start of radicle emergence was followed by the epicotyl development (Fig. 2.7b).

HRH desiccation at 4 °C and desiccation on a gelled medium also improved the germination frequencies (average of 66 and 62 %, respectively) but were less effective in promoting germination than HRH desiccation at 25 °C. Desiccation in an empty plate and cold-wet stratification resulted in no or poor germination (average of 1 %).

Table 2.2 Water content and germination frequency of Japanese black pine somatic embryos after slow desiccation at HRH condition

Desiccation period	Water content	Germination frequency (%) after			
(days)	(%)	7 days	14 days	21 days	28 days
0	72.0 ± 1.5	2	10	18	21
21	62.4 ± 2.7	66	71	71	71

Water content values represent the mean ± SE, $n = 3$ replicates of 25 somatic embryos. Germination frequencies were calculated after 6 weeks of culture from 530 to 784 somatic embryos for the desiccation period of 0 and 21 days at 25 °C, respectively

Table 2.3 Effect of different PMT on germination frequency of Japanese black pine somatic embryos

Somatic embryo line	Control	Slow desiccation at HRH condition		Desiccation on gelled medium	Desiccation in an empty plate	Cold-wet stratification
		25 °C	4 °C			
T216-401	2 (4/200)	85 (170/201)	59 (78/132)	63 (72/115)	0 (0/100)	1 (1/100)
T205-401	0 (0/100)	70 (105/151)	61 (61/100)	55 (55/100)	0 (0/100)	0 (0/100)
T054-820	4 (10/235)	93 (293/315)	72 (152/211)	68 (68/100)	2 (2/100)	2 (2/100)

Values in parentheses represent germinated somatic embryos/total somatic embryos tested. Germination frequencies were calculated after 6 weeks of culture

Fig. 2.7 Somatic embryo germination in Japanese black pine with no PMT (**a**) and with HRH desiccation at 25 °C (**b**). *Bars* 1 cm

Table 2.4 Effect of slow desiccation (HRH at 25 °C) on germination and conversion frequencies of somatic embryos in six different families of Japanese black pine

Family	Germination frequency (%)		Conversion frequency (%)	
	Control	Slow desiccation	Control	Slow desiccation
T0216	15 (113/730)	74 (726/985)	15 (106/730)	72 (707/985)
T0205	30 (250/840)	81 (655/812)	24 (205/840)	76 (619/812)
T0064	6 (12/200)	93 (296/320)	1 (2/200)	90 (288/320)
T0054	2 (16/847)	84 (1098/1303)	0 (4/847)	80 (1040/1303)
T0090	5 (5/100)	94 (188/200)	1 (1/100)	88 (176/200)
T9037	8 (8/100)	77 (289/374)	2 (2/100)	76 (285/374)

Values in parentheses represent germinated or converted somatic embryos/total somatic embryos tested. Germination and conversion frequencies were calculated after 6 and 12 weeks of culture, respectively

2.6.3 Effect of Slow Desiccation on the Germination and Conversion Frequencies of Somatic Embryos

Based on the previous results, the most successful PMT achieved (HRH desiccation at 25 °C) was further performed to evaluate their effect on germination and conversion frequencies in somatic embryos originated from 6 different families of Japanese black pine. As shown in Table 2.4, desiccation of somatic embryos at HRH at 25 °C resulted not only in a marked increment on germination frequencies but also in the subsequent improvement of plant conversion rates in all families tested. The total average in terms of the germination and conversion frequency of somatic embryos was improved by more than fivefold (14.3–81.4 %) and more than sixfold (11.4–78.0 %), respectively, compared with the control.

2.7 *In Vitro* Growth and Acclimatization of Somatic Plants

Regenerated plantlets were transferred into 300-ml flasks containing 100 ml of fresh germination medium supplemented with 30 g l^{-1} sucrose, 5 g l^{-1} AC, and 10 g l^{-1} agar (Wako Pure Chemical Industries, Osaka, Japan) (Fig. 2.8a) or into Magenta® vessels (Sigma, St. Louis, USA) containing Florialite® (Nisshinbo Industries, Inc., Tokyo, Japan) irrigated with a 0.1 % (v/v) Hyponex® 6-10-5 plant-food solution (Hyponex Japan Co., Ltd., Osaka, Japan) and kept under the same conditions as described above for 16–20 weeks prior to ex vitro acclimatization (Fig. 2.8b).

The developed somatic plants were transplanted into plastic pots filled with Kanuma soil and acclimatized inside a growth chamber at 25 °C and 80 % relative humidity. During the first 2 weeks, the plants were kept in plastic boxes with transparent covers (Assist No.2, Shinkigosei Co., Ltd., Tokyo, Japan) and were irrigated with tap water one time per week. The covers were gradually opened during the next 2 weeks and completely removed 1 month after transplantation. The pots were fertilized using a modified nutrient solution (Nagao 1983), which included 143 mg l^{-1} NH_4NO_3, 55.1 mg l^{-1} $NaH_2PO_4 \cdot 2H_2O$, 47.1 mg l^{-1} KCl, 52.5 mg l^{-1} $CaCl_2 \cdot 2H_2O$, 61 mg l^{-1} $MgSO_4 \cdot 7H_2O$, 25 mg l^{-1} Fe(III) EDTA, 0.1 mg l^{-1} Cu EDTA, 0.1 mg l^{-1} Mn EDTA, 0.1 mg l^{-1} Zn EDTA, 1.5 mg l^{-1} H_3BO_3, 0.01 mg l^{-1} KI, 0.005 mg l^{-1} $CoCl_2 \cdot 6H_2O$, and 0.005 mg l^{-1} MoO_3. The regenerated somatic plants were successfully acclimatized (Fig. 2.9a) and are growing in the field (Fig. 2.9b).

2.8 Discussion and Concluding Remarks

Although the ability of Japanese black pine to produce mature embryos varied according to genotype, the somatic embryo production in 75 % of genotypes tested (six of eight lines) was

Fig. 2.8 *In vitro* growth of Japanese black pine emblings into flasks containing medium (**a**) and into Magenta boxes containing Florialite (**b**). *Bars* 1 cm

Fig. 2.9 Acclimatized somatic plants of Japanese black pine (**a**). Somatic plants growing in the field (**b**). *Bars* 1 cm (**a**), 1 m (**b**)

over 100 cotyledonary embryos per plate. A range from 35 to 949 embryos per plate were produced. The maturation frequency and the quality of embryos produced are the most important criteria for the optimization of a somatic embryogenesis protocol. The quality of an embryo is related to its morphology and the ability to produce normal plants. A high plant conversion frequency is essential for somatic embryogenesis protocols in commercial mass production, breeding programs, and genetic engineering (Maruyama et al. 2005a).

In order to improve plant conversion protocols, the desiccation of somatic embryos after PEG-mediated maturation has been recommended for conifer species. For most species, desiccation presumably acts to terminate developmental processes and to initiate those metabolic processes necessary to prepare the seeds for germination and growth (Kermode and Bewley 1985).

For Japanese black pine, the comparison of PMT indicated that the slow desiccation of somatic embryos under HRH was the most effective in promoting germination (Maruyama and Hosoi 2012). Similarly, Roberts et al. (1990) indicated that slow drying at HRH promoted germination up to 90 % in somatic embryos of the interior spruce. Slow desiccation at 4 °C and desiccation on a gelled medium also considerably improved the germination and conversion rate when compared with the control, but were only mildly effective when compared with the best result. Both treatments improved the quality of germinants and decreased the time required for germination, but did not increase the total number of germinants (Maruyama and Hosoi 2012). These results suggest that although the quality of germinating somatic embryos was improved by desiccation treatments, the germination frequency was affected by the rate of desiccation in each condition (data not shown).

Fast desiccation treatments resulted in increased damage and death of Japanese black pine somatic embryos. Desiccation in a laminar flow cabinet and into empty plates was found to be ineffective in stimulating germination (Maruyama and Hosoi 2012). For conifer somatic embryos, the desiccation tolerance generally decreased with increasing rapidity of desiccation (Bomal and Tremblay 1999). Roberts et al. (1990) showed that humidities of 81 % and lower were lethal to the somatic embryos of interior spruce, whereas germination was enhanced following treatments at humidities greater than 95 % relative to untreated controls. These observations indicate that desiccation at HRH causes physiological changes in somatic embryos which promote germination (Kong and Yeung 1992). This notwithstanding, the effect of desiccation on somatic embryos germination is not a simple process, due to it being affected by complex factors such as the developmental stage of embryo maturation, accumulation of storage reserves and endogenous hormones, and desiccation tolerance of somatic embryos (Roberts et al. 1990;

Beardmore and Charest 1995; Jones and van Staden 2001; Bonga et al. 2010; Kharenko et al. 2011; Lu et al. 2011; Abrahamsson et al. 2011).

In summary, based on our results, desiccation treatments of the cotyledonary somatic embryo of Japanese black pine after PEG-mediated maturation were found to be effective in promoting germination. The beneficial effect of desiccation treatment on germination frequency was dependent not only on the decrease in the water content of somatic embryos but also on the desiccation rate. In the present study, the embryo drying rates for some treatments may have been more rapid than desirable or there is evidence that a shorter drying period may have been more beneficial. The mechanism of desiccation tolerance in somatic embryos is complex and requires more research. However, despite the doubts to be clarified, we can conclude that the germination frequencies, the synchronization of root emergence, and subsequent plant conversion of somatic embryos of Japanese black pine were dramatically improved by slow desiccation at HRH.

The improved regeneration system represents a promising tool for the efficient propagation and for future genetic engineering approaches to develop resistant clones to the pine wilt disease.

Acknowledgments We express our gratitude to the Ibaraki Prefectural Government Forestry Technology Center for the generous supply of seeds.

References

Abrahamsson M, Valladares S, Larsson E, Clapham D, von Arnold S (2011) Patterning during somatic embryogenesis in relation to polar auxin transport and programmed cell death. Plant Cell Tissue Org Cult doi:10.1007/s11240-011-0103-8

Beardmore T, Charest PJ (1995) Black spruce somatic embryo germination and desiccation tolerance. I. Effects of abscisic acid, cold, and heat treatments on the germinability of mature black spruce somatic embryos. Can J For Res 25:1763–1772

Becwar MR, Nagmani R, Wann SR (1990) Initiation of embryogenic cultures and somatic embryo development in loblolly pine (*Pinus taeda*). Can J For Res 20:810–817

Bomal C, Tremblay M (1999) Effect of desiccation to low moisture content on germination, synchronization of root emergence, and plant regeneration of black spruce embryos. Plant Cell Tissue Org Cult 56:193–200

Bonga JM, Klimaszewska KK, von Aderkas P (2010) Recalcitrance in clonal propagation, in particular of conifers. Plant Cell Tissue Org Cult 100:241–254

Charest PJ, Devantier Y, Lachance D (1996) Stable genetic transformation of *Picea mariana* (black spruce) via particle bombardment. In Vitro Cell Dev Biol Plant 32:91–99

Ellis DD, McCabe DE, McInnis S, Ramachandran R, Russell DR, Wallace KM, Martinell BJ, Roberts DR, Raffa KF, McCown BH (1993) Stable transformation of *Picea glauca* by particle acceleration. Bio/Technology 11:84–89

Garin E, Isabel N, Plourde A (1998) Screening of large numbers of seed families of *Pinus strobus* L. for somatic embryogenesis from immature and mature zygotic embryos. Plant Cell Rep 18:37–43

Hay EI, Charest PJ (1999) Somatic embryo germination and desiccation tolerance in conifers. In: Jain SM, Gupta PK, Newton RJ (eds) Somatic embryogenesis in woody plants, vol 4. Kluwer Academic, Dordrecht, pp 61–96

Hosoi Y, Ishii K (2001) Somatic embryogenesis and plantlet regeneration in *Pinus armandii* var. *amamiana*. In: Morohoshi N, Komamine A (eds) Molecular breeding of woody plants. Elsevier Science, Amsterdam, pp 313–318

Hosoi Y, Maruyama TE (2012) Plant regeneration from embryogenic tissue of Pinus luchuensis Mayr, an endemic species in Ryukyu Island, Japan. Plant Biotechnol 29:401–406

Igasaki T, Sato T, Akashi N, Mohri T, Maruyama E, Kinoshita I, Walter C, Shinohara K (2003) Somatic embryogenesis and plant regeneration from immature zygotic embryos of *Cryptomeria japonica* D. Don. Plant Cell Rep 22:239–243

Ishii K, Maruyama E (1999) Regeneration from somatic embryos of three Japanese conifers. Biologia 16:54–57

Ishii K, Maruyama E, Hosoi Y (2001) Somatic embryogenesis of Japanese conifers. In: Morohoshi N, Komamine A (eds) Molecular breeding of woody plants. Elsevier Science, Amsterdam, pp 297–304

Jain SM, Gupta PK, Newton RJ (eds) (1995) Somatic embryogenesis in woody plants Vol. 3, Gymnosperms. Kluwer Academic, Dordrecht

Jones NB, van Staden J (1999) Somatic embryogenesis in *Pinus patula* Scheide et Deppe. In: Jain SM, Gupta PK, Newton RJ (eds) Somatic embryogenesis in woody plants, vol 4. Kluwer Academic, Dordrecht, pp 431–447

Jones NB, van Staden J (2001) Improved somatic embryo production from embryogenic tissue of *Pinus patula*. In Vitro Cell Dev Biol Plant 37:543–549

Kanetani S, Akiba M, Nakamura K, Gyokusen K, Saito A (2001) The process of decline of an endangered tree species, *Pinus armandii* Franch. var. *amamiana* (Koidz.) Hatusima, on the southern slope of Mt. Hasadake in Yaku-shima Island. J For Res 6:307–310

Kermode AR, Bewley D (1985) The role of maturation drying in the transition from seed development to germination. I. Acquisition of desiccation-tolerance and germinability during development of *Ricinus communis* L. seeds. J Exp Bot 12:1906–1915

Kharenko OA, Zaharia LI, Giblin M, Cekic V, Taylor DC, Palmer CD, Abrams SR, Loewen MC (2011) Abscisic acid metabolism and lipid accumulation of a cell suspension culture of *Lesquerella fendleri*. Plant Cell Tissue Org Cult 105:415–422

Kim YW, Moon HK (2007) Regeneration of plant by somatic embryogenesis in *Pinus rigida* × *P. taeda*. In Vitro Cell Dev Bio Plant 43:335–342

Kishi Y (1995) The pine wood nematode and the Japanese pine sawyer. Thomas, Tokyo

Kiyohara T, Tokushige Y (1971) Inoculation experiments on a nematode, *Bursaphelenchus* sp., onto pine trees. J Jpn For Soc 53:210–218 (in Japanese with English summary)

Klimaszewska K, Cyr DR (2002) Conifer somatic embryogenesis: I. Dev Dendrobiol 48:31–39

Klimaszewska K, Devantier Y, Lachance D, Lelu M-A, Charest PJ (1997) *Larix laricina* (tamarack): somatic embryogenesis and genetic transformation. Can J For Res 27:538–550

Klimaszewska K, Park JS, Overton C, Maceacheron I, Bonga JM (2001) Optimized somatic embryogenesis in *Pinus strobus* L. In Vitro Cell Dev Biol Plant 37:392–399

Klimaszewska K, Trontin JF, Becwar MR, Devillard C, Park YS, Lelu-Walter MA (2007) Recent progress in somatic embryogenesis of four *Pinus* spp. Tree For Sci Biotechnol 1:11–25

Kong L, Yeung E (1992) Development of white spruce somatic embryos: II. Continual shoot meristem development during germination. In Vitro Cell Dev Biol 28P:125–131

Lelu M-A, Bastien C, Drugeault A, Gouez M-L, Klimaszewska K (1999) Somatic embryogenesis and plantlet development in *Pinus sylvestris* and *Pinus pinaster* on medium with and without growth regulators. Physiol Plant 105:719–728

Lu J, Vahala J, Pappinen A (2011) Involvement of ethylene in somatic embryogenesis in Scots pine (*Pinus sylvestris* L.). Plant Cell Tissue Org Cult 107:25–33

Mamiya Y (1983) Pathology of the pine wilt disease by *Bursaphelenchus xylophilus*. Annu Rev Phytopathol 21:201–220

Mamiya Y, Enda N (1972) Transmission of *Bursaphelenchus lignicolus* (Nematoda: Aphelenchoididae) by *Monochamus alternus* (Coleoptera: Cerambycidae). Nematologica 18:159–162

Maruyama TE, Hosoi Y (2012) Post-maturation treatment improves and synchronizes somatic embryo germination of tree species of Japanese pines. Plant Cell Tissue Org Cult 110:45–52

Maruyama E, Tanaka T, Hosoi Y, Ishii K, Morohoshi N (2000) Embryogenic cell culture, protoplast regeneration, cryopreservation, biolistic gene transfer and plant regeneration in Japanese cedar (*Cryptomeria japonica* D. Don). Plant Biotechnol 17:281–296

Maruyama E, Hosoi Y, Ishii K (2002) Somatic embryogenesis in Sawara cypress (*Chamaecyparis pisifera* Sieb. et Zucc.) for stable and efficient plant regeneration, propagation and protoplast culture. J For Res 7:23–34

Maruyama E, Hosoi Y, Ishii K (2005a) Somatic embryo production and plant regeneration of Japanese black pine (*Pinus thunbergii*). J For Res 10:403–407

Maruyama E, Hosoi Y, Ishii K (2005b) Propagation of Japanese red pine (*Pinus densiflora* Zieb. et Zucc.). Prop Ornam Plants 4:199–204

Maruyama E, Ishii K, Hosoi Y (2005c) Efficient plant regeneration of Hinoki cypress (*Chamaecyparis obtusa* Sieb. et Zucc.) via somatic embryogenesis. J For Res 10:73–77

Maruyama E, Hosoi Y, Ishii K (2007) Somatic embryogenesis and plant regeration in yakutanegoyou, *Pinus armandii* Franch. var. *amamiana* (Koidz.) Hatusima, an endemic and endangered species in Japan. In Vitro Cell Dev Biol Plant 43:28–34

Miguel C, Gonçalves S, Tereso S, Marum L, Maroco J, Olivera MM (2004) Somatic embryogenesis from 20 open-pollinated families of Portuguese plus trees of maritime pine. Plant Cell Tissue Org Cult 76:121–130

Nagao A (1983) Differences of flower initiation of *Cryptomeria japonica* under various alternating temperatures. J Jpn For Soc 65:335–338 (in Japanese)

Park YS, Bonga JM, Cameron SI, Barrett JD, Forbes K, DeVerno LL, Klimaszewska K (1999) Somatic embryogenesis in jack pine (*Pinus banksiana* Lamb). In: Jain SM, Gupta PK, Newton RJ (eds) Somatic embryogenesis in woody plants, vol 4. Kluwer Academic, Dordrecht, pp 491–504

Park YS, Lelu-Walter MA, Harvengt L, Trontin JF, MacEacheron I, Klimaszewska K, Bonga JM (2006) Initiation of somatic embryogenesis in *Pinus banksiana*, *P. strobus*, *P. pinaster*, and *P. sylvestris* at three laboratories in Canada and France. Plant Cell Tissue Org Cult 86:87–101

Pullman GS, Namjoshi K, Zhang Y (2003) Somatic embryogenesis in loblolly pine (*Pinus taeda* L.): improving culture initiation with abscisic acid and silver nitrate. Plant Cell Rep 22:85–95

Roberts DR, Sutton BCS, Flinn BS (1990) Synchronous and high frequency germination of interior spruce somatic embryos following partial drying at high relative humidity. Can J Bot 68:1086–1090

Sato H, Sakuyama T, Kobayashi M (1987) Transmission of *Bursaphelenchus xylophilus* (Steiner et Buhrer) Nickle (Nematoda, Aphelenchoididae) by *Monochamus saltuarius* (Gebler) (Coleoptera, Cerambycidae). J Jpn For Soc 69:492–496 (in Japanese with English summary)

Shin D-I, Podila GK, Huang YH, Karnosky DF (1994) Transgenic larch expressing genes for herbicide and insect resistance. Can J For Res 24:2059–2067

Smith DR (1996) Growth medium. United States patent # 5,565,355

Stasolla C, Loukanina N, Ashihara H, Yeung EC, Thorpe TE (2001) Purine and pyrimidine metabolism during the partial drying treatment of white spruce (*Picea glauca*) somatic embryos. Physiol Plant 111:93–101

Taniguchi T (2001) Plant regeneration from somatic embryos in *Pinus thunbergii* (Japanese black pine) and *Pinus densiflora* (Japanese red pine). In: Morohoshi N, Komamine A (eds) Molecular breeding of woody plants. Elsevier Science, Amsterdam, pp 318–324

Walter C, Grace LJ, Wagner A, White DWR, Walden AR, Donaldson SS, Hinton H, Gardner RC, Smith DR (1998) Stable transformation and regeneration of transgenic plants of *Pinus radiata* D. Don. Plant Cell Rep 17:460–468

Somatic Embryogenesis in Sawara Cypress (*Chamaecyparis pisifera* Sieb. *et* Zucc.)

3

Yoshihisa Hosoi and Tsuyoshi E. Maruyama

Abstract

Seeds containing immature zygotic embryos were used as explants for embryogenic culture initiation. Embryogenic culture was maintained and proliferated by 2–3-week-interval subcultures. Embryogenic tissue suspended in 2–3 ml of medium was plated on 90-mm-diameter plates containing maturation medium. Mature cotyledonary somatic embryos were collected from the maturation medium and transferred to the germination medium. After 2–4 weeks of culture, most of the somatic embryos germinated and were converted into plantlets. Protoplasts were isolated from embryogenic tissue maintained for about 1 year. Proliferated embryogenic cells from cultured protoplasts were transferred to maturation medium, and after about 6 weeks of culturing, the cotyledonary somatic embryos were produced. Matured somatic embryos germinated and then converted into plantlets.

Keywords

Tissue culture • Protoplast culture • Suspension culture • Single cell • Suspensor cell • Embryogenic cell • Polyethylene glycol • Abscisic acid • Maltose

3.1 Introduction

Sawara cypress (*Chamaecyparis pisifera* Sieb. *et* Zucc.), a native tree species in Japan, is one of the six species in the genus *Chamaecyparis* found worldwide. This species grows 30-m tall with 1-m diameter at breast height and is mainly used for making barrels, buckets, traditional bathtubs, furniture, coffin, and as interior joinery material. In addition, it is prized as ornamental tree in parks and gardens and as one of the most popular house hedges. Moreover, this species is an important genetic resource for hybridization with a related hinoki cypress (*Chamaecyparis obtusa* Sieb. *et* Zucc.) which is one of the most important commercial timber trees in Japan, rep-

Y. Hosoi (✉) • T.E. Maruyama
Department of Molecular and Cell Biology, Forestry and Forest Products Research Institute, Matsunosato 1, Tsukuba 305-8687, Japan
e-mail: yh2884@ffpri.affrc.go.jp

A. Mujib (ed.), *Somatic Embryogenesis in Ornamentals and Its Applications*,
DOI 10.1007/978-81-322-2683-3_3

resenting about 25 % of the plantation area in the country (Fukuhara 1989).

A stable and efficient plant regeneration system is important not only for the effective propagation of selected trees but also for genetic transformation and somatic hybridization to develop disease-resistant hybrids. Somatic embryogenesis is the most attractive techniques for mass propagation and conservation of genetic resources by cryopreservation of the embryogenic tissues without changing its genetic makeup and the loss of juvenility (Park et al. 1998).

Since somatic embryogenesis and the plantlet regeneration of gymnosperm woody species were first reported in Norway spruce (Hakman et al. 1985; Chalupa 1985; Hakman and von Arnold 1985), successful studies in other conifers have been reported (Tautorus et al. 1991; Jain et al. 1995; Stasolla and Yeung 2003). However, for many species, the maturation of large numbers of somatic embryos and their subsequent plant conversion is sometimes difficult, and effective utilization remains problematic.

In this chapter, we describe a stable and efficient plant regeneration system in sawara cypress *via* somatic embryogenesis. The initiation of embryogenic cultures, their maintenance and proliferation, maturation of somatic embryos, germination and plant conversion, ex vitro acclimatization and growth in the field, and the protoplast culture are described in subsequent sections.

3.2 Embryogenic Culture Initiation

Immature open-pollinated cones (Fig. 3.1a) of sawara cypress were collected from mother trees in late June to early July. The collected cones were subsequently disinfected by 1-min immersion in 99.5 % ethanol and dried in the laminar flow cabinet before dissection. The excised seeds (Fig. 3.1b) were disinfected with 1 % (w/v available chlorine) sodium hypochlorite solution for 15 min and then rinsed five times with sterile distilled water. Seeds containing immature zygotic embryos were used as explants for embryogenic culture initiation.

The explants were cultured in 24-well plates (one per well) containing 1/2 MS medium (Murashige and Skoog 1962). MS medium with basal salts reduced to half the standard concentration (but replacing all NH_4NO_3 with 1000 mg l^{-1} glutamine) was supplemented with 10 μM 2,4-dichlorophenoxyacetic acid (2,4-D), 5 μM 6-benzylaminopurine (BA), and 10 g l^{-1} sucrose. The pH of the medium was adjusted to 5.8 prior to autoclaving for 15 min at 121 °C. The cultures were kept in darkness at 25 °C. The presence or absence of the distinct early stages of embryos characterized by an embryonal head with a suspensor system from the explants was observed under an inverted microscope weekly for up to 3 months.

Fig. 3.1 Collected open-pollinated cones (**a**) and excised seeds (**b**) of sawara cypress. *Bars* 5 mm

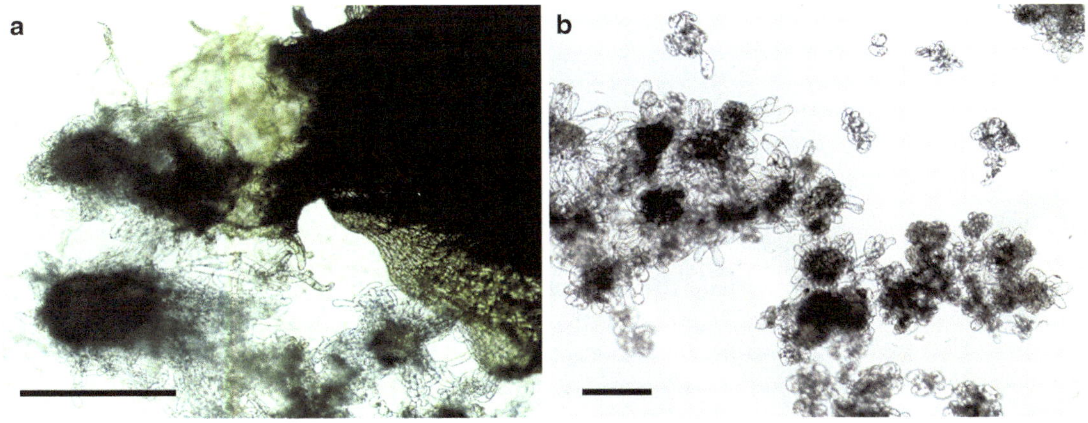

Fig. 3.2 Embryogenic tissue initiation (**a**) and proliferation (**b**) from seed of sawara cypress. *Bars* 500 μm

Fig. 3.3 Embryogenic response of four open-pollinated seed families of sawara cypress

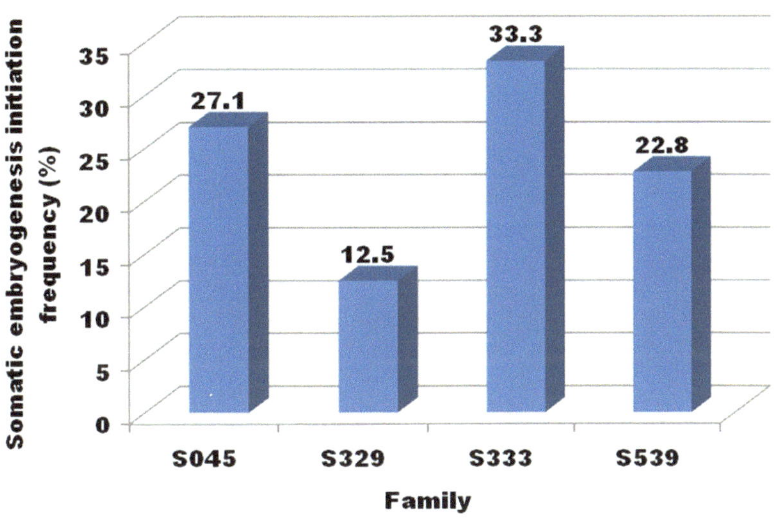

Embryogenic tissues (ET) extruding from the micropylar ends of explants appeared mostly after 2–4 weeks of culture (Fig. 3.2a), and the proliferation of ET was clearly confirmed after 4–6 of culture (Fig. 3.2b). About 7 weeks after culturing, the somatic embryogenesis initiation frequency varied from 12.5 to 33.3 % (Fig. 3.3). This result represents a relatively high somatic embryogenesis initiation frequency in comparison with the initiation range of 14.5–17.2 % reported in the related hinoki cypress (Maruyama et al. 2005).

3.3 Maintenance and Proliferation of Embryogenic Cultures

The maintenance and proliferation of embryogenic cultures (EC) were possible in several media containing a combination of 2,4-D (1–10 μM) plus BA (0–3 μM). The principal characteristics of these media were the reduction in the concentration of inorganic components from the standard and the addition of filter-sterilized glutamine as an organic nitrogen source. In our culture routines, EC were maintained and proliferated by 2–3-week-interval subcultures on 90-mm-diameter plates containing 1/2 MS medium or 1/2 LP medium (Aitken-Christie and Thorpe 1984) supplemented with 30 g l^{-1} sucrose, 3 μM 2,4-D, 1 μM BA, and 1.5 g l^{-1} glutamine. ET proliferated readily and retained their original translucent and mucilaginous appearance. In general, solid medium was used for the maintenance routine and liquid medium for rapid proliferation of the cultures. The low density in subculture routine helped maintain suitable conditions for EC (densely embryonal head with a distinct suspensor system) in the suspension culture. Before the maturation step, about 10–20 mg fresh weight (FW) of ET from the solid medium were transferred to a 100 ml flask containing 30–40 ml of medium (of a composition equivalent to that used for the maintenance and proliferation but without gellan gum) and cultured for about 2 weeks on a rotary shaker at 50–70 rpm, in darkness at 25 °C.

3.4 Maturation of Somatic Embryos

About 100 mg FW of ET suspended in 2–3 ml of medium was plated on 70-mm-diameter filter paper disks over 90-mm-diameter plates containing 30–40 ml of maturation medium. Maturation medium containing basal salts and vitamins from the original EM medium (Maruyama et al. 2000), sucrose or maltose, abscisic acid (ABA), activated charcoal (AC), polyethylene glycol 4000 (PEG), and amino acids (Smith 1996) (g l^{-1}: glutamine 7.3, asparagine 2.1, arginine 0.7, citrulline 0.079, ornithine 0.076, lysine 0.055, alanine 0.04, and proline 0.053). The plates were sealed with Novix-II film (Iwaki Glass Co., Ltd., Chiba, Japan) and kept in darkness at 25 °C for 6–12 weeks.

3.4.1 Effect of Kind and Concentration of Sugar

Figure 3.4 shows the effect of two different kinds of sugar on the maturation of somatic embryos. At the tested sugar concentrations, optimal results were achieved by using maltose as a carbohydrate source. Although 30 and 50 g l^{-1} did not result in a significant difference in terms of cotyledonary embryos per plate, the peak embryo maturation frequency resulted from the medium containing 50 g l^{-1} maltose with an average of 372 mature embryos. In contrast, when sucrose was used, 50 g l^{-1} resulted in a decrease of maturation frequency. Maltose has been considered a better carbohydrate source than sucrose or glucose for embryo maturation (Uddin et al. 1990; Uddin 1993). Similarly, a medium containing maltose as a carbohydrate source was reported as effective to enhance somatic embryo maturation in the loblolly pine (Li et al. 1998). This report inferred that about a tenfold enhancement was achieved by using maltose to replace sucrose and that the morphology of cotyledonary embryos was improved. In our results, although significant differences were achieved in terms of the somatic embryo maturation efficiency, the morphology of cotyledonary embryos induced on the medium with sucrose or maltose was similar.

3.4.2 Effects of ABA and AC

Figure 3.5 shows the beneficial effect of increased ABA content in medium supplemented with AC on the maturation of sawara cypress somatic embryos. The best result was achieved with 100 μM ABA in the presence of AC, obtaining an average of 348 cotyledonary embryos per plate. The higher the ABA concentration, the greater the number of mature embryos. A similar result was reported in *Pinus strobus* (Klimaszewska and Smith 1997), *Picea glauca-engelmannii* complex (Roberts et al. 1990a), and *P. glauca* (Dunstan et al. 1991). The addition of AC into the media notably

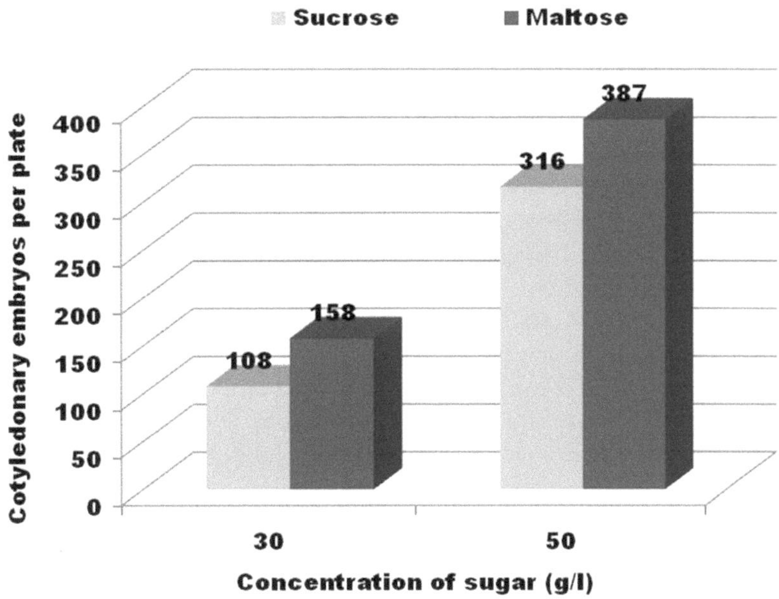

Fig. 3.4 Effect of kind of sugar on maturation of sawara cypress somatic embryos

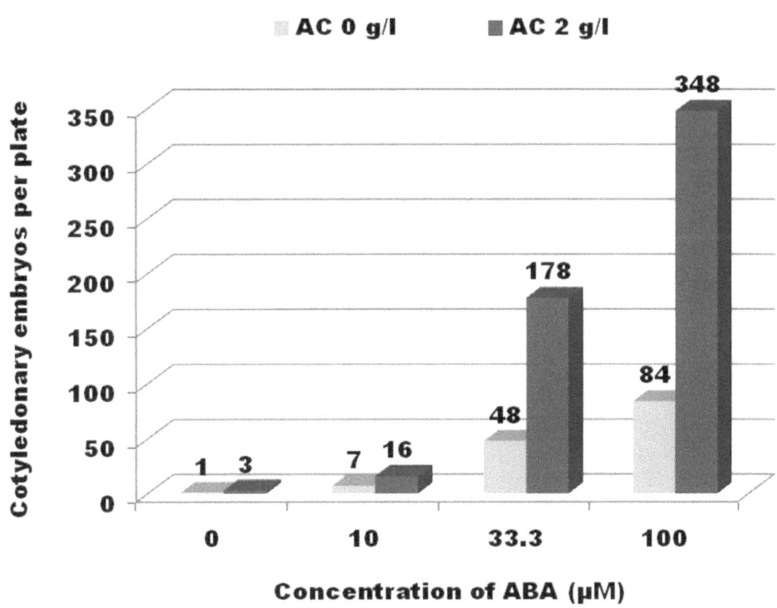

Fig. 3.5 Effect of ABA and AC on maturation of sawara cypress somatic embryos

enhanced the maturation efficiency, with around a fourfold enhancement achieved by using 33.3–100 μM in combination with 2 g l^{-1} AC. Pullman and Gupta (1991) reported further improved loblolly pine embryo development using a combination of ABA and AC, while Gupta et al. (1995) reported further improved quality of cotyledonary embryos of Douglas fir (*Pseudotsuga menziesii*) by a combination of ABA, AC, and PEG. Similarly, Lelu-Walter et al. (2006) indicated that coating the cells with AC reduced ET proliferation and significantly enhanced the maturation of maritime pine somatic embryos. AC is widely used in tissue culture media, where it is believed to function as an adsorbent for toxic metabolic products and residual hormones (von Aderkas et al. 2002; Pullman and Gupta 1991).

ABA-free medium or those supplemented with a low concentration (10 μM) failed to stimulate appropriately embryo maturation, producing only a few cotyledonary embryos (Fig. 3.5). ET on medium without ABA did not develop beyond the embryo stage 1, as described elsewhere (von Arnold and Hakman 1988). Most of the proembryos showed arrested development, whereas the proliferation of ET was evident. Lelu et al. (1999) reported that mature embryos of *Pinus sylvestris* and *P. pinaster* were produced in far higher numbers and that the development of cotyledonary somatic embryos versus abnormal shooty ones was enhanced with the addition of 60 μM ABA in comparison with media without ABA. Somatic embryos of the hybrid larch (*Larix* × *leptoeuropaea*) developed normally on a medium supplemented with 60 μM ABA but abnormally on a medium with no ABA (Gutmann et al. 1996). Most of the studies on somatic embryogenesis in conifers have reported ABA as a key hormone in embryo development and that the number and quality of embryo produced were vastly reduced in its absence (Durzan and Gupta 1987; von Arnold and Hakman 1988; Attree and Fowke 1993; Dunstan et al. 1998).

Several authors have suggested that the role of ABA in somatic embryogenesis is to inhibit cleavage polyembryony with the consequent development of individual somatic embryos (Durzan and Gupta 1987; Boulay et al. 1988; Krogstrup et al. 1988; Gupta et al. 1991), to stimulate the accumulation of nutrients, lipids, proteins, and carbohydrates (Hakman and von Arnold 1988) and suppress precocious germination (Roberts et al. 1990a). In addition, Gupta et al. (1993) reported improved desiccation tolerance to less than 10 % water content with 80–90 % germination rates in Norway spruce embryos produced with a combination of ABA and AC. The use of ABA for somatic embryo maturation in gymnosperms is extensively reported in the compilation of Jain et al. (1995).

3.4.3 Effect of PEG

As shown in Fig. 3.6, the addition of PEG stimulated the cotyledonary embryo production of sawara cypress, with a higher concentration of PEG in the medium resulting in a higher maturation frequency. The best result was obtained at a concentration of 150 g l^{-1} with an average number of 1043 cotyledonary embryos collected per plate, in comparison with 382, 215, and 13 embryos per plate at concentrations of 75, 50, and 0 g l^{-1}, respectively. In the absence of PEG, most of the proembryos did not develop into cotyledonary embryos. ET proliferation was evi-

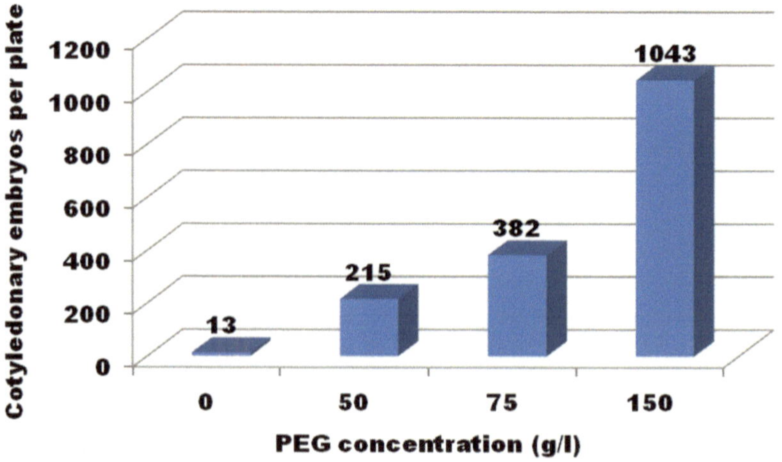

Fig. 3.6 Effect of PEG on maturation of sawara cypress somatic embryos

dent, and most of them developed into structures consisting of small embryonal heads from which elongated suspensors extended (stage 1 somatic embryos).

The use of PEG in combination with ABA has become routine for stimulating somatic embryo maturation in many gymnosperms. The beneficial effect of PEG on embryo maturation may be related to a water stress induction similar to that generated by desiccation (Attree and Fowke 1993); to an increase in the accumulation of storage reserves, such as storage proteins, lipids, and polypeptides (Roberts et al. 1990a; Attree et al. 1992; Misra et al. 1993); and to a tolerance to water loss (Attree et al. 1991). In contrast, some authors have reported that PEG promotes maturation but inhibits the further development of *Picea glauca* (Kong and Yeung 1995) and *P. abies* somatic embryos (Bozhkov and von Arnold 1998). In recent years, several studies have reported promotion of the maturation of somatic embryos by the addition of ABA into medium solidified with a high concentration of gellan gum in the absence of PEG (Klimaszewska and Smith 1997; Lelu et al. 1999). However, in our study, a high concentration of gellan gum in the absence of PEG was not effective in promoting the somatic embryo maturation of sawara cypress (data not shown).

3.5 Germination and Plant Conversion

Mature cotyledonary somatic embryos were collected from the maturation medium and transferred to the germination medium (1/2 LP or a 1/2 EM without PGR and supplemented with 2 g l^{-1} AC and 10 g l^{-1} agar). Cultures were kept at 25 °C under a photon flux density of about 65 μmol $m^{-2}s^{-1}$ with cooling and fluorescent lamps for 16 h daily.

The start of germination was observed as early as 3–5 days after transfer to the germination medium, and after 2–4 weeks of culture, most of the somatic embryos germinated and were converted into plantlets. As shows in

Table 3.1 Effect of PEG concentration in maturation medium on production, size, germination, and plant conversion of sawara cypress somatic embryos

PEG (g/l)	Embryos per plate	Size range (mm)	Germination frequency (%)	Conversion frequency (%)
0	13	3–10	97	92
50	215	2–8	98	93
75	382	2–6	97	93
150	1043	1–3	97	92

Table 3.1, the germination frequencies and plant conversion rates of sawara cypress were similar in somatic embryos derived from different PEG-treated media. Morphological difference among somatic embryos obtained on media supplemented with different concentrations of PEG was restricted to the size of them. The higher the PEG concentration, the smaller the size of mature embryos (Fig. 3.7). Nevertheless, the embryo size was not found to be influential in germination and subsequently plant conversion. Cotyledonary embryos germinated and converted into plants at high frequencies independent of their size (Table 3.1).

Regenerated plants were transferred to 300 ml flasks containing 100 ml of fresh medium (same composition used for the germination but with 30 g l^{-1} sucrose and 5 g l^{-1} AC) and kept under the same conditions described above for 8–12 weeks before ex vitro acclimatization.

3.6 Ex Vitro Acclimatization and Growth of Somatic Plants

The developed plants were transplanted into plastic pots filled with vermiculite and acclimatized inside a growth chamber at 25 °C in 80 % relative humidity. During the first 2 weeks, plants were kept under high relative humidity in plastic boxes with transparent plastic covers and irrigated with tap water one time per week. Subsequently, the cover was gradually opened, and the pots were fertilized with a nutrient solution modified from Nagao (1983) containing in

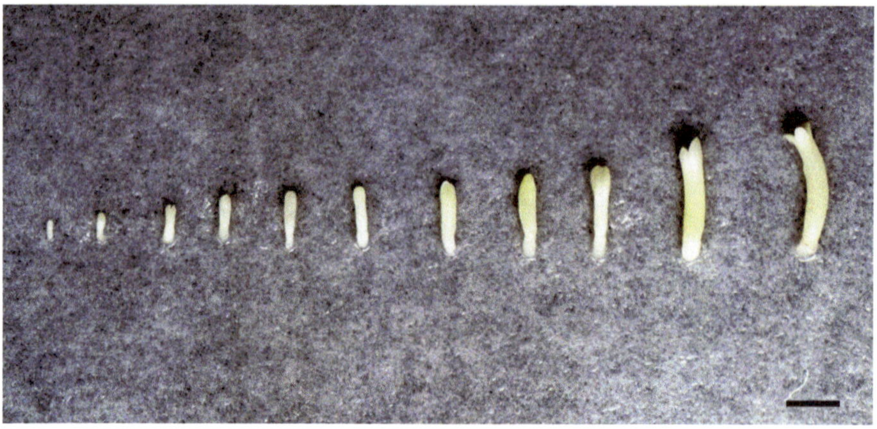

Fig. 3.7 Different somatic embryo sizes of sawara cypress in function to the concentration of PEG in maturation media. *Bar* 5 mm

Fig. 3.8 Somatic plants of sawara cypress growing in the field. *Bar* 1 m

mg l^{-1}: NH_4NO_3 143, $NaH_2PO_4 \cdot 2H_2O$ 55.1, KCl 47.1, $CaCl_2 \cdot 2H_2O$ 52.5, $MgSO_4 \cdot 7H_2O$ 61, Fe(III) EDTA 25, Cu EDTA 0.1, Mn EDTA 0.1, Zn EDTA 0.1, H_3BO_3 1.5, KI 0.01, $CoCl_2 \cdot 6H_2O$ 0.005, and MoO_3 0.005. The covers were completely removed about 4 weeks after transplanting.

The acclimatized plants were transferred to a greenhouse and grown under controlled conditions for 6–8 months before transplanting them to the growth test field. One hundred somatic plants and ten seedlings were planted in the field at a 1.0 × 1.0 m spacing. To calculate the mean growth, the height of 30 somatic trees and 7 seedlings was measured every 2 years during the age of 4–10 years (Fig. 3.8). Although no significant difference was observed in the mean height until the age of 4 years, the growth of the seedlings was markedly superior to that of somatic trees after 6 years of age (Fig. 3.9).

The reason for that was not explored in this study; however, we speculate that the difference in growth can be attributed to the culture conditions of somatic plants. Evidences that

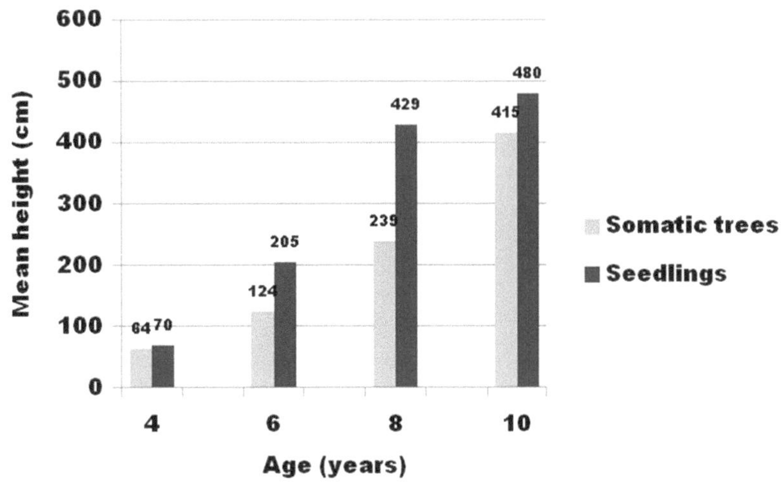

Fig. 3.9 Increment in the mean height of somatic trees versus seedlings of sawara cypress growing in the field during 4–10 years of growth

culture conditions during somatic embryo development may reduce the growth of somatic plants were reported for several conifers (Högberg et al. 2003). The use of PEG and long-term contact with ABA during maturation substantially reduced the growth of regenerated Norway spruce plants (Bozhkov and von Arnold 1998; Högberg et al. 2001). Similarly, inferior growth increment in somatic plants versus that in seedlings has reported in Douglas fir (Dean et al. 2008) and Scots pine (Niskanen et al. 2008).

3.7 Protoplast Culture

Protoplasts were isolated from ET maintained for about 1 year. Cells from 14- to 21-day-old embryogenic suspension cultures were collected on 100 μm nylon Falcon cell strainer (Becton, Dickinson & Co., Massachusetts, USA) and incubated in an enzyme solution containing 1 % (w/v) Cellulase Onozuka RS (Yakult Honsha Co., Ltd., Tokyo, Japan) (RS), 0.1 % (w/v) Pectolyase Y-23 (Kikkoman Corp., Tokyo, Japan) (P), and 0.6 M mannitol for 5 h at 25 °C. The digest was filtered through 40-μm cell strainer and then centrifuged at 100 × g for 3 min. The supernatant was removed, and the protoplasts were resuspended in 0.6 M mannitol solution and centrifuged once more. The yield of protoplasts was counted with hemocytometer, and their viability was verified using fluorescein diacetate (FDA). Protoplasts were cultured at a density of 2×10^3 protoplasts per ml in 96-well plates containing 50 μl of 0.6 M mannitol proliferation media supplemented with 2,4-D (1, 3, 10, 30, and 100 μM). Plates containing protoplast cultures in wells and sterile distilled water in outspaces among wells (to maintain relative humidity) were sealed with Novix-II film and kept at 28 °C in darkness. The plating efficiency percentages (number of colonies/initial plating density × 100) were determined after 6 weeks of culturing.

The protoplast viability examination by FDA staining showed that up to more than 90 % of protoplasts survived after isolation. This result indicated that the combination of RS (1 %) and P (0.1 %) in the presence of 0.6 M mannitol as osmotic agent was effective for protoplast isolation (Fig. 3.10a) from embryogenic cells of sawara cypress. Subsequently, cell division (Fig. 3.10b) was mostly observed 1–2 weeks after isolation, and after 3–4 weeks of culture in media under the conditions described above, colony formation (Fig. 3.10c) was achieved in all concentrations of 2,4-D but with most efficiency at concentrations of 10 μM (Fig. 3.11). About 1–2 weeks after, the initiation of elongated suspensor cells was observed (Fig. 3.12a). The formation of vacuolated suspensor cells increased

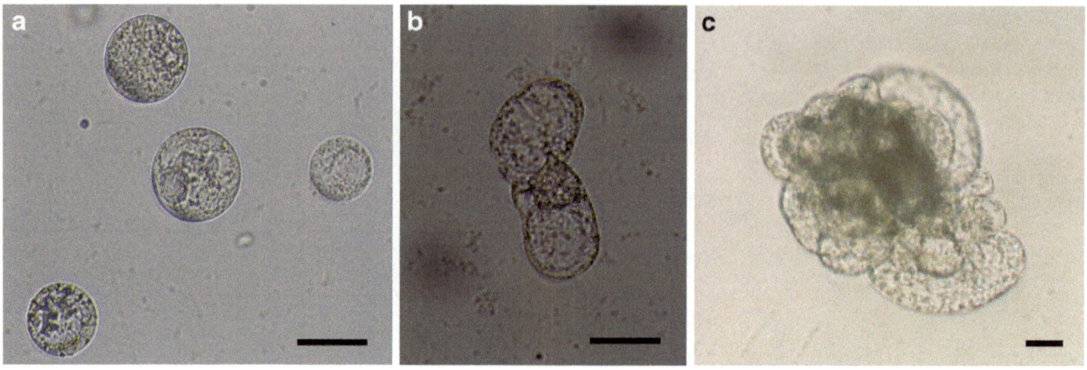

Fig. 3.10 Isolation (**a**), cell division (**b**), and colony formation (**c**) in cultured protoplast of sawara cypress. *Bars* 50 μm

Fig. 3.11 Effect of 2,4-D concentration on the plating efficiency percentages of sawara cypress protoplasts

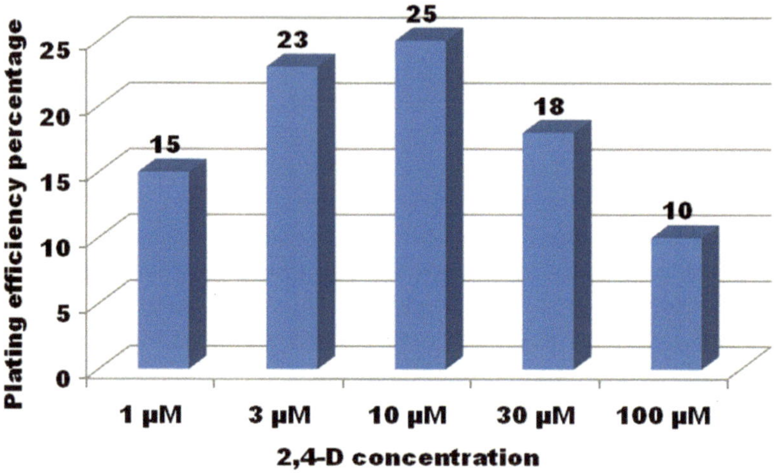

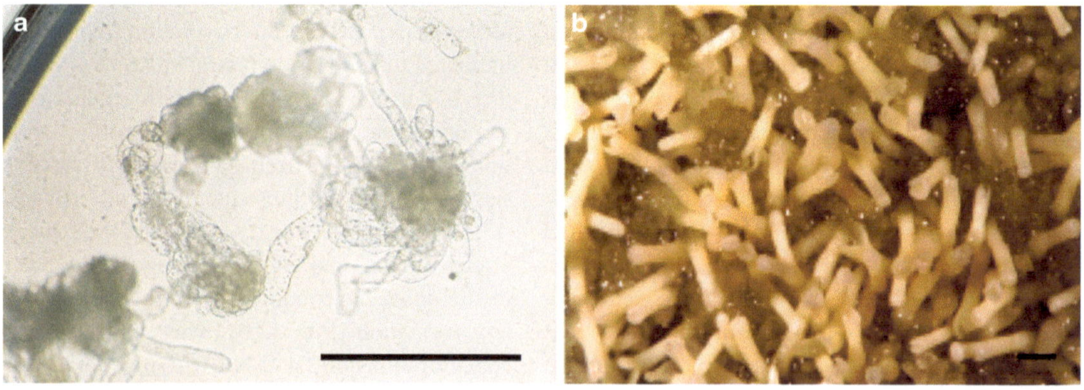

Fig. 3.12 Embryogenic cell proliferation (**a**) and somatic embryos production (**b**) from cultured protoplast of sawara cypress. *Bars* 1 mm

Fig. 3.13 Germination (**a**) and plantlet conversion (**b**) of somatic embryos produced from cultured protoplasts of sawara cypress. *Bars* 1 cm

over time, and after 6–7 weeks of culturing, the proliferation of embryogenic cells was evident resembling that of the original ET.

Subsequently, proliferated embryogenic cells from cultured protoplasts were transferred to maturation medium, and after about 6 weeks of culturing, the cotyledonary somatic embryos were produced (Fig. 3.12b). Matured somatic embryos germinated (Fig. 3.13a) and then converted into plantlets (Fig. 3.13b) about 1 week and 1 month after transfer to PGR-free medium, respectively.

Establishment of an efficient protoplast culture system represents an important tool in fundamental research on the synthesis of cell walls, membrane transports, and the cytoskeleton in relation to the cell cycle and division (Klimaszewska 1995). In addition, the single cell culture system represents a powerful tool for studies on the physiology of different cell types, analysis of differentiation programs, genetic manipulation of plant cells, and cell-cell interactions (Spangenberg et al. 1986; Schweiger et al. 1987).

3.8 Concluding Remarks

An effective plant regeneration system has been achieved for sawara cypress *via* somatic embryogenesis. A high somatic embryo maturation efficiency was obtained when ET were cultured on the medium supplemented with a combination of maltose, PEG, ABA, and AC. More than 1000 cotyledonary embryos per plate were produced on a maturation medium supplemented with 150 g l^{-1} PEG (Maruyama et al. 2002). In addition to high somatic embryo maturation efficiency, the subsequent high germination and plant conversion frequencies attained demonstrated the high quality of the somatic embryos produced. Cotyledonary somatic embryos readily germinated after transfer to a PGR-free medium without any kind of post-maturation treatment, as was previously reported as necessary to promote the germination of somatic embryos of some other species (Roberts et al. 1990b; Roberts et al. 1991; Kong and Yeung 1992; Kong and Yeung 1995; Jones and van Staden 2001). Thus, almost all mature somatic embryos germinated and developed normal plants.

On the other hand, the achieved differences in growth performance between somatic trees and seedlings represent an important parameter for improving the culture protocol. Post-effects of culture conditions could be avoided by modifying the procedure of somatic embryogenesis (Högberg et al. 2001). The epicotyl length and the presence of lateral roots as combined selection criteria at ex vitro transfer could be used to identify somatic plants having height growth characteristics comparable to those of seedlings. Early selection according to these criteria improves clonal performance and reduces intraclonal variation in somatic plants of Norway spruce (Högberg et al. 2003).

In addition, a simple procedure for protoplast culture to develop an efficient technique for individual culture of single cell was described. This technique offers significant advantages for improvement of tissue culture system by identification and selection of desirable cells, as well as for genetic manipulation through direct gene transfer such by microinjection and electroporation.

The described somatic embryogenesis system should permit, in the near future, the mass propagation of selected trees and the genetic engineering of this species.

References

Aitken-Christie J, Thorpe TA (1984) Clonal propagation: gymnosperms. In: Vasil IK (ed) Cell culture and somatic cell genetics of plants, vol 1. Academic, San Diego, pp 82–959

Attree SM, Fowke LC (1993) Somatic embryogenesis and synthetic seeds of conifers. Plant Cell Tiss Org Cult 35:1–35

Attree SM, Moore D, Sawhney VK, Fowke LC (1991) Enhanced maturation and desiccation tolerance of white spruce [*Picea glauca* (Moench) Voss] somatic embryos: effects of a non-plasmolysing water stress and abscisic acid. Ann Bot 68:519–525

Attree SM, Pomeroy MK, Fowke LC (1992) Manipulation of conditions for the culture of somatic embryos of white spruce for improved triacylglycerol biosynthesis and desiccation tolerance. Planta 187:395–404

Boulay MP, Gupta PK, Krogstrup P, Durzan DJ (1988) Development of somatic embryos from cell suspension cultures of Norway spruce (*Picea abies* Karst.). Plant Cell Rep 7:134–137

Bozhkov PV, von Arnold S (1998) Polyethylene glycol promotes maturation but inhibits further development of *Picea abies* somatic embryos. Physiol Plant 104:211–224

Chalupa V (1985) Somatic embryogenesis and plantlet regeneration from cultured immature and mature embryos of *Picea abies* (L.). Karst. Commun Inst For Cech 14:57–63

Dean CA, Welty DE, Herold GE (2008) Performance and genetic parameters of somatic and zygotic progenies of Coastal Douglas-fir at 7.5-years across Washington and Oregon, USA. Silvae Genet 58:212–219

Dunstan DI, Bethune TD, Abrams SR (1991) Racemic abscisic acid and abscisyl alcohol promote maturation of white spruce (*Picea glauca*) somatic embryos. Plant Sci 76:219–228

Dunstan DI, Bekkaoui F, Pilon M, Fowke LC, Abrams SR (1998) Effects of abscisic acid and analogues on the maturation of white spruce (*Picea glauca*) somatic embryos. Plant Sci 58:77–84

Durzan DJ, Gupta PK (1987) Somatic embryogenesis and polyembryogenesis in Douglas-fir cell suspension cultures. Plant Sci 52:229–235

Fukuhara N (1989) Fertility in interspecific-crossing between hinoki (*Chamaecyparis obtusa* Endl.) and Sawara (*C. pisifera* Endl.) and identification of the hybrids. Bull For For Prod Res Inst 354:1–38

Gupta PK, Timmis R, Pullman G, Yancey M, Kreitinger M, Carlson W, Carpenter C (1991) Development of an embryogenic system for automated propagation of forest trees. In: Vasil IK (ed) Cell culture and somatic cell genetics of plants, vol 8. Academic, San Diego, pp 75–93

Gupta PK, Pullman G, Timmis R, Kreitinger M, Carlson W, Grob J, Welty E (1993) Forestry in the 21st century: the biotechnology of somatic embryogenesis. Bio/Technology 11:454–459

Gupta PK, Timmis R, Timmis KA, Carlson WC, Welty EDE (1995) Somatic embryogenesis in Douglas-fir (*Pseudotsuga menziesii*). In: Jain SM et al (eds) Somatic embryogenesis in woody plants, vol 3. Kluwer Academic, Dordrecht, pp 303–313

Gutmann M, von Aderkas P, Label P, Lelu M-A (1996) Effects of abscisic acid on somatic embryo maturation of hybrid larch. J Exp Bot 35:1905–1917

Hakman I, Arnold S (1988) Somatic embryogenesis and plant regeneration from suspension cultures of *Picea glauca* (white spruce). Physiol Plant 72:579–587

Hakman I, von Arnold S (1985) Plantlet regeneration through somatic embryos in *Picea abies* (Norway spruce). J Plant Physiol 121:149–158

Hakman I, Fowke LC, von Arnold S, Eriksson T (1985) The development of somatic embryos in tissue cultures initiated from immature embryos of *Picea abies* (Norway spruce). Plant Sci 38:53–59

Högberg KA, Bozhkov P, Grönroos R, von Arnold S (2001) Critical factors affecting ex vitro performance of somatic embryo plants of *Picea abies*. Scand J For Res 16:295–304

Högberg KA, Bozhkov P, von Arnold S (2003) Early selection improves clonal performance and reduces intraclonal variation of Norway spruce plants propagated by somatic embryogenesis. Tree Physiol 23:211–216

Jain SM, Gupta PK, Newton R (1995) Somatic embryogenesis in woody plants, vol 3. Kluwer Academic, Dordrecht

Jones NB, van Staden J (2001) Improved somatic embryo production from embryogenic tissue of *Pinus patula*. In Vitro Cell Dev Biol Plant 37:543–549

Klimaszewska K (1995) Somatic embryogenesis in *Picea mariana* (Mill.). In: Jain SM et al (eds) Somatic Embryogenesis in Woody Plants, vol 3. Kluwer Academic, Dordrecht, pp 67–79

Klimaszewska K, Smith DR (1997) Maturation of somatic embryos of *Pinus strobus* is promoted by a high concentration of gellan gum. Physiol Plant 100:949–957

Kong L, Yeung E (1992) Development of white spruce somatic embryos: II. Continual shoot meristem development during germination. In vitro Cell Dev Biol 28P:125–131

Kong L, Yeung E (1995) Effects of silver nitrate and polyethylene glycol on white spruce (*Picea glauca*) somatic embryo development: enhancing cotyledonary embryo formation and endogenous ABA content. Physiol Plant 93:298–304

Krogstrup P, Eriksen EN, Moller JD, Roulund H (1988) Somatic embryogenesis in Sitka spruce (*Picea sitchensis* (Bong.) Carr.). Plant Cell Rep 7:594–597

Lelu M-A, Bastien C, Drugeault A, Gouez M-L, Klimaszewska K (1999) Somatic embryogenesis and plantlet development in *Pinus sylvestris* and *Pinus pinaster* on medium with and without growth regulators. Physiol Plant 105:719–728

Lelu-Walter MA, Bernier-Cardou M, Klimaszewska K (2006) Simplified and improved somatic embryogenesis for clonal propagation of *Pinus pinaster*. Plant Cell Rep 25:767–776

Li XY, Huang FH, Murphy JB, Gbur EEJR (1998) Polyethylene glycol and maltose enhance somatic embryo maturation in loblolly pine (*Pinus taeda* L.). In vitro Cell Dev Bio Plant 34:22–26

Maruyama E, Tanaka T, Hosoi Y, Ishii K, Morohoshi N (2000) Embryogenic cell culture, protoplast regeneration, cryopreservation, biolistic gene transfer and plant regeneration in Japanese cedar (*Cryptomeria japonica* D. Don). Plant Biotechnol 17:281–296

Maruyama E, Hosoi Y, Ishii K (2002) Somatic embryogenesis in Sawara cypress (*Chamaecyparis pisifera* Sieb. et Zucc.) for stable and efficient plant regeneration, propagation and protoplast culture. J For Res 7:23–34

Maruyama E, Ishii K, Hosoi Y (2005) Efficient plant regeneration of Hinoki cypress (*Chamaecyparis obtusa*) via somatic embryogenesis. J For Res 10:73–77

Misra S, Attree SM, Leal I, Fowke LC (1993) Effects of abscisic acid, osmoticum and desiccation on synthesis of proteins during the development of white spruce somatic embryos. Ann Bot 71:11–22

Murashige T, Skoog F (1962) A revised medium for rapid growth and bioassays with tobacco cultures. Physiol Plant 15:473–497

Nagao A (1983) Differences of flower initiation of *Cryptomeria japonica* under various alternating temperatures. J Jpn For Soc 65:335–338 (in Japanese)

Niskanen AM, Stenvall N, Pakkanen A, Pulkkinen P (2008) Comparison of growth and stem form characters of *Pinus sylvestris* clones and seedlings of the same origin in a 10-year field trial. Scand J For Res 23:484–490

Park YS, Barrett JD, Bonga JM (1998) Application of somatic embryogenesis in high-value clonal forestry: deployment, genetic control, and stability of cryopreserved clones. In vitro Cell Dev Biol Plant 34:231–239

Pullman GS, Gupta PK (1991) Method for reproducing coniferous plants by somatic embryogenesis using absorbent materials in the development stage media. U.S. Patent No. 5,034,326

Roberts DR, Flinn BS, Webb DT, Webster FB, Sutton BCS (1990a) Abscisic acid and indole-3-butyric acid regulation of maturation and accumulation of storage proteins in somatic embryos of interior spruce. Physiol Plant 78:355–360

Roberts DR, Sutton BCS, Flinn BS (1990b) Synchronous and high frequency germination of interior spruce somatic embryos following partial drying at high relative humidity. Can J Bot 68:1086–1090

Roberts DR, Lazaroff WR, Webster FB (1991) Interaction between maturation and high relative humidity treatments and their effects on germination of Sitka spruce somatic embryos. J Plant Physiol 138:1–6

Schweiger H-G, Dirk J, Koop H-U, Kranz E, Neuhaus G, Spangenberg G, Wolff D (1987) Individual selection, culture and manipulation of higher plant cells. Theor Appl Genet 73:769–783

Smith DR (1996) Growth medium. U. S. Patent No. 5,565,455

Spangenberg G, Koop H-U, Lichter R, Schweiger HG (1986) Microculture of single protoplasts of *Brassica napus*. Physiol Plant 66:1–8

Stasolla C, Yeung EC (2003) Recent advances in conifer somatic embryogenesis: improving somatic embryo quality. Plant Cell Tiss Org Cult 74:15–35

Tautorus TE, Fowke LC, Dunstan DI (1991) Somatic embryogenesis in conifers. Can J Bot 69:1873–1899

Uddin M. (1993) Somatic embryogenesis in Gymnosperms. U.S. Patent No. 5,187,092

Uddin MR, Dinus RJ, Webb DT (1990) Effects of different carbohydrates on maturation of *Pinus taeda* somatic embryos. Abstracts VII International Congress on Plant Tissue and Cell Culture, Amsterdam, Netherlands, p 272

von Aderkas P, Label P, Lelu MA (2002) Charcoal effects on early developmental and hormonal levels of somatic embryos of hybrid larch. Tree Physiol 22:431–434

von Arnold S, Hakman I (1988) Regulation of somatic developments in *Picea abies* by abscisic acid (ABA). J Plant Physiol 132:164–169

Optimizing Factors Affecting Somatic Embryogenesis in Cineraria

4

Iyyakkannu Sivanesan and Byoung Ryong Jeong

Abstract

We established a reproducible protocol for somatic embryogenesis and plant regeneration of cineraria. 2,4-Dichlorophenoxy acetic acid (2,4-D) had a significant effect on somatic embryo formation. Addition of cytokinins such as 2-isopentenyladenine (2-iP), 6-benzyladenine (BA), and thidiazuron (TDZ) to the 2,4-D containing medium enhanced the frequency of somatic embryo induction and average number of somatic embryos per explant. However, the nature of SE varied depending on combination of 2,4-D and cytokinins. Cotyledon and leaf explants developed somatic embryos on Murashige and Skoog (MS) medium supplemented with 3.0 mg l^{-1} 2,4-D and 1.0 mg l^{-1} BA. Among the two explants, leaves were found to be the most effective for somatic embryogenesis and subsequent plant regeneration. Most of the embryos developed from the cotyledon explants showed precocious germination. Furthermore, somatic embryos obtained from the cotyledon explants developed hyperhydric shoots. Thus, induction and development of SE in cineraria is also affected by the age of the explants. Globular embryos developed into normal plantlets through heart, torpedo, and cotyledonary stages, similar to zygotic embryos, when cultured on MS medium supplemented with gibberellic acid (GA_3). The in vitro-developed plantlets were successfully acclimatized in the greenhouse with 98 % survival.

Keywords

Ammonium nitrate • Auxin • Benzyladenine • Gibberellic acid • Ornamental plants • Plant regeneration • Temperature

I. Sivanesan (✉)
Present address: Department of Molecular Biotechnology, Konkuk University,
1, Hwayang-dong, Gwangjin-gu, Seoul 143 701, Republic of Korea
e-mail: isivanesan@yahoo.com

B.R. Jeong
Institute of Agriculture and Life Science, Department of Horticulture, Division of Applied Life Science (BK21 Plus), Gyeongsang National University, Jinju 660 701, Republic of Korea

A. Mujib (ed.), *Somatic Embryogenesis in Ornamentals and Its Applications*,
DOI 10.1007/978-81-322-2683-3_4

4.1 Introduction

Somatic embryogenesis is the process of histodifferentiation for regeneration of complete plants by haploid or diploid somatic cells which resemble zygotic embryos (Williams and Maheswaran 1986). The process of histodifferentiation of a somatic cell to an embryogenic one is mediated by abiotic stress imposed by culture medium composition and subculture interval (Feher 2008). A number of factors including type and composition of plant growth regulators (PGRs) (Sivanesan et al. 2011), light conditions, explant source (Sharma and Rajam 1995), and genotype (Kim et al. 2003) can influence the process of successful somatic embryogenesis. Several types of stress, such as wounding, heavy metal ions, osmotic stress, high salt concentrations, and high levels of PGRs, have been suggested to induce the somatic embryo (SE) in plants (Zavattieri et al. 2010). Thus, somatic embryogenesis is a process that is normally suppressed in plants, and stress treatments are derepressing the somatic embryogenesis process. Somatic embryogenesis is a multistep regeneration process in plants starting with proembryogenic masses that progress to the next stages maintaining the bipolar pattern of cell. Auxins and cytokinins are essential to maintain proembryogenic proliferation (Xu et al. 2013), whereas embryo formation is triggered by the withdrawal of PGRs (Bozhkov et al. 2002). The further development of early somatic embryos and maturation requires abscisic acid (Rai et al. 2011; Su et al. 2012). Initiation of SE is strongly controlled by genetic additive effects, and as a result plants of some families' respond well in propagation, while others are difficult to propagate by somatic embryogenesis (Salvo et al. 2014).

Cineraria are attractive pot plants belonging to the family Asteraceae. *Senecio cruentus* is an ornamental species; it originates in Canary Islands and is widely cultivated for its colorful flowers. Cineraria can be easily propagated by seeds, but seedlings are heterozygous producing flowers of different colors (Malueg et al. 1994). It is propagated vegetatively by cuttings; however, shortage of plant materials has often been affecting the large-scale commercial propagation (Sivanesan and Jeong 2012). Thus, propagation of new selections could be achieved more rapidly using in vitro propagation methods. In vitro clonal propagation of ornamental plants is an important technique that can be applied for large-scale production of select plants or cultivars. Clonal propagation through somatic embryogenesis has become an essential method for the improvement of most economically important plants. Plantlet regeneration through somatic embryogenesis is also a potential tool for the production of synthetic seeds and genetic transformation. In addition, this process is also considered as a most suitable system for physiological and morphological studies in plant during morphological development and conservation of desired genotypes through mass propagation (Lelu-Walter et al. 2013). Effective procedures for the initiation of somatic embryogenesis from mature plants permit more rapid propagation of elite plants, thus increasing genetic gains in each breeding generation (Dey et al. 2015). Direct somatic embryogenesis reduces the time required for plant propagation, which may be beneficial to minimize culture-induced genetic changes (Sivanesan et al. 2011). Somatic embryogenesis and plant regeneration have been reported in *Senecio* species (Malueg et al. 1989, 1994; Nam et al. 2005), but low frequency of embryo induction and conversion, abnormal embryos, and hyperhydricity hinder the application of these protocols for large-scale commercial propagation.

The addition of auxin to the culture medium is required to induce somatic embryos in most plant species. However, in some plant species, the inclusion of cytokinin alone is sufficient to induce somatic embryos. In cineraria, addition of both auxin and cytokinin to the culture medium is required for the induction of SE (Malueg et al. 1994; Nam et al. 2005). The maturation, germination, and conversion of SE into plantlets are other important steps during somatic embryogenesis. Several chemical and physical factors affect the development and conversion of SE such as abscisic acid (ABA), gibberellic acid (GA_3), polyethylene glycol (PEG), sucrose, light intensity, humidity, and temperature. We optimized the conditions for somatic embryo induction and studied the effect of ABA, PEG, and sucrose concentration on embryo maturation and germination (Nam et al. 2005). In this chapter, we describe an efficient procedure for somatic embryogenesis and plant regeneration from cotyledon and leaf explants in *S. cruentus*.

4.2 Somatic Embryogenesis

Seeds of *S. cruentus* 'Tokyo Daruma' (Sakata Seed Co., Yokohama, Japan) were sterilized with 70 % (v/v) ethanol for 30 s and 2.0 % (v/v) sodium hypochlorite for 10 min. Each treatment was followed by four rinses with sterile distilled water, and the seeds were germinated on Murashige and Skoog (MS) medium containing 3 % (w/v) sucrose and 0.8 % (w/v) agar. The pH of the medium was adjusted to 5.8 before autoclaving at 121 °C for 20 min. The cultures were maintained at 25 ± 2 °C under a16 h photoperiod with 15 μmol m^{-2} s^{-1} photosynthetic photon flux density (PPFD). Cotyledon (0.5 cm) and leaf (0.5–1.0 cm) explants were taken from 7- to 21-day-old seedlings, respectively. For SE induction, the explants were cultured on MS supplemented with various concentrations of auxins [2,4-dichlorophenoxyacetic acid (2,4-D), indole-3-butyric acid (IBA), or α-naphthaleneacetic acid (NAA)] alone or in combination with cytokinins (2-isopentenyladenine (2-iP), 6-benzyladenine (BA), or thidiazuron (TDZ)] (Table 4.1). The cultures were maintained for 2 weeks at 25 ± 2 °C in the dark and then exposed to light of 45 μmol m^{-2} s^{-1} PPFD provided by cool white fluorescent light with a light/dark cycle of 16/8 h for 4 weeks. To study the effects of ammonium nitrate (NH_4NO_3) and temperature on SE induction, cotyledon and leaf explants were cultured on optimal SE induction medium (MS + 3.0 mg l^{-1} 2,4-D + 1.0 mg l^{-1} BA)

Table 4.1 Effects of plant growth regulators on somatic embryogenesis from cotyledon and leaf explants of *S. cruentus* after 6 weeks of culture

PGRs (mg l^{-1})						Somatic embryo induction (%)		No. of somatic embryos/explant	
2,4-D	IBA	NAA	2iP	BA	TDZ	Cotyledon	Leaf	Cotyledon	Leaf
0						0.0 ± 0.0	0.0 ± 0.0	0.0 ± 0.0	0.0 ± 0.0
1.0						13.3 ± 2.2	18.0 ± 3.0	14.0 ± 1.6	28.3 ± 1.7
2.0						37.7 ± 1.9	39.5 ± 2.4	26.7 ± 1.2	39.0 ± 2.2
3.0						52.8 ± 3.5	66.8 ± 3.7	32.0 ± 1.7	45.0 ± 1.6
3.0			1.0			84.7 ± 1.2	87.6 ± 2.6	63.6 ± 3.3	84.3 ± 4.6
3.0				1.0		93.6 ± 3.4	97.0 ± 1.4	87.3 ± 2.1	135.3 ± 5.7
3.0					1.0	92.0 ± 2.4	95.6 ± 1.0	77.3 ± 1.7	101.6 ± 2.8
	1.0					Root	Root	Root	Root
	2.0					Root	Root	Root	Root
	3.0					Callus	Callus	Callus	Callus
	3.0		1.0			84.7 ± 1.2	92.0 ± 2.0	5.3 ± 0.9	7.3 ± 0.5
	3.0			1.0		84.7 ± 1.2	92.0 ± 2.0	7.0 ± 1.4	11.0 ± 0.8
	3.0				1.0	84.7 ± 1.2	92.0 ± 2.0	9.0 ± 0.8	13.7 ± 1.7
		1.0				Root	Root	Root	Root
		2.0				Callus + root	Callus + root	Callus + root	Callus + root
		3.0				Callus	Callus	Callus	Callus
		3.0	1.0			84.7 ± 1.2	92.0 ± 2.0	6.0 ± 1.6	19.6 ± 1.3
		3.0		1.0		84.7 ± 1.2	92.0 ± 2.0	12.0 ± 1.6	23.7 ± 2.6
		3.0			1.0	87.2 ± 4.6	92.0 ± 2.0	15.3 ± 0.9	21.6 ± 1.2

Values are mean ± standard deviation (SD) of results

Table 4.2 Effects of ammonium nitrate concentration and temperature on somatic embryogenesis from cotyledon and leaf explants of *S. cruentus*

NH_4NO_3		Somatic embryo induction (%)		No. of somatic embryos/explant	
(mg l^{-1})	Temperature (°C)	Cotyledon	Leaf	Cotyledon	Leaf
0		18.2 ± 1.8	27.7 ± 3.3	2.3 ± 0.9	4.6 ± 1.7
412.5	15	29.6 ± 1.7	36.7 ± 1.7	5.6 ± 0.9	7.7 ± 0.5
825.0		52.7 ± 3.7	58.0 ± 1.6	6.3 ± 1.7	9.3 ± 0.9
1650		62.3 ± 3.7	67.0 ± 1.6	9.7 ± 0.5	14.0 ± 0.8
0		38.7 ± 2.1	40.0 ± 1.6	5.6 ± 1.2	6.3 ± 1.2
412.5	20	62.3 ± 2.5	71.6 ± 2.1	13.7 ± 1.2	18.0 ± 1.0
825.0		76.3 ± 1.7	84.7 ± 1.2	17.0 ± 0.8	23.3 ± 2.5
1650		87.0 ± 2.2	91.3 ± 1.6	21.7 ± 3.3	29.7 ± 1.3
0		44.0 ± 1.6	39.3 ± 2.7	12.3 ± 1.7	16.0 ± 2.2
412.5	25	63.0 ± 2.2	75.3 ± 1.2	34.0 ± 1.6	55.6 ± 2.6
825.0		86.3 ± 3.8	90.6 ± 2.5	53.0 ± 3.7	76.6 ± 2.5
1650		93.6 ± 3.4	97.0 ± 1.4	87.3 ± 2.1	135.3 ± 5.7

Values are mean ± standard deviation (SD) of results

with 0, 412.5, 825.0, or 1650 mg l^{-1} NH_4NO_3, and the cultures were maintained for 2 weeks at 15, 20, or 25 ± 2 °C in the dark and then exposed to light of 45 μmol m^{-2} s^{-1} PPFD. Globular stage embryos were transferred to the MS containing various concentrations of GA_3 for SE maturation and germination (Table 4.2).

4.2.1 Role of Plant Growth Regulators (PGRs) on Somatic Embryo Induction

PGRs play crucial roles in the induction and development of SEs. Cotyledon and leaf explants failed to develop SE on MS medium devoid of PGRs, and their color gradually turned from green to brown exhibiting necrosis. Somatic embryogenesis was also induced on auxin-free media in several plants. Thus, the presence of PGRs in the culture medium was essential for the induction of somatic embryos in cineraria. Auxin is often required for the induction of callus, root, or SE from various explants, and the process of somatic embryogenesis is generally initiated on the culture medium supplemented with high concentrations of auxins. In alfalfa, supplementation of low concentration of 2,4-D induced callus from leaf protoplast, while at ten times higher levels of 2,4-D provoked in forming embryo-like structures (Feher et al. 2002). However, somatic embryogenesis was induced in several plants on auxin-free media (Jimenez 2001). Among the auxins, only 2,4-D promoted direct somatic embryogenesis in both explants of cineraria, whereas IBA and NAA induced callus and roots from the cut ends and surface of the explants. 2,4-D is one of the most widely used herbicides in the world, and Feher et al. (2003) suggested that it functions as a stress chemical as well as an auxin. Besides, auxinic herbicides have been shown to interact with ethylene and ABA, thereby increasing the cellular levels of these stress hormones (Feher et al. 2003). The seedling explants of carrot formed SEs when the culture medium was supplemented with ABA (Nishiwaki et al. 2000). Increasing concentration of sucrose or addition of NaCl or CdCl to the PGR-free medium induces SEs in carrot (Zavattieri et al. 2010). The stress caused by in vitro culture conditions and media composition can also induce SE formation. 2,4-D has been proven to be the most potent auxin and effective for SE induction in several ornamental plants (Tanaka et al. 2000; Sivanesan et al. 2011, 2012).

Incubating explants under a short period of darkness may be beneficial for in vitro morphogenesis by reducing blackening and chlorosis and preventing the formation of some growth inhibitors. Dennis (2000) reported that culture in the darkness suppressed unwanted tissue differentiation in explants tissues, for example, by limiting the development of plastids into chloroplasts. SEs were induced under light conditions, but culturing the explants 14 days in the dark followed by 4 weeks under light conditions resulted in high frequency of SE induction. This is in agreement with our previous study (Nam et al. 2005). Similar result was also found in other ornamental species such as *Campanula punctata* (Sivanesan et al. 2011), *Crocus vernus* (Sivanesan et al. 2012), and *Dendranthema grandiflorum* (Tanaka et al. 2000). The tissue sensitivity of the explants to the PGRs may have been altered by the dark treatment, thereby resulting in a higher frequency of embryo formation (Zobayed and Saxena 2003). In geranium continuous exposure of light treatment significantly reduced the level of endogenous plant growth hormones (Hutchinson et al. 2000). Globular embryos induced directly on the surface of the explants after weeks of culture, without a callus phase, and the embryos were yellow, compact, and pale. Of the various concentrations of 2,4-D tested, the highest frequency of somatic embryo induction (66.8 %) and mean number of SE (45.0) per leaf explant were obtained on MS supplemented with 3.0 mg l^{-1} 2,4-D (Table 4.1). In contrast, SE did not induce from the cotyledon explants of cineraria 'Jester Pink' on MS ammended with 2,4-D, while 'Early Blue' formed SE on the medium containing 2,4-D (Nam et al. 2005). Thus, somatic embryogenesis in cineraria is cultivar dependent. Somatic embryogenesis has been described as being genetically determined. Cineraria seeds are heterozygous in nature; therefore, further studies on genetic analysis in various cultivars of cineraria may help to find out the genes involved in somatic embryogenesis in cineraria. However, SEs were initiated from all cultivars of cineraria on MS containing auxin plus cytokinins. In several plants, a combination of auxin and cytokinin stimulates the formation of SEs.

The frequency of SE induction and average number of SEs per explant significantly increased when the optimal concentration of 2,4-D (3.0 mg l^{-1}) was combined with cytokinins. However, the nature of SEs varied depending on combination of 2,4-D and cytokinins. The explants developed fused SEs on MS supplemented with 2,4-D and 2iP after 6 weeks of culture (Fig. 4.1a). Auxin polar transport was reported to be essential for bilateral symmetry during early plant embryogenesis. Liu et al. (1993) reported that TIBA, trans-cinnamic acid, and 9-hydroxyfluorene-9-carboxylic acid induced the formation of fused cotyledons in Indian mustard. Cytokinin influences cell-to-cell auxin transport by modification of expression of several auxin transport compounds. The addition of ABA to the induction medium significantly decreased the proportion of abnormal embryos in several plants. In soybean, the supplementation of ABA to the culture medium containing 2,4-D, adenine, and kinetin increased the number of SEs which grew from the globular to heart stage (Phillips and Collins 1981). SEs with abnormal cotyledons failed to convert into normal plantlets (Todd and Yeung 1993). Normal SE formed two meristematic centers and then organized normal two cotyledons. The addition of ABA induces dormancy of SEs, which makes embryos to form two suitable meristematic centers (Ammirato 1985). Thus, incorporation of ABA to the SE induction medium may reduce the formation of abnormal embryos. When the medium was supplemented with 2,4-D and BA, a few embryos germinated precociously (Fig. 4.1b). The inclusion of high molecular weight PEG to the SE induction medium significantly improves the synchrony of development and the quality of mature SEs, particularly in terms of storage reserves, and prevents precocious germination (Attree et al. 1992). Nam et al. (2005) reported that addition of 7.8 μM ABA or 3.0 % PEG or 6.0 % sucrose to the culture medium prevented precocious germination in cineraria. On medium containing 2,4-D and TDZ, the explants formed SE and also callus on the surface after 6 weeks of culture (Fig. 4.1c). The callus formation may be due to the high activity of TDZ. Thidiazuron, a cotton defoliant, has been shown both auxin and cytokinin effects (Guo et al. 2011). Ferreira et al. (2006) reported that treatment of TDZ significantly increases the

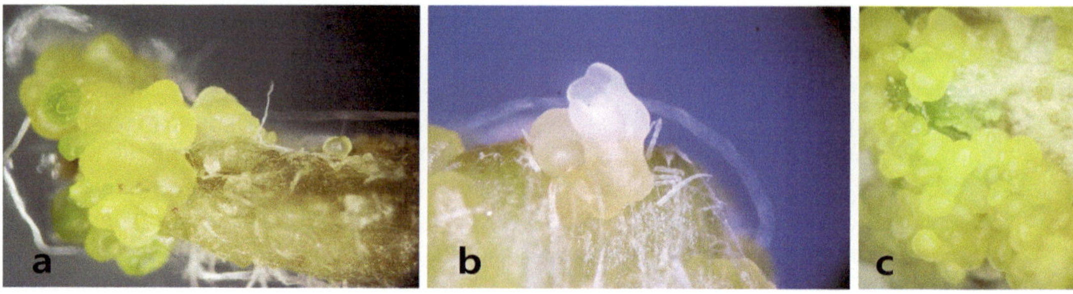

Fig. 4.1 Effects of combination of 2,4-D and cytokinins on somatic embryogenesis from explants of *S. cruentus* after 6 weeks of culture. (**a**) 2,4-D+2iP. (**b**) 2,4-D+BA. (**c**) 2,4-D+TDZ

endogenous cytokinins and IAA levels of the explants of *Dendrobium*. Thus, reducing the concentration of TDZ in the induction medium can minimize the callus formation and the number of SEs. Both explants formed SE and callus on MS medium containing IBA or NAA in combination with cytokinins after 6 weeks of culture. Of the various combinations of auxins and cytokinins studied, the highest frequency of SE induction (97.0 %) and mean number of SE (135.3) per leaf explant were obtained on MS supplemented with 3.0 mg l^{-1} 2,4-D and 1.0 mg l^{-1} BA. Similar result has also been reported in cineraria 'Hansa' (Malueg et al. 1994) and 'Jester Pink' (Nam et al. 2005). The use of 2,4-D and BA for SE induction has been reported for other members of Asteraceae (Filho et al. 1993; Correa et al. 2009).

4.2.2 Influence of Explants on Somatic Embryo Induction

Somatic embryogenesis in cineraria has been achieved using various explants such as cotyledons (Malueg et al. 1994; Nam et al. 2005), hypocotyls (Nam et al. 2005), and leaves (Malueg et al. 1994). The ability of cineraria cultivars to develop SEs is influenced by the type of explant. Nam et al. (2005) reported that cotyledon explants were found to be the best for SE induction than hypocotyls. In this study, cotyledon and leaf explants produced SEs on MS supplemented with 3.0 mg l^{-1} 2,4-D and 1.0 mg l^{-1} BA (Table 4.1). However, leaf explants were more efficient than cotyledon in SE induction. This may be due to variation in the endogenous levels of PGRs in the explants. SE began to appear mostly on the cut end of cotyledon within 2 weeks of culture (dark condition), while SEs were formed on the cut ends and surface of the leaf explants after 3 weeks of culture. Most of the embryos developed from the cotyledon explants showed precocious germination after 6 weeks of culture (Fig. 4.2a). Addition of ABA, PEG or sucrose to the culture medium prevents precocious germination, but most of the SE-derived plantlets exhibited hyperhydricity (Nam et al. 2005). Therefore, the normal process of SE development is closely coupled with the osmotic environment surrounding the SEs (Yeung and Claudio 2000). On the other hand, compact yellow globular embryos were formed on the leaf explants after 6 weeks of culture. Thus, induction and development of SE in cineraria is also affected by the age of the explants. In most plant species, younger explants were more responsive than the older explants for SE induction. However, mature leaf explants found to be the best for SE induction and plant regeneration in cineraria.

4.2.3 Effects of Different Concentrations of NH_4NO_3 and Temperature on Somatic Embryo Induction

Growth and morphogenesis in tissue cultures are greatly influenced by the availability of nitrogen

Fig. 4.2 Effects of combination of 2,4-D and BA on somatic embryogenesis from explants of *S. cruentus* after 6 weeks of culture. (**a**) Cotyledonary explant. (**b**) Leaf explant

and the form in which it is used (Sivanesan and Jeong 2012). Ammonium nitrate concentrations have been shown to be important for the induction of somatic embryogenesis in many plants (Smith and Krikorian 1989; Choi et al. 1998; Greer et al. 2009). Gertsson (1988) reported that a low number of adventitious shoots were obtained from petiole explants of *Senecio hybridus* when the total nitrogen in MS was increased to 75 mM and that more shoots were produced when the total nitrogen was reduced to 30 mM. In contrast, decreasing the levels of NH_4NO_3 had a negative effect on adventitious shoot induction in *S. cruentus* (Sivanesan and Jeong 2012). In this study, the frequency of SE induction and mean number of SEs per explant gradually decreased with decrease in the level of NH_4NO_3 concentration from 1650 to 0 mg l^{-1} (Table 4.2). The quality of the globular stage embryos did not differ markedly in used treatments. Morphogenesis including embryogenesis in culture is also influenced by temperature. The explants were maintained at three different temperatures in order to study the effect of temperature on somatic embryogenesis. Decreasing the culture temperature from 25 to 15 °C decreased the frequency of SE induction. The explants maintained under 25 °C developed maximum number of SEs, while those maintained under 15 °C produced less number of SEs. The results suggest that NH_4NO_3 concentration and temperature had a significant effect on somatic embryogenesis in cineraria. The morphology of globular embryos was affected by NH_4NO_3 concentration and temperature. The globular embryos developed on SE induction medium containing low levels of NH_4NO_3 (412.5, 825.0 mg l^{-1}) at 20 °C were purple in color (Fig. 4.3a). Thus, anthocyanin synthesis could be stimulated in cineraria in vitro cultures by reducing NH_4NO_3 levels and temperature. Variegated plantlets were produced from the globular embryos when the embryos were transferred to the germination medium devoid of NH_4NO_3 and at 15 °C (Fig. 4.3b). Somaclonal variation has also been observed in cineraria (Sivanesan and Jeong 2012) when the shoot cultures are maintained at 15 °C. Anthocyanins are flavonoid metabolites that contribute to the coloration of various plant tissues and in vitro cultured cells and tissues. The application of ammonium has been found to decrease the level of anthocyanins in petals of *Gerbera hybrida* (Huang et al. 2008), while nitrogen deficiency increased the levels of anthocyanins in *Arabidopsis thaliana* (Scheible et al. 2004). Temperature also affects anthocyanin synthesis in various plants. Low temperature enhances anthocyanin synthesis in *A. thaliana* (Leyva et al. 1995) and *Zea mays* (Christie et al. 1994).

Fig. 4.3 Effects of different concentrations of NH_4NO_3 and temperature on somatic embryogenesis from explants of *S. cruentus*. (**a**) Pigmented somatic embryos developed on MS medium containing 412.5 mg l^{-1} NH_4NO_3 at 20 °C. (**b**) Variegated shoot developed from the SE

4.3 Maturation and Germination of Somatic Embryos

Embryo maturation and simultaneous conversion to plantlets is one of the steps in somatic embryogenesis, which partially depends on embryo quality (Sivanesan et al. 2012). Somatic embryogenesis is often initiated on the culture medium supplemented with PGRs; however, development beyond the globular stage was inhibited by maintaining the same induction medium. Thus, for promoting further growth and development of SEs, it is necessary to subculture the embryos to the PGR-free medium. In cineraria, globular embryos initiated on MS containing 2,4-D and BA matured and converted into plantlets on MS with activated charcoal (Malueg et al. 1994). Activated charcoal is composed of carbon arranged in a quasi-graphitic form in small particle size. It is often used in medium to improve growth and development of cell, tissue, and organ (Thomas 2008). Activated charcoal has the potential to absorb some inorganic ions, auxins, cytokinins, and phenolics. The positive effect of activated charcoal on embryo maturation and conversion was probably by adsorption of PGRs (von Aderkas et al. 2002), ethylene (Johansson 1983), and growth inhibitory substances like 5-hdroxymetylfurfural in the culture medium (Weatherhead et al. 1978), alteration of culture medium pH to an optimum level for development (Owen et al. 1991), and release of substances naturally present in or adsorbed by activated charcoal (Pan and van Staden 1998). The addition of activated charcoal to the maturation medium has also been proven beneficial for development and conversion of SEs in *Campanula punctata* (Sivanesan et al. 2011) and *Crocus vernus* (Sivanesan et al. 2012). On the other hand, ABA and osmotic stress (PEG, sucrose) have been reported beneficial for embryo maturation in cineraria (Nam et al. 2005). The combination of ABA and PEG is also used to stimulate SE maturation in *Corydalis yanhusuo* (Sagare et al. 2000) and *Panax ginseng* (Langhansova et al. 2004). Treatment of 3.8 μM ABA or 6.0 % sucrose was best for embryo maturation. ABA and sucrose are known to be important factors for seed maturation in higher plants. The authors reported that treatment of ABA was the best in promoting the conversion ratio than PEG or sucrose treatment. ABA is often used in in vitro culture to promote the maturation of SEs and to store carbohydrate reserves in embryos during maturation. The influence of ABA on somatic embryo maturation and germination has been reported in many plants (Rai et al. 2011). The conversion ratio from SEs to plantlets in cineraria was 30 % (Nam et al. 2005). Thus, it is necessary to improve the conversion ratio of somatic

Fig. 4.4 Somatic embryogenesis and plant regeneration from cotyledon and leaf explants of *S. cruentus*. (**a**) Globular stage. (**b**) Heart stage. (**c**) Torpedo stage. (**d**) Cotyledonary stage. (**e**) Germination and root formation. (**f**) SE-derived plantlets. (**g**) and (**h**) SE germination and conversion. (**i**) Hyperhydric shoots developed from the SE obtained from cotyledon explants

Table 4.3 Effects of different concentrations of GA_3 on germination of somatic embryos after 5 weeks of culture

GA_3 (mg l^{-1})	Germination (%)
0	29.7 ± 5.9
0.5	42.7 ± 5.0
1.0	65.0 ± 5.1
2.0	84.3 ± 3.1
4.0	71.3 ± 3.3

Values are mean ± SD of results

embryos to be of practical use. In this study, globular embryos completed germination through heart, torpedo, and cotyledonary stages when the embryos were cultured on MS basal medium (Fig. 4.4a–d). However, the frequency of conversion was very low. Furthermore, SE having malformed cotyledons occurred during the embryo development. The low frequency of SE conversion to plantlets is due to the carryover effects of PGRs. GA_3 had a significant effect on SE maturation and conversion (Table 4.3). Gibberellins are known as growth-promoting hormones, being involved in several processes during plant development, like shoot elongation, flower development, breaking dormancy, and seed germination (Linkies and Leubner-Metzger 2012). The frequency of SE maturation and conversion were 29.7, 42.7, 65.0, 84.3, or 71.3 % when MS was supplemented with 0, 0.5, 1.0, 2.0, or 4.0 mg l^{-1} GA_3, respectively. The fact that GA_3 stimulates the maturation and conversion of SE is well known. However, SE obtained from the cotyledon explants developed hyperhydric shoots (Fig. 4.4). This is in agreement with our previous report (Nam et al. 2005). The results suggest that somatic embryogenesis and plant regeneration in cineraria largely depends on the type of explant. Further, understanding physiological and molecular nature of the explants will be useful for accessing the quality of the SE. Plantlets developed from the SE with shoots and roots were separated and grew further on hormone-free MS medium for 4 weeks. The in vitro-developed plantlets were successfully acclimatized in the greenhouse with 98 % survival.

4.4 Conclusion

We developed a simple and reproducible procedure for somatic embryogenesis and plant regeneration of cineraria. 2,4-D and BA had a positive effect on SE induction in both explants. Among the two explants, leaf was found to be the most effective for somatic embryogenesis and subsequent plant regeneration. GA_3 had a positive effect on SE maturation and conversion. This protocol could be utilized for genetic transformation and large-scale commercial propagation of cineraria. Most of the SEs were attached to the explants; these must be separated from the explants and embryo clusters for artificial seed production. Further studies are needed to confirm the genetic homogeneity of the in vitro-regenerated plantlets.

References

Ammirato PV (1985) Patterns of development in culture. In: Henke RR (ed) Tissue culture in forestry and agriculture. Plenum, New York, pp 9–29

Attree SM, Pomeroy MK, Fowke LC (1992) Manipulation of conditions for the culture of somatic embryos of white spruce for improved triacylglycerol biosynthesis and desiccation tolerance. Planta 187:395–404

Bozhkov PV, Filonova LH, von Arnold S (2002) A key developmental switch during Norway spruce somatic embryogenesis is induced by withdrawal of growth regulators and is associated with cell death and extracellular acidification. Biotechnol Bioeng 77:658–667

Choi YE, Yang DC, Choi KT (1998) Induction of somatic embryos by macrosalt stress from mature zygotic embryos of *Panax ginseng*. Plant Cell Tissue Organ Cult 52:177–181

Christie PJ, Alfenito MR, Walbot V (1994) Impact of low-temperature stress on general phenylpropanoid and anthocyanin pathways: enhancement of transcript abundance and anthocyanin pigmentation in maize seedlings. Planta 194:541–549

Correa CM, de Oliveira GN, Astariata LV, Santarem ER (2009) Plant regeneration through somatic embryogenesis of yacon [*Smallanthus sonchifolius* (Poepp. and Endl.) H. Robinson]. Braz Arch Biol Technol 52:549–554

Dennis JG (2000) Nonzygotic embryogenesis. In: Robert NT, Dennis JG (eds) Plant tissue culture concepts and laboratory exercises. CRC Press, Boca Raton, pp 175–189

Dey T, Saha S, Ghosh PD (2015) Somaclonal variation among somatic embryo derived plants- evaluation of agronomically important somaclones and detection of genetic changes by RAPD in *Cymbopogon winterianus*. South Afr J Bot 96:112–121

Feher A (2008) The initiation phase of somatic embryogenesis: what we know and what we don't. Acta Biol Szeged 52:53–56

Feher A, Pasternak T, Otvos K, Miskolczi P, Dudits D (2002) Induction of embryogenic competence in somatic plant cells: a review. Biologia 57:5–12

Feher A, Pasternak TP, Dudits D (2003) Transition of somatic plant cells to an embryogenic state. Plant Cell Tissue Organ Cult 74:201–228

Ferreira WM, Kerbauy GB, Kraus JE, Pescador R, Suzuki RM (2006) Thidiazuron influences the endogenous levels of cytokinins and IAA during the flowering of isolated shoots of *Dendrobium*. J Plant Physiol 163:1126–1134

Filho JCB, Hashimoto JM, Vieira LG (1993) Induction of somatic embryogenesis from leaf explants of *Stevia rebaudiana*. Braz J Plant Physiol 5:51–53

Gertsson UE (1988) Large-scale in vitro propagation of Senecio X hybridus Hyl. J Hortic Sci 63:131–136

Guo B, Abbasi BH, Zeb A, Xu LL, Wei YH (2011) Thidiazuron: a multi-dimensional plant growth regulator. Afr J Biotechnol 10:8984–9000

Huang Z, Liang M, Peng J, Xing T, Wang X (2008) Exogenous ammonium inhibits petal pigmentation and expansion in *Gerbera hybrida*. Physiol Plant 133:254–265

Hutchinson MJ, Sanaratna T, Sahi SV, Saxena PK (2000) Light mediates endogenous plant growth substances in thidiazuron-induced somatic embryogenesis in geranium hypocotyls cultures. J Plant Biochem Biotechnol 9:1–6

Jimenez VM (2001) Regulation of in vitro somatic embryogenesis with emphasis on to the role of endogenous hormones. Rev Bras Fisiol Veg 13:196–223

Johansson L (1983) Effects of activated charcoal in anther cultures. Physiol Plant 59:397–403

Kim CK, Chung JD, Jee SO, Oh JY (2003) Somatic embryogenesis from in vitro grown leaf explants of *Rosa hybrid*a L. J Plant Biotechnol 5:161–164

Langhansova L, Konradova H, Vanek T (2004) Polyethylene glycol and abscisic acid improve maturation and regeneration of *Panax ginseng* somatic embryos. Plant Cell Rep 22:725–730

Lelu-Walter MA, Thompson D, Harvengt L, Sanchez L, Toribio M, Paques LE (2013) Somatic embryogenesis in forestry with a focus on Europe: state-of-the-art, benefits, challenges and future direction. Tree Genet Genomes 9:883–899

Leyva A, Jarillo JA, Salinas J, Martinez-Zapater JM (1995) Low temperature induces the accumulation of phenylalanine ammonia-lyase and chalcone synthase mRNAs of *Arabidopsis thaliana* in a light-dependent manner. Plant Physiol 108:39–46

Linkies A, Leubner-Metzger G (2012) Beyond gibberellins and abscisic acid: How ethylene and jasmonates control seed germination. Plant Cell Rep 31:253–270

Liu CM, Xu ZH, Chua NH (1993) Polar transport is essential for the establishment of bilateral symmetry during early plant embryogenesis. Plant Cell 5:621–630

Malueg KR, McDaniel GL, Trigiano RN (1989) Somatic embryogenesis of florists' cineraria. HortSci 24:744

Malueg KR, McDaniel GL, Graham ET, Trigiano RN (1994) A three media transfer system for direct somatic embryogenesis from leaves of *Senecio x hybridus* Hyl. (Asteraceae). Plant Cell Tissue Organ Cult 36:249–253

Nam EY, Kim GH, Jeong BR (2005) Direct somatic embryogenesis and plant regeneration in cineraria (*Senecio cruentus*). J Korean Soc Hortic Sci 46:210–216

Nishiwaki M, Fujino K, Koda Y, Masuda K, Kikuta Y (2000) Somatic embryogenesis induced by the simple application of abscisic acid to carrot (*Daucus carota* L.) seedlings in culture. Planta 211:756–759

Owen HR, Wengerd D, Miller AR (1991) Culture medium pH is influenced by basal medium, carbohydrate source, gelling agent, activated charcoal, and medium storage method. Plant Cell Rep 10:583–586

Pan MJ, Van Staden J (1998) The use of charcoal in in vitro culture-a review. Plant Growth Regul 26:155–163

Phillips GC, Collins GB (1981) Induction and development of somatic embryos from cell suspension cultures of soybean. Plant Cell Tissue Organ Cult 1:123–129

Rai MK, Shekhawat NS, Harish GAK, Phulwaria M, Ram K, Jaiswal U (2011) The role of abscisic acid in plant tissue culture: a review of recent progress. Plant Cell Tissue Organ Cult 106:179–190

Sagare AP, Lee YL, Lin TC, Chen CC, Tsay HS (2000) Cytokinin induced somatic embryogenesis and plant regeneration in *Corydalis yanhusuo* (Fumariaceae)- a medicinal plant. Plant Sci 160:139–147

Salvo SAGD, Hirsch CN, Buell CR, Kaeppler SM, Kaeppler HF (2014) Whole transcriptome profiling of maize during early somatic embryogenesis reveals altered expression of stress factors and embryogenesis-related genes. PLoS One 9, e111407

Scheible W-R, Morcuende R, Czechowski T, Fritz C, Osuna D, Palacios-Rojas N, Schindelasch D, Thimm O, Udvardi MK, Stitt M (2004) Genome-wide reprogramming of primary and secondary metabolism, protein synthesis, cellular growth processes, and the regulatory infrastructure of *Arabidopsis* in response to nitrogen. Plant Physiol 136:2483–2499

Sharma P, Rajam MV (1995) Genotype, explant and position effects on organogenesis and somatic embryogenesis in eggplant (*Solanum melongena* L.). J Exp Bot 46:135–141

Sivanesan I, Jeong BR (2012) Identification of somaclonal variants in proliferating shoot cultures of *Senecio cruentus* cv. Tokyo Daruma. Plant Cell Tissue Organ Cult 111:247–253

Sivanesan I, Lim MY, Jeong BR (2011) Somatic embryogenesis and plant regeneration from leaf and petiole explants of *Campanula punctata* Lam. var. *rubriflora* Makino. Plant Cell Tissue Organ Cult 107:365–369

Sivanesan I, Son MS, Jana S, Jeong BR (2012) Secondary somatic embryogenesis in *Crocus vernus* (L.) Hill. Propag Ornam Plants 12:163–170

Smith DL, Krikorian AD (1989) Release of somatic embryogenic potential from excised zygotic embryos of carrot and maintenance of proembryonic cultures in hormone-free medium. Am J Bot 76:1832–1843

Su YH, Su YX, Liu YG, Zhang XS (2012) Abscisic acid is required for somatic embryo initiation through mediating spatial auxin response in Arabidopsis. Plant Growth Regul 69:167–176

Tanaka K, Kanno Y, Kudo S, Suzuki M (2000) Somatic embryogenesis and plant regeneration in chrysanthemum (*Dendranthema grandiflorum* (Ramat.) Kitamura). Plant Cell Rep 19:946–953

Thomas TD (2008) The role of activated charcoal in plant tissue culture. Biotechnol Adv 6:618–631

Todd CN, Yeung EC (1993) Failure to establish a functional shoot meristem may be a cause of conversion failure in somatic embryos of *Daucus carota* (Apiaceae). Am J Bot 80:1284–1291

von Aderkas P, Label P, Lelu MA (2002) Charcoal affects early development and hormonal concentrations of somatic embryos of hybrid larch. Tree Physiol 22:431–434

Weatherhead MA, Burdon J, Henshaw GG (1978) Some effects of activated charcoal as an additive to plant tissue culture media. Z Pflanzenphysiol 89:141–147

Williams EG, Maheswaran G (1986) Somatic embryogenesis: factors influencing coordinated behaviour of cells as an embryogenic group. Ann Bot 57:443–462

Xu Z, Zhang C, Zhang X, Liu C, Wu Z, Yang Z, Zhou K, Yang X, Li F (2013) Transcriptome profiling reveals auxin and cytokinin regulating somatic embryogenesis in different sister lines of cotton cultivar CCRI24. J Integr Plant Biol 55:631–642

Yeung EC, Claudio S (2000) Somatic embryogenesis–apical meristems and embryo conversion. Korean J Plant Tissue Cult 27:299–307

Zavattieri MA, Frederico AM, Lima M, Sabino R, Arnholdt-Schmitt B (2010) Induction of somatic embryogenesis as an example of stress-related plant reactions. Electron J Biotechnol 13:1–9. doi:10.2225/vol13-issue1-fulltext-4

Zobayed SMA, Saxena PK (2003) In vitro regeneration of *Echinacea purpurea* L.: enhancement of somatic embryogenesis by indolebutyric acid and dark pre-incubation. In Vitro Cell Dev Biol Plant 39:605–612

Advances in Proteomics of Somatic Embryogenesis

5

Magdalena I. Tchorbadjieva

Abstract

Plants are sessile organisms and, as such, have evolved a remarkable developmental plasticity allowing them to cope with the adverse effects of numerous biotic and abiotic factors from the environment. One of the most intriguing examples of this plasticity is somatic embryogenesis during which somatic cells dedifferentiate into cells that are capable to form embryos. The latter are morphologically similar to zygotic embryos and can regenerate whole plants. The transition of somatic cells into embryogenic ones is the most intriguing and the part of somatic embryogenesis least understood. To better understand the mechanisms of somatic embryogenesis, comparative proteomic approaches have been used, and in recent years, hundreds of proteins have been identified in embryogenic and nonembryogenic cultures, in somatic and zygotic embryos, and during distinct stages of somatic embryogenesis in both angiosperms and gymnosperms.

This review provides a comprehensive analysis of current advances in the proteomics of SE, focusing on the presence and putative function of differentially expressed proteins involved in processes such as stress response, protein synthesis and processing, cell proliferation, signal transduction and energy metabolism, and their potential use as embryogenic markers.

Keywords

Differentially expressed proteins • Embryogenic and non-embryogenic callus • Proteomics • PR protein • Heat shock protein • Actin • Annexin • Signal transduction • Storage protein • Secretome • Programmed cell death • Up-regulation of genes • Somatic embryogenesis

M.I. Tchorbadjieva (✉)
Department of Biochemistry, Faculty of Biology, Sofia University,
8 Dragan Tsankov str., 1164, Sofia, Bulgaria
e-mail: magd@biofac.uni-sofia.bg

A. Mujib (ed.), *Somatic Embryogenesis in Ornamentals and Its Applications*,
DOI 10.1007/978-81-322-2683-3_5

5.1 Introduction

Somatic embryogenesis (SE) represents a unique phenomenon in the plant kingdom. This developmental pathway is one of the most striking examples of the plant cell developmental plasticity (Fehér 2008; Fehér 2014). It includes a series of characteristic events such as dedifferentiation of somatic cells, activation of cell division, and reprogramming of their metabolism and gene expression patterns. The transition of somatic cells into embryogenic ones is the most intriguing and the part of somatic embryogenesis least understood (Fehér 2005; Karami et al. 2009; Elhiti et al. 2013). It is still not known why and how somatic cells regain totipotency and embryonic cell fate giving rise to somatic embryos that are morphologically similar to zygotic embryos (Fehér 2014). More than two decades ago, SE was proposed to be a developmental stress response (Dudits et al. 1991), and now it is widely accepted that stress and hormones play a crucial role in collectively inducing cell dedifferentiation and initiation of embryogenic program in plants with responsive genotype (Fehér et al. 2003; Ikeda-Iwai et al. 2003; Rose and Nolan 2006). Still, the underlying mechanisms are hardly known justifying the listing of the question "How does a single somatic cell become a whole plant?" among the current 125 most important scientific questions (Vogel 2005).

Efforts have been made to elucidate the patterns of gene expression that may play a critical role in the process of SE and especially in the somatic-to-embryogenic developmental transition. Several gene classes associated with SE including SERK, LEAFY COTYLEDON, BABY BOOM, WUSCHEL, and PICKLE have been identified (reviewed by Karami et al. 2009; Yang and Zhang 2010; Fehér 2014). However, the lack of a clear correlation between mRNA and protein abundance due to the variation in mRNA stability, translatability, and protein stability sets some limits on mRNA expression profiling. Furthermore, protein structure, activity, and function can be altered and regulated by subcellular localization, interaction with other molecules, and posttranslational modifications that would not be detected by mRNA analysis (Rose et al. 2004). Consequently, there is a growing recognition that this approach should be complemented with profiling methods of the final gene products or proteins themselves.

Proteomics is a powerful approach aimed at systematic studies of protein structure, function, interaction, and dynamics. Protein changes occurring as a response to developmental and environmental states, abundances of proteins, posttranslational modifications, and protein isoforms can also be investigated (Valledor and Jorrin 2011). Within the last decade, improvements in the high-resolution two-dimensional gel electrophoresis (2-DE) and mass spectrometry (MS) technique have allowed the identification of proteins linked to SE competence and development in diverse plant species such as *Manihot esculenta, Catharanthus roseus, Phoenix dactylifera, Cyclamen persicum* (reviewed by Aslam et al. 2013), *Quercus suber* (Gomez-Garay et al. 2013), *Vitis vinifera* (Marsoni et al. 2008; Zhang et al. 2009), *Vigna unguiculata* (Nogueira et al. 2007), *Acca sellowiana* (Cangahuala-Inocente et al. 2009), *Citrus sinensis* (Pan et al. 2009), *Coffea arabica* (Tonietto et al. 2012), *Crocus sativus* (Sharifi et al. 2012), *Araucaria angustifolia* (Jo et al. 2014), *Elaeis guineensis* (Silva Rde et al. 2014), *Medicago truncatula* (de Jong et al. 2007; Almeida et al. 2012), *Zea mays* (Sun et al. 2013; Varhaníková et al. 2014), and *Pinus pinaster* (Morel et al. 2014). These reports included studies on protein expression changes during the transitional state from somatic to embryogenic cells, identified differentially expressed proteins in nonembryogenic and embryogenic calli and in somatic and zygotic embryos, compared specific proteins that accumulate during different stages of somatic embryogenesis, and discovered putative protein markers for somatic embryogenesis.

5.2 Proteomics of Somatic Embryogenesis: A Ten-Year Survey (2005–2015)

Transcriptomics and proteomics approaches have been applied to investigate global gene expression during somatic embryo formation in several plant species, including angiosperms and gymnosperms (see Table 5.1). SE can be divided into

Table 5.1 Overview of plant proteomics studies of somatic embryogenesis (2005–2015)

Plant species	Common name	Source	Proteomics techniques	References
Acca sellowiana	Feijoa	Somatic embryo development	2-DE/MALDI-TOF MS	Cangahuala-Inocente et al. (2009)
Acca sellowiana (O Berg.) Burret	Feijoa	Somatic embryo regeneration	2D DIGE/Q-TOF MS/MS	Fraga et al. (2013)
Araucaria angustifolia (Bert.) O. Ktze	Brazilian pine	Embryogenic vs. nonembryogenic cell line	2-DE/MALDI-TOF/TOF MS/MS	Jo et al. (2014)
Carica papaya L.	Papaya	Somatic embryo development	2-DE/MALDI-TOF/TOF MS/MS	Vale Ede et al. (2014)
Citrus sinensis Osbeck	Valencia sweet orange	Somatic embryo development	2-DE/MALDI-TOF MS	Pan et al. (2009)
Coffea arabica	Coffee	Somatic embryo development	2-DE/MALDI-TOF/TOF MS/MS	Tonietto et al. (2012)
Crocus sativus L.	Saffron	Embryogenic vs. nonembryogenic calli	2-DE/MALDI-TOF/TOF MS/MS	Sharifi et al. (2012)
Cyclamen persicum	The Persian cyclamen	Embryogenic vs. nonembryogenic cell lines	2D DIGE/ MALDI-TOF MS	Lyngved et al. (2008)
Cyclamen persicum	The Persian cyclamen	Somatic vs. zygotic embryos	2-DE/nano-LC-MS/MS	Winkelmann et al. (2006)
Cyclamen persicum	The Persian cyclamen	Somatic vs. zygotic embryos	2-DE/LC-MS/MS	Rode et al. (2011)
Cyclamen persicum	The Persian cyclamen	Somatic embryo development and maturation	2-DE/LC-MS/MS	Rode et al. (2012)
Cyclamen persicum	The Persian cyclamen	Somatic embryo development and nonembryogenic callus	2-DE/MALDI-TOF/TOF MS/MS	Bian et al. (2010)
Cyphomandra betacea	Tamarillo	Embryogenic vs. nonembryogenic calli	2-DE/LC-MS/MS	Correia et al. (2012)
Elaeis guineensis Jacq.	Oil palm	Somatic embryo development	2-DE/MALDI-TOF/TOF MS/MS	Silva Rde et al. (2014)
Eruca sativa Mill	Rucola	Embryogenic vs. nonembryogenic calli	1-DE/MALDI-TOF MS	Chen et al. (2012)
Gossypium hirsutum L.	Cotton	Somatic embryo development	iTRAQ/LC-ESI-MS/MS	Ge et al. (2015)
Larix principis-rupprechtii Mayr	Prince Rupprecht's larch	Embryogenic vs. nonembryogenic calli	iTRAQ/LC-MS/MS	Zhao et al. (2015)
Larix × eurolepis	Hybrid larch	Somatic vs. zygotic embryo maturation	1-DE/LC-MS/MS 2-DE/LC-MS/MS	Teyssier et al. (2014)
Manihot esculenta Crantz	Cassava	Secondary somatic embryogenesis	2-DE/MALDI-TOF/TOF MS/MS	Baba et al. (2008)
Manihot esculenta Crantz	Cassava	Somatic embryos, plantlets, and roots	1-DE/LC-ESI-MS/MS	Li et al. (2010)
Medicago truncatula	Barrel clover	Embryogenic (2HA) vs. nonembryogenic cell lines (Jemalong)	2-DE/MALDI-TOF MS/MS	Imin et al. (2005)
Medicago truncatula	Barrel clover	Embryogenic (M9-10A) vs. nonembryogenic (M9) cell lines	2-DE/MALDI-TOF/TOF MS/MS	Almeida et al. (2012)
Percea americana	Avocado	Embryogenic vs. nonembryogenic cell lines	2-D DIGE/MALDI-TOF/TOF MS/MS	Guzmán-García et al. (2013)
Phoenix dactylifera L.	Date palm	Somatic vs. zygotic embryos	2-DE/MALDI-TOF/TOF MS/MS	Sghaier-Hammami et al. (2009)

(continued)

Table 5.1 (continued)

Plant species	Common name	Source	Proteomics techniques	References
Picea balfouriana	Balfour spruce	Somatic embryo development	iTRAQ/LC-MS/MS	Li et al. (2015)
Picea glauca	White spruce	Somatic embryo maturation	2-DE/LC-MS/MS	Lippert et al. (2005)
Pinus pinaster Ait.	Maritime pine	Somatic vs. zygotic embryo maturation	1-DE/LC-MS/MS 2-DE/LC-MS/MS	Morel et al. (2014)
Quercus suber	Cork oak	Somatic embryo development	2-D DIGE/MALDI-TOF/TOF MS/MS	Gomez-Garay et al. (2013)
Theobroma cacao L.	Cacao	Somatic vs. zygotic embryos	2-DE/LC-MS/MS	Noah et al. (2013)
Vanilla planifolia Andrews	Vanilla orchid	Callus regeneration	2-DE/MALDI-TOF/TOF MS/MS	Tan et al. (2013)
Vigna unguiculata	Cowpea	Embryogenic cell suspensions	2-DE/MALDI-TOF/TOF MS/MS	Nogueira et al. (2007)
Vitis vinifera cv. Thompson seedless	Grape vine	Embryogenic vs. nonembryogenic calli	2-DE/LC-ESI-MS/MS	Marsoni et al. (2008)
Vitis vinifera L. cv. Cabernet sauvignon	Grape vine	Embryogenic vs. nonembryogenic calli	2-DE/MALDI-TOF/TOF MS/MS	Zhang et al. (2009)
Zea mays L. H99 inbred cell line	Maize	Embryogenic vs. nonembryogenic calli	2-DE/MALDI-TOF MS	Sun et al. (2013)
Zea mays L. A19 inbred cell line	Maize	Embryogenic vs. nonembryogenic calli	2-DE/Q-TOF MS/MS	Varhaníková et al. (2014)

two phases: induction and expression. During the induction phase, somatic cells are undifferentiated, acquire embryogenic competence, and proliferate as embryogenic formations. Cellular reprogramming takes place which is associated with the expression of specific genes. In the expression phase, the embryogenic cells differentiate to form somatic embryos and after that plants. Somatic embryo development encompasses key stages of ZE: the heart and torpedo stages in the case of dicotyledonous species; the globular, scutellar, and coleoptilar stages in the case of monocotyledonous species; and early and late embryogenesis in the case of gymnosperm species (Rocha and Dornelas 2013). Proteomic studies related to SE have focused mainly on (1) early stages of embryogenesis, when the embryogenic competence is acquired; (2) comparison of embryogenic and nonembryogenic calli; and (3) comparison of zygotic and somatic embryos and have revealed a plethora of differentially expressed proteins.

However, the results obtained from these studies are not entirely identical, perhaps due to the different plant species, organs, methods, and conditions used. And still, numerous proteins involved in fundamental cellular processes of SE were found to be identical in the different plant species. The biological significance of key differentially expressed proteins and their associated cellular roles is discussed in this section, stressing on the perspective of their SE or in vitro morphogenesis relevance. At a functional level, proteins identified in this study were derived from a broad variety of cellular processes, which emphasize the range of changes associated with embryo development. Suggested protein markers for SE offer the possibility of determining the embryogenic potential of plant cells in culture long before any morphological changes have taken place and of gaining further information on the molecular basis of induction and differentiation of plant cells (Tchorbadjieva 2005).

The proteins were classified into the following functional categories based on their primary biological process: (1) stress response; (2) protein synthesis, processing, and fate; (3) cell proliferation, cell wall biogenesis, and cell elongation; (4) transcription and signal transduction; (5) metabolism and energy state; and (6) storage proteins.

5.2.1 Proteins Involved in Stress/ Defense Response

5.2.1.1 Proteins Involved in Oxidative Stress Response

Recent proteomic studies have strongly emphasized the role of stress proteins during SE (Takáč et al. 2011).

The oxidative stress in plant tissue culture is well documented (Zavattieri et al. 2010). A burst in ROS is indispensable in modulating cell division and the reprogramming of cell metabolism as an adaptation response to stress. The initiation of embryogenic cultures (EC) often involves wounding of anthers, leaf explants, hypocotyls, etc. and placing them on callus induction medium. The latter contains the synthetic auxin 2,4-D, known by its herbicide effect triggering considerable oxidative stress in plant cells. ROS has also been implicated as a second messenger during auxin and stress-induced embryogenesis (Maraschin et al. 2005). Uncontrolled production of ROS could severely damage cellular proteins and membranes. Therefore, plant cells regulate ROS levels through sophisticated mechanisms such as scavenging with antioxidant defense proteins. Peroxidases, catalases, and superoxide dismutases (SOD) function as key players in the modulation of ROS levels and ROS-mediated stress signaling (Apel and Hirt 2004). Thus, it is not surprising that enzymes involved in detoxification have been detected in all studies regarding the early stages of somatic embryogenesis.

Ascorbate peroxidase (APX) plays a critical role in oxidative stress response through reactive oxygen species (ROS) scavenging. APX isozymes differ with respect to structure, substrate specificity, and tissue distribution, and their expression is highly responsive to environmental conditions and stress. A differential accumulation of APX has been reported in *Vitis vinifera* embryogenic (EC) and embryogenic calli (NEC) – two acidic isoforms were detected in EC only, while two basic isoforms were specific to NEC (Marsoni et al. 2008; Zhang et al. 2009). The proteomic profiles of the embryogenic phases of *Elaeis guineensis* revealed an APX present in all stages of somatic embryo development and a peroxidase present in primary callus and a putative secretory peroxidase unique to proembryogenic callus (Silva Rde et al. 2014). APX was found more abundant in the EC of two inbred lines of *Zea mays* L. – A19 (Varhaníková et al. 2014) and H99 (Sun et al. 2013) – as well as in callus of saffron (Sharifi et al. 2012). This might suggest that maintaining low levels of hydrogen peroxide is important for cell reprogramming towards SE. An increased APX was detected in mature embryos of cork oak as compared to proliferating embryos, with a slight decrease in intermediate stages as compared to the initial stage (Gomez-Garay et al. 2013).

In *Cyclamen persicum*, catalase was upregulated in embryogenic suspensions (ES) (Lyngved et al. 2008). Rode et al. (2012) found high levels of catalases of which the more acidic forms were specific for callus and early stages of SE. In contrast, the more basic forms had a higher abundance in the late embryos. These catalases are therefore candidates for stage-specific isoforms, probably affected by different phosphorylation levels. In mature embryos of oak, catalase and APX with diminished levels in proliferating embryos and increased abundance in the mature ones were detected (Gomez-Garay et al. 2013).

Glutathione-S-transferase (GST) proteins are involved in several processes, including protection from oxidative stress and detoxification of xenobiotics (Dixon et al. 2002). Winkelmann et al. (2006) found that GST is accumulated in embryogenic tissue and somatic embryos and may also have a possible role in detoxifying excessive amounts of auxin (Fehér et al. 2003). In *Citrus*, three different isoforms of GST with differential expression were detected – one was upregulated in late stages of SE, while the other two were only transiently downregulated in the first weeks after callus induction (Pan et al. 2009). Several examples have been reported in which members of the GST gene family were upregulated during auxin-induced somatic embryogenesis (Imin et al. 2005; Lyngved et al. 2008; Guzmán-García et al. 2013). Consistently, GST accumulation has been reported in somatic embryos of *Cyclamen persicum* (Winkelmann et al. 2006) and *Vitis vinifera* (Marsoni et al.

2008). A differential expression of GST was observed in *Medicago truncatula* – it consistently increased in both embryogenic and nonembryogenic lines from day 0 to 14 d reflecting the explant response to in vitro stress. When the first embryos started to differentiate, it dropped to its initial level in the embryogenic line M9-10a, suggesting that the M9-10a explants were able to cope with imposed oxidative stress (Almeida et al. 2012). Similar results were obtained in *Citrus sinensis* (Pan et al. 2009).

Thus, it can be concluded that oxidative stress may stimulate cell differentiation to promote somatic embryo formation, whereas other antioxidative proteins may serve to protect cells from toxicity caused by long-term in vitro culturing.

Superoxide dismutase (SOD) converts superoxide radicals to hydrogen peroxide, thus providing the first line of defense against oxidative stress. Mn-SOD is mainly localized in the mitochondrial matrix and the peroxisomes. SOD accumulation has been reported to be involved in triggering SE and is thought to be required for embryo germination. In *Cyclamen persicum*, SODs were abundant in both somatic and zygotic embryos (Winkelmann et al. 2006), but in *Quercus*, a significantly higher abundance of this enzyme was detected in somatic versus zygotic embryos (Gomez-Garay et al. 2013). A Mn-SOD was upregulated to much higher levels in EC than in NEC of *Vitis* (Zhang et al. 2009) and was overexpressed in cork oak (Gomez-Garay et al. 2013) and cotton cotyledonary and mature stages of somatic embryos (Ge et al. 2015).

During the scavenging of ROS, scavenger recovery proteins such as thioredoxin and glutaredoxin are necessary to regenerate scavengers. Imin et al. (2005) identified a thioredoxin H (Trx) which appeared early in explant cultures of the embryogenic cell line 2HA and became undetectable at later stages of cell proliferation suggesting that Trx H plays a significant role during early stages of commitment from the vegetative stage to a pathway of cellular differentiation and proliferation. The Trx H group of proteins is involved in a broad scope of biological functions, acting as cofactors, transcription regulators, protein-binding regulators, protein folding catalysts, growth factors, and antioxidants. In cotton cotyledonary embryos, three scavenger recovery proteins, namely, glutaredoxin, glutaredoxin-related proteins, and thiol–disulfide isomerase/thioredoxins, showed various abundance profiles, indicating that these proteins might regenerate scavengers for response to ROS in cotton embryo growth (Ge et al. 2015). In oak torpedo somatic embryos, Trx showed increased level upon maturation (Gomez-Garay et al. 2013). Peroxiredoxin (1-Cys Prx) showed considerably elevated levels of expression in later stages of SE development in the highly embryogenic cell lines 2HA (Imin et al. 2005) and M9-10a (Almeida et al. 2012) of *Medicago sativa*. Another peroxiredoxin – 2-Cys Prx – was highly expressed in *Quercus* torpedo somatic embryos (Gomez-Garay et al. 2013) and somatic embryos of *Cyclamen* (Rode et al. 2011). Peroxiredoxins are thiol-dependent antioxidants containing one (1-Cys) or two (2-Cys) conserved Cys residues that protect lipids, enzymes, and DNA against reactive oxygen species (Imin et al. 2005). A high abundance of these enzymes suggests the occurrence of stress within the somatic embryos tissue beyond that due to embryogenesis.

Cyclophilins constitute a subgroup of a large family of proteins called immunophilins. Most cyclophilins display peptidylprolyl isomerase (PPIase) enzymatic activity and active role in protein folding which render them molecular chaperones. Several of these members have also been directly linked to multiple stresses (Kumari et al. 2012). A higher expression of cyclophilin in the early stages of embryogenesis and response to various stresses has been reported as well as its involvement in the control of ROS (Ruan et al. 2011). The upregulation of cyclophilin in the globular and cotyledonary phases of *Coffea arabica* (Noah et al. 2013) may be related to the control of ROS levels. It is possible that a higher accumulation of ROS occurs in the globular stage and may account for the higher expression of cyclophilin to control ROS levels in the cell.

These results show that a dynamic expression of oxidative stress-related proteins occurs during SE. Somatic embryogenic lines show higher levels of GST, APX, and SOD proteins. Assuming that the generation of a significant amount of stress and ROS is a prerequisite to induce SE, it is possible that SE lines may have a better ability of controlling oxidative stress by regulation of the ROS-scavenging system and hence maintaining ROS homeostasis during somatic embryo growth and maturation. In this way, ROS can act as signaling molecules playing an important role during auxin-induced SE. As a whole, this sustains the hypothesis that SE is a cellular stress-adaptive response to in vitro culture conditions (Dudits et al. 1991; Pasternak et al. 2002).

5.2.1.2 Pathogenesis-Related Proteins

Another group of proteins involved in stress/defense response is the pathogenesis-related proteins (PR). PR proteins have been observed mostly during early stages of embryogenesis. A chitinase and β-1,3-glucanase considerably increased during the formation of proembryogenic masses (PEMs) in oil palm (Silva Rde et al. 2014). An increased expression of endochitinase was observed during the induction phase of SE in the highly embryogenic M9-10A of *M. truncatula* (Almeida et al. 2012). Chitinase IV was significantly more expressed in EC of *Larix principis* (Zhao et al. 2015). One of the most abundant proteins in embryogenic suspension cultures of cowpea was identified as chitinase from the PR-4 family of PR proteins (Nogueira et al. 2007). During SE in *Picea glauca*, the transcript of a β-1,3-glucanase gene was highly abundant in embryogenic tissues and gradually decreased during the induction of somatic embryos with the lowest abundance occurring during the globular embryo stage (Dong and Dunstan 1997). Five types of β-1,3-glucanase were detected at the globular stage in *M. truncatula* embryogenic cultures (Imin et al. 2004).

Chitinases catalyze the hydrolytic cleavage of the β-1,4-glycoside bond between N-acetylglucosamine residues, mainly present in chitin. β-1,3-glucanases are enzymes that catalyze the hydrolysis of the fungal cell wall polymer β-1,3-glucan. Chitinase induction is often coordinated with the expression of specific β-1,3-glucanases and other PR proteins in response to pathogen attack, as well as abiotic and biotic stresses. Despite their typical involvement in the defense response of plants, chitinases and glucanases are also expressed in healthy plants in an organ-specific and developmentally regulated pattern, suggesting a nondefensive role in plant development as well (Kasprzewska 2003). The accumulation of chitinases at a much higher level in the medium of embryo cultures than in nonembryogenic lines suggested that chitinase could be related to embryogenic competence for SE. A frequent occurrence of chitinases in embryogenic tissues has been associated with enzymatic activity on arabinogalactan proteins that enable the control or maintenance of embryogenic cell fate (van Hengel et al. 2001).

In both embryogenic and nonembryogenic cell lines of *Medicago truncatula*, two of the most abundant proteins were an abscisic acid (ABA)-responsive protein with homology to the pathogenesis-related protein PR10-1 and PR10-1 itself. Interestingly, they changed little throughout the 8 weeks of culture, suggesting a general role for ABA-responsive proteins and PR10 proteins in cell maintenance or cell defense (Imin et al. 2005). In EC of cowpea, a highly abundant protein was identified as PR10 (Nogueira et al. 2007). As some classes of PR-10 proteins are inducible by auxin, it is possible that this protein mediates the cell responses to growth regulators in cowpea. Also, PR10 proteins were found to be upregulated in NEC (Marsoni et al. 2008; Cangahuala-Inocente et al. 2009; Zhang et al. 2009) and downregulated in EC of *Vitis* (Zhang et al. 2009) and *Zea mays* inbred line H99 (Sun et al. 2013). PR10 proteins belong to an intracellular defense-related protein family and show homology to ribonucleases. In addition, PR10 genes are highly expressed in response to biotic and abiotic stresses (van Loon et al. 2006). The downregulation of PR10 in EC appears correlated with a better ability of controlling oxidative stress in EC cells. Thaumatin (a PR5 protein) was identified with maximum abundance at initial embryo stage of *Quercus suber* (Gomez-Garay et al. 2013). Thaumatins are subject to complex expression profiles regulated by environmental factors and developmental stages (Liu et al. 2010).

PR protein synthesis during in vitro cultures may be associated with the adaptation of plant cells to new environmental conditions, resulting in the activation of defense mechanisms that are likely not directly related to a specific morphogenic pathway such as SE. One hypothesis is that plant cells may activate signaling pathways that trigger cellular events in response to environmental stress, leading to the formation of embryonal structures (Van Loon et al. 2006). These findings indicate that PR-type proteins can have a developmental role and, through their enzymatic activities, may generate signal molecules that could act as endogenous elicitors in morphogenesis. Such elicitors could also play a role in activating other types of defensive responses.

5.2.2 Protein Synthesis, Processing, Folding, and Fate

5.2.2.1 Heat Shock Proteins (HSPs)

SE is a complete cell reprogramming process, and the initial stage of dedifferentiation of somatic cells to embryo-like structure needs global change in gene expression and protein complement. It requires the assembly and stabilization of newly synthesized proteins, as well as the modification and removal of peptides (Fehér et al. 2003).

The consideration of SE as a specific form of the stress response is supported by experimental findings that show the involvement of the heat shock systems in this developmental reprogramming. Many heat shock proteins (HSPs) are molecular chaperones, which are formed in response to stress and also are developmentally regulated. They assist in the correct folding of nascent and misfolded polypeptides by preventing their aggregation. Some of them are responsible for assembly translocation and degradation. Their function may be of increased importance under stress conditions, where misfolding of polypeptides occurs more commonly (Wang et al. 2004). Although HSPs are referred to as stress-responsive proteins, however, many of them are expressed in the absence of stress, during normal cell growth aiding in protein folding and subcellular sorting. HSPs are involved in SE even without being triggered by external signals.

Members of the HSP family HSP60 and HSP70 have been reported to be expressed at higher levels in EC than in NEC (Correia et al. 2012; Guzmán-García et al. 2013; Teyssier et al. 2014; Vale Ede et al. 2014; Zhao et al. 2015). Two different studies performed in *V. vinifera* reported the use of HSP70 as a possible marker for the embryogenic capacity based on their higher levels in embryogenic callus than in non-embryogenic callus (Marsoni et al. 2008; Zhang et al. 2009).

The ER luminal binding protein (BiP), a member of the HSP70 family, was overexpressed in EC of *Larix principis* (Zhao et al. 2015) and *Carica papaya* (Vale Ede et al. 2014). BiP localizes to the endoplasmic reticulum (ER). It acts in the translocation of proteins through the ER and assists in the proper folding and maturation of newly synthesized proteins entering the organelle (Vale Ede et al. 2014). HSP70 and luminal binding proteins have also been reported to be more abundant during the first stages of the development of SE in *M. truncatula* (Imin et al. 2005) and *Crocus sativus* (Sharifi et al. 2012). The expression level of these proteins was the highest at 5 weeks after the induction of SE from leaf explants, coinciding with the generation of calli. However, the chaperones showed a decrease in the 8-week cultures, when somatic embryos appeared. Similar results were reported in *P. glauca* (Lippert et al. 2005). In oil palm, an HSP90 was identified in all stages of SE with a higher intensity in callus after 14 days on induction medium. HSP90 seems an interesting candidate to be further investigated regarding its ability to induce embryogenic competence in oil palm (Silva Rde et al. 2014).

An ATP-binding protein is another HSP that showed increased abundance in the initial proliferation embryo stage (PSE) of cork oak. This protein controls protein folding during cell reorganization from somatic plant cells to the embryogenic pathway (Gomez-Garay et al. 2013).

This may imply that a higher level of expression of the chaperones is required for the maintenance of cells during early culture, playing a protective function in response to the stress conditions that characterize in vitro growth.

Different HSP levels were found to be stage specific during embryogenesis in *Cyclamen*. High levels of HSP20 and HSP70 were representative for differentiated embryos marking late stages of embryogenesis, while increased abundances of HSP60 and HSP101 were typical for earlier stages and callus. These proteins help newly synthesized proteins to fold and minimize protein aggregation upon stresses imposed during initiation of SE from somatic cells (Rode et al. 2012).

Besides at early stages of SE, HSPs were more abundant at later stages of embryo development, highly increasing during somatic embryo maturation of cork oak (Gomez-Garay et al. 2013), hybrid larch (Teyssier et al. 2014), and somatic and zygotic embryos of *C. persicum* (Winkelmann et al. 2006). Probably their high expression is necessary to prepare the embryos for desiccation, consistent with the high level of proteins synthesized at this stage. In coffee SE, an HSP70 was more abundant in the cotyledonary phase when compared to the torpedo stage and was not observed in the globular stage (Tonietto et al. 2012).

Small heat shock proteins (sHSPs) of 15–30 kDa have been reported in high abundance during the late stages of SE. Small HSPs are not only crucial components of the plant heat shock response but also play important roles in adaptation to various other stresses and development (Waters 2011). One sHSP was much more expressed in EC than in NEC of *Vitis* (Zhang et al. 2009). An sHSP was found to accumulate during maturation of oak somatic embryos (Gomez-Garay et al. 2013). Small HSPs have been also found to be upregulated from early to mature stages of SE development in *P. glauca* (Lippert et al. 2005) and *P. abies* (Businge et al. 2013). In *Pinus pinaster*, two sHSPs were found to be overexpressed in cotyledonary somatic embryos (two isoforms of HSP 18.2), and one in cotyledonary zygotic embryos (class II HSP 17.6) together with HSP60 and HSP70 has been overexpressed in both types of embryos (Morel et al. 2014). In *Phoenix dactylifera*, stress-related proteins of the HSP family (17.6 and 70 kDa) were detected in zygotic embryos only. These proteins are especially abundant at the late stages of embryo maturation (Sghaier-Hammami et al. 2009). The presence of sHSP 17.6 in zygotic embryos only makes it a potential marker of zygotic embryo maturity in *Pinus pinaster* (Morel et al. 2014) and *Phoenix dactylifera*. HSP 18.2 and HSP 17.6 belong to a small HSP family (HSP 20) which has been detected in differentiated embryos in *Cyclamen* (Rode et al. 2012), but their function is not yet known.

The data suggest that the increased chaperone proteins may play a fundamental role in SE possibly by alleviating stresses associated with global reprogramming during somatic to embryogenic transition. In addition, HSP proteins all contribute to the preparation of the embryo for subsequent desiccation and germination phases.

5.2.2.2 Protein Processing and Fate

Cellular reprogramming ultimately requires the synthesis, protein folding, and posttranslational modification of newly synthesized proteins, as well as the removal of proteins that are no longer needed. The ubiquitin/26S proteasome pathway is a major factor in controlled proteolysis. The ubiquitin–proteasome pathway can be regulated at the level of ubiquitination or the level of proteasome activity (Glickman and Ciechanover 2002). In plants, the ubiquitin–proteasome system can control nearly every aspect of growth and development, such as the cell cycle, embryogenesis, defense, environmental responses, and hormone signaling (Vierstra 2009).

Consistently, 26S proteasome regulatory particle triple-A ATPase subunits α and β were found to be upregulated in EC of *C. persicum* (Lyngved et al. 2008), *V. vinifera* (Zhang et al. 2009), *C. sativus* (Sharifi et al. 2012), and *Z. mays* (Sun et al. 2013). High expression levels of the proteasome subunits in EC imply the possible role of proteasome machinery in callus establishment through removal of explant-associated proteins that are no longer needed. Subunits of the 26S proteasome and ubiquitin were highly abundant throughout SE in *C. persicum* and *C. sativus*, but especially noticeable in the late torpedo stages (Sharifi et al. 2012; Rode et al. 2012). It is well known that proteasome activity is closely aligned with cell proliferation processes. Rapid cell division marking the early stages of SE is accompanied by an increase in proteasome subunits. Thus, high levels of enzymes involved in the 26S proteasome-dependent proteolysis pathway seem to be important for the switch from callus to globular embryo as well as from globular to torpedo-shaped embryo in these species. A decrease in proteasome subunit abundance was observed during cotyledonary embryo development in cork oak (Gomez-Garay et al. 2013) and cotton (Ge et al. 2015). In cotyledonary embryos, the need for saving energy slows down protein synthesis, whereas reduction in protein degradation also facilitates nutrient storage for germination and plantlet regeneration (Ge et al. 2015). Also, proteasome activity decreases when cells are stimulated to differentiate as is the case. In a previous study on early somatic embryo development in *Picea glauca*, it has been reported that three proteasome proteins were simultaneously downregulated as the culture medium containing abscisic acid stimulates embryo maturation and concomitant differentiation, thereby reducing cell division (Lippert et al. 2005). Hence, the hypothesis that a decrease or absence of proteasome subunits might be an appropriate marker for tracking proper embryonic development has been widely accepted.

To summarize, controlled proteolysis executed by the 26S proteasome–ubiquitin system is needed in order for the somatic embryogenesis to succeed.

Proteomic studies of SE revealed the presence of proteinases and proteinase inhibitors in several plant species (Nogueira et al. 2007; Marsoni et al. 2008; Sharifi et al. 2012; Noah et al. 2013). A leucine aminopeptidase was found to be unique to the PEG treatment of embryogenic cultures of *Carica papaya* improving somatic embryo production thus being indicative of better control of embryonic development under this treatment (Vale Ede et al. 2014). An exceptional high accumulation of trypsin inhibitors and aspartic proteinase was observed in somatic and zygotic embryos of *Theobroma cacao* L., respectively (Noah et al. 2013). Regarding the crucial role of aspartic proteinase, which is responsible for the initial breakdown of proteins during germination, its relatively early accumulation in zygotic embryo in the torpedo stage might be an indication of the onset of maturation. Recently, Guilloteau et al. (2005) found trypsin inhibitor to be part of an active aspartic proteinase complex in cacao seeds. They suggested this trypsin inhibitor subunit to protect storage proteins from precocious hydrolysis that might result from the high aspartic proteinase content in cacao seeds.

Besides their well-known function to degrade damaged, misfolded, and harmful nonfunctional proteins, proteases indeed play key roles in the maturation of cell wall proteins and the generation of active peptides (van der Hoorn 2008). The endogenous cysteine proteinase inhibitors – the cystatins – form a tight, reversible complex with cysteine proteases, thus exerting a fine control of their activity during embryogenesis, organogenesis, programmed cell death, and tolerance to abiotic and biotic stresses (Benchabane et al. 2010; Turk et al. 2012). Thus, the precise control of the proteolytic processes is of utmost importance for the correct plant growth and development (Novinec and Lenarčič 2013).

5.2.2.3 Protein Synthesis

Protein metabolism is a key factor in somatic embryogenesis induction. In line with this, one of the largest functional groups found in ECs in *Cyphomandra betacea* was that of proteins involved in protein biosynthesis, namely, ribosomal proteins. In addition, pentatricopeptide repeat-containing proteins, known by their role in protein–protein interactions related to a large variety of functions (chaperoning, transcription, etc.), were exclusively found in EC (Correia et al. 2012). Accordingly, three protein synthesis-related proteins [histone H2B.2; 40S ribosomal protein, protein disulfide isomerase (PDI)], were detected in EC of maize which is in agreement with the high number of proteins synthesized during the formation of calli (Sun et al. 2013). Ten differentially expressed ribosomal proteins were downregulated in cotyledonary embryos, indicating that biosynthesis of proteins was decreased in cotyledonary embryos (Ge et al. 2015).

To summarize, a global change in protein metabolism takes place during the somatic-to-embryogenic cell transition. This is reflected in the overexpression of molecular chaperones such as HSPs and cyclophilins, proteasome subunits to process unnecessary proteins, a duo of proteinases and their inhibitors to fine-tune developmental processes, as well as proteins involved in intense protein synthesis.

5.2.3 Cell Proliferation and Cell Wall Metabolism

The induction of SE includes dedifferentiation in somatic plant cells and establishment of embryogenic competence. These processes require reactivation of the cell division in somatic plant cells. Embryogenesis is accompanied by a morphogenesis process during which the embryo differentiates through several distinct stages – globular, heart-shaped, torpedo, and cotyledon stages. The morphogenesis is based on coordinated cell elongation and division (Pan et al. 2009).

Tubulin and actin are ubiquitous components of the eukaryotic cell. Actin microfilaments and tubulin microtubules comprise the cytoskeleton. Alpha- and β-tubulins are assembled coordinately, forming microtubules in response to various intracellular and extracellular signals and participating in many different functions in eukaryotic cells. Tubulins are known to be associated with cell division and cell elongation and played a significant role in the separation of the organelles and daughter chromosomes (mitosis) (Pan et al. 2009).

Although tubulin is considered a housekeeping protein and has been widely used as a constitutive standard in gene expression, its upregulated differential expression was observed during SE until the globular stage in *C. sinensis* (Pan et al. 2009), but gradually downregulated upon maturation in white spruce (Lippert et al. 2005) and cork oak (Gomez-Garay et al. 2013). Also, some tubulins exhibited higher expression in the EC compared to the NEC (Zhang et al. 2009; Sun et al. 2013; Zhao et al. 2015).

In plants, the actin cytoskeleton plays significant roles in the definition of cell polarity and orientation of cell division, cell elongation, cell wall development, transport processes, positioning of membrane receptors, and in PCD. A study on the role of actin isoforms in SE in Norway spruce demonstrated that actin isoforms were expressed predominantly in suspensor cells (Schwarzerova et al. 2010).

Actin is also a common reference gene in gene quantitative expression and was recently found to be involved in programmed cell death (PCD) and cell defense mechanisms against biotic and abiotic stress. Depolymerization of the actin cytoskeleton acts as a downstream regulatory factor in the plant stress adaptation networks and participates in triggering the execution of PCD (Malerba et al. 2008). In accordance, different actin forms were found in embryonic masses of cassava (Baba et al. 2008) and *C. persicum* (Lyngved et al. 2008). Two actin isoforms were observed at higher levels in EC of *Larix principis* (Zhao et al. 2015).

Proliferating cell nuclear antigen (PCNA) is an evolutionarily well-conserved protein found in all eukaryotic species. It is able to interact with multiple partners in several metabolic pathways such as DNA repair, translation DNA synthesis, DNA methylation, chromatin remodeling, and cell growth and apoptosis, suggesting a function in the regulation of the cell cycle (Maga and Hubscher 2003). In accordance, the involvement of PCNA in the SE process was first shown in *Vitis vinifera* (Marsoni et al. 2008) and Zhao et al. (2015) detected an elevated level of PCNA in EC of *Larix principis*.

Annexins have been identified in cotyledons of cassava somatic embryos undergoing secondary SE (Baba et al. 2008). Annexins occurred at more than double the abundance in EC of *Larix principis* (Zhao et al. 2015) and were found highly expressed during SE of oil palm (Silva Rde et al. 2014) and in globular and cotyledonary embryos of coffee (Tonietto et al. 2012). Annexins are multifunctional proteins expressed throughout the life cycle that appear capable of linking Ca^{2+}, redox reactions, and lipid signaling to coordinate development in response to the biotic and abiotic environment. Strong evidence also indicates that annexins are involved in cell division during the cell cycle (Laohavisit and Davies 2011).

Accepting the fact that the initiation of somatic embryogenesis is closely linked to hormone-induced cell divisions, its molecular characterization can also be based on genes with cell cycle-dependent expression (Dudits et al. 1995). One protein that is involved in cell growth and cell division through microtubule stabilization, the translationally controlled tumor protein homolog (TCTP), was unique to ES (Lyngved et al. 2008). This protein has previously been identified in embryogenic cell suspensions of cowpea (Nogueira et al. 2007).

The plant cell wall is a highly dynamic entity whose structure and composition changes dramatically during plant growth and cellular differentiation; it serves as the first mediator in cell-to-cell communications and protects cells from biotic and abiotic stresses (Wolf et al. 2012). Cell wall and membrane formation are enhanced during embryogenesis. In line with this, α-1,4-glucan protein synthase, a glycosyltransferase, was upregulated in EC of *Larix principis* (Zhao et al. 2015), and this is consistent with observations in *C. persicum* (Lyngved et al. 2008). In addition, higher abundance of this protein was found in mature somatic embryos during the SE process (Teyssier et al. 2014). The enzyme builds up α-1,4-glucan chains covalently bound to protein and is involved in UDP forming and the synthesis of polysaccharides in the cell wall. A cell wall development protein containing a leucine-rich repeat (LRR) and an extension domain, a LRR family protein/extension family protein which is involved in cell expansion and growth, was identified in embryogenic cultures of *Cyclamen*. The presence of high levels of this protein in both globular embryos and torpedo embryos is consistent with the active embryogenic cell division that occurs during these developmental stages of SE (Bian et al. 2010).

Enzymes involved in cell wall and membrane synthesis were upregulated in embryogenic suspension cultures (ES). Acyl-[acyl-carrier protein] desaturase (unique to ES) takes part in fatty acid biosynthesis and is a key determinant of the overall level of unsaturated fatty acids in the cell (Lyngved et al. 2008). Several proteins involved in cell wall metabolism were found upregulated (pectin methylesterase, pectate lyase, cinnamyl alcohol dehydrogenase) and others downregulated (pectin acetylesterase and beta-expansin 1a precursor) in cotyledonary embryos of cotton. The downregulated pectin acetylesterase and upregulated pectin methylesterases promote cell and organelle elongation in cotyledonary embryos. Overall, cell wall synthesis and the loosening of cross-links are likely to regulate cell expansion in somatic embryos (Ge et al. 2015).

To summarize, higher levels of some proteins involved in the cell cycle, cell wall biogenesis, and expansion were found in EC than in NEC as well as throughout the process of SE in agreement with an active embryogenic cells division, cell wall plasticity, cell growth, and differentiation.

5.2.4 Signal Transduction

Levels of some proteins involved in signaling were found to be higher in ES than in NES in *C. persicum* (Lyngved et al. 2008) and *Larix principis* (Zhao et al. 2015). The 14-3-3-like protein GF14-D, similar to a protein found in embryogenic cell suspensions by Nogueira et al. (2007), belongs to the 14-3-3 proteins, which are highly conserved phosphoserine-/phosphothreonine-binding proteins. 14-3-3 proteins regulate a broad range of target proteins in all eukaryotes through phosphorylation and may allow the growth and development of cells to be coordinated with the metabolic status of the plant as well as environmental stresses (reviewed in Denison et al. 2011). The phosphorylation-dependent binding of 14-3-3 proteins regulate negatively the activity of mitochondrial and chloroplast ATP synthases that may suggest a mechanism for plant cells to adapt to environmental changes such as nutrient supply, and especially exogenous plant growth regulators, during the initiation step of SE. In addition, these proteins participate in the regulation of various biochemical processes during seed development (Zhao et al. 2015). Such proteins were found in embryogenic cultures of *C. papaya* (Vale Ede et al. 2014). Two 14-3-3 proteins were more abundant in proliferating embryos than in their mature counterparts in oak (Gomez-Garay et al. 2013).

Protein phosphatase 2A (PP2A) 65-kDa regulatory subunit has been previously associated with the embryogenesis process by Marsoni et al. (2008). PP2A is a ubiquitous and conserved serine/threonine phosphatase with a multifunctional regulatory activity in plants. The A subunit, in particular, is essential for auxin transport, functioning as a positive regulator of the PP2A holoenzyme, and is involved in differential cell elongation responses. Plant PP2As also participate in biotic and abiotic stress signaling pathways. Recent functional analysis of PP2As revealed that they are key components of stress signal transduction pathways, playing positive and dynamic roles in stress signaling (reviewed in Pais et al. 2009). Dudits et al. (1995) reported a marked increase in phosphorylation of defined proteins in embryogenic cells compared to the nonembryogenic cells. Accordingly, the embryogenic suspensions of *C. persicum* expressed higher levels of PP2A and a 14-3-3-like protein, both of which play important roles in protein phosphorylation (Lyngved et al. 2008). In addition, G proteins and calreticulin have been classified as promising candidates for involvement in signal transduction in *C. persicum* and *Larix principis* embryogenic cultures (Lyngved et al. 2008; Zhao et al. 2015).

The higher abundance of PP2A regulatory subunit and 14-3-3 proteins in embryogenic cultures and at early stages of embryogenesis suggests that they might play a protective role against the stress from in vitro culturing in addition to the stress accompanying the cellular reprogramming.

5.2.5 Proteins Involved in the Metabolism

During SE, plant cells undergo reprogramming of their metabolism and energy consumption, especially carbon and nitrogen metabolism (Fehér et al. 2003). The fast growth and high cell division rate of callus and developing somatic embryos require high amounts of energy. This explains why the largest class of differentially expressed proteins in EC and during SE usually are proteins involved in energy metabolism. Several glycolytic enzymes, as well as proteins involved in energy metabolism most frequently detected in the proteomic studies of different plant species, are discussed below.

The glycolytic enzyme, triosephosphate isomerase, plays a significant role in glycolysis and is essential for effective energy generation. It catalyzes the reversible interconversion of the triosephosphate isomers dihydroxyacetone phosphate and D-glyceraldehyde 3-phosphate. In comparison with zygotic embryos, higher accumulation of triosephosphate isomerase occurred in somatic embryos in *C. persicum* (Winkelmann et al. 2006) and in EC of *Larix principis* (Zhao et al. 2015) and cowpea (Nogueira et al. 2007). However, in EC of *Vitis* (Marsoni et al. 2008), triosephosphate isomerase was downregulated. A similar result was found in *Cyclamen* torpedo embryos (Bian et al. 2010). Ito et al. (2003) proposed that triosephosphate isomerase was among the key enzymes of the regulation mechanism that slowed down glycolysis and the tricarboxylic acid cycle rate under oxidative stress in order to lower the deleterious production of reactive oxygen species (Marsoni et al. 2008).

The overexpression of transketolase, one enzyme of the pentose pathway, could be correlated with the active proliferation of somatic embryos that need biosynthetic intermediate and NADPH (Marsoni et al. 2008; Lyngved et al. 2008).

Fructokinase (FRK) is a primary enzyme of glycolysis and is involved in starch synthesis. FRK was one of the most abundant proteins in the proteome reference map generated from embryogenic cell cultures in *M. truncatula* (Imin et al. 2005). This enzyme was also detected during the SE of *Citrus sinensis*, with the highest expression level in early stages of embryogenic callus (Pan et al. 2009). This protein has been reported previously as exclusively increased in developed calli compared with corm explant (Sharifi et al. 2012). In oil palm, fructokinase was detected first at the earliest stage of SE induction and was also observed in the next stages of development indicating that starch synthesis occurred during the first 2 weeks of SE induction. The deposition of starch in the cortex of the embryogenic callus has been observed in *C. sativa* and is mobilized and used as an energy source by the meristematic cells during intense cell division and differentiation (Silva Rde et al. 2014).

Enolase that catalyzes the conversion of 2-phosphoglycerate (2-PG) to phosphoenolpyruvate (PEP) in the glycolytic pathway was found highly abundant in somatic embryos of *C. persicum* (Rode et al. 2012), *Eruca sativa* (Chen et al. 2012), and oil palm (Silva Rde et al. 2014). Lippert et al. (2005) detected a high expression of this protein only in the torpedo stage of *Picea glauca* and therefore suggested that enolase could be an interesting candidate to be used as a molecular marker of embryogenesis maturation. Similarly, enolase was also expressed only in the torpedo stage of coffee somatic embryos, indicating that this protein could be used as a molecular marker of the torpedo stage in different plants (Tonietto et al. 2012). A highly abundant enolase in EC was also confirmed by two independent proteomic studies in *Z. mays* (Sun et al. 2013; Varhaníková et al. 2014).

Glyceraldehyde-3-phosphate dehydrogenase (GAPDH) is an essential enzyme of glycolysis. GAPDH has been found in higher amounts in somatic embryos compared to zygotic embryos (Winkelmann et al. 2006). Four GAPDH proteins were unique to ES in *Cyclamen* (Lyngved et al. 2008). Nogueira et al. (2007) also identified several proteins as GAPDH in proembryogenic masses of cowpea. GAPDH was highly expressed in the globular phase and showed a reduced expression in the other SE stages of coffee (Tonietto et al. 2012), while GAPDH peaked at oak cotyledonary somatic embryo suggesting that this stage is more demanding in terms of energy and precursors for the synthesis of primary metabolites such as amino acids and fatty acids (Gomez-Garay et al. 2013). The distinct abundance patterns of GAPDH could reflect regulation along the glycolysis pathway at different developmental stages.

Smith and Krikorian (1990) have shown that low pH is essential for maintaining the proembryogenic stage in carrot, while high average intracellular pH favors regeneration. At low pH, the ATPase H^+ pump is stimulated. Accordingly, vacuolar ATP synthase subunit B2, ATP synthase subunit beta, and ATPase alpha F1 were highly upregulated in ES compared to NES in *Cyclamen* (Lyngved et al. 2008). The 14-3-3 proteins known as positive regulators of H^+-ATPase activity (Chen et al. 2006) were upregulated in ES, too. The abundance of ATPase in EC may explain the proembryogenic state of the culture, but may also mean that the embryogenic cells need more energy for their metabolic changes (Lyngved et al. 2008). ATP is a ubiquitous energy source that can also act as a signaling molecule in cellular metabolism. In *Abies alba* Mill., an increase in the levels of adenosine triphosphate (ATP) is associated with the maturation of SEs (Petrussa et al. 2009). In addition, mitochondrial activities like ATP catabolism influence plant cell death, which is considered essential for correct SE maturation (Bozhkov et al. 2005). A beta subunit of mitochondrial ATPase in mature ZEs and embryogenic R cell line during the proliferation phase was detected in *A. angustifolia* (Silveira et al. 2008; Jo et al. 2014) defining the mitochondrial ATPase as the marker for selection of EC lines responsive to maturation treatments. In plants (Vianello et al. 2007), the energy status, particularly ATP levels, is crucial to the start of PCD. Thus, it is possible that the proper development of SEs in the R cell line could be linked with an adequate PCD mechanism, which was absent in the nonembryogenic B cell line (Jo et al. 2014). The presence of this protein suggests a high energetic metabolism in ECs of the cell line. However, downregulation of the β subunit of ATP synthase was observed in EC of *Vitis* (Zhang et al. 2009) and *Zea mays* (Sun et al. 2013) possibly indicating considerable damage to this subunit.

The upregulation of S-adenosyl methionine (SAM) synthase has been observed in various conifers at stages from early to late embryogenesis of both somatic and zygotic embryos (Lippert et al. 2005; Balbuena et al. 2009; Jo et al. 2014; Teyssier et al. 2014; Morel et al. 2014). SAM synthase likely contributes to enhanced amino acid metabolism (Teyssier et al. 2014) as well as to alterations in both polyamine content and ethylene biosynthesis (Morel et al. 2014) and might be essential for the transition of PEM to somatic embryos. In *Picea glauca*, SAM synthase has been characterized as a biochemical marker of early somatic embryo development (Lippert et al. 2005). SAM synthase converts the amino acid methionine into SAM, which is the donor of methyl groups to the DNA methylation system by SAM-dependent methyltransferase. High levels of methylation during plant embryogenesis are associated with chromatin modeling, selective gene expression, and growth of somatic embryos (Morel et al. 2014).

Interestingly, energy metabolism is more active in somatic embryos compared to zygotic embryos, as previously reported by Sghaier-Hammami et al. (2009), Teyssier et al. (2014), Morel et al. (2014), and Noah et al. (2013) in mature somatic embryos from *Phoenix dactylifera* L., *Larix* × *eurolepsis*, *Pinus pinaster*, and *Theobroma cacao* L., respectively. The higher abundance of glycolytic, citrate cycle, and ATP synthesis enzymes in somatic embryos than in zygotic embryos indicates a more active energy metabolism and ATP demand in the former than in the latter and may explain why matured somatic embryos are capable of entering the germination phase without undergoing the desiccation-induced dormancy program observed for zygotic embryos.

SE is a complex developmental process, and the data discussed above indicate that it is extensively based on carbohydrate metabolism ensuring the heavy energy demand required for metabolic processes that occur during cell division and elongation.

5.2.6 Storage Proteins

The differential accumulation of specific storage proteins during different stages of somatic and zygotic embryo development has been reported, and they may be indicators of embryo developmental stages. Proteomic studies reveal that synthesis of storage proteins starts with the globular stage of somatic and zygotic embryos reaching its peak at the maturation stage. 7S globulin has been identified in *Cyclamen* (Winkelmann et al. 2006; Rode et al. 2011) and in *Pinus pinaster* (Morel et al. 2014) zygotic and somatic embryos, with higher content in zygotic embryos. The accumulation of 11S globulin in the torpedo phase of coffee somatic embryos could be to guarantee the energy necessary during embryo maturation in the cotyledonary phase (Tonietto et al. 2012). In oil palm, glutelin accumulation could be used as a biological marker of an early stage of SE since it was only observed in the first stage of SE (Silva Rde et al. 2014).

Legumin precursor was highly abundant in oak globular and cotyledonary embryos (Gomez-Garay et al. 2013) and has been found upregulated in the mature in vitro derived embryo tissue of *Picea glauca* (Lippert et al. 2005). In some species, the level of this type of reserve proteins has been used to distinguish between zygotic and somatic embryos due to a lower accumulation of reserve proteins in somatic embryos. Three storage proteins, namely, seed storage protein vicilin A and B and 2S albumin storage protein, were upregulated in cotton cotyledonary embryos (Ge et al. 2015). Legumin-like and vicilin-like proteins were highly expressed in both somatic and zygotic cotyledonary embryos of hybrid larch (Teyssier et al. 2014). A glutelin in *Phoenix* (Sghaier-Hammami et al. 2009) and a legumin-like protein in *Pinus pinaster* (Morel et al. 2014) were downregulated in somatic embryos relative to zygotic embryos, supporting the hypothesis of a lower accumulation of reserve proteins in somatic embryos. The putative novel storage proteins in *Cyclamen*, the small enolases (Rode et al. 2011), were present in all developmental stages, but their abundance increased significantly towards later stages of embryogenesis in *Cyclamen* embryos (Rode et al. 2012).

The synthesis of large amounts of storage proteins is regarded as a marker for embryo maturation, and the storage proteins themselves potentially represent excellent markers of embryo quality in gymnosperms as well as in angiosperms. These proteins all contribute to the preparation of the embryo for subsequent desiccation and germination phases. The comparative proteomic studies indicate that the somatic embryo lacks/accumulates fewer reserve proteins compared to the zygotic embryo and there are possibilities of improving the storage protein content which depends on the SE culture medium as was observed in date palm. The addition of abscisic acid to the culture medium of date palm somatic embryos induced the accumulation of a 22 kDa glutelin enabling the embryos to accumulate the necessary reserves for normal germination (Sghaier et al. 2009). Studies at the proteomic level may help to improve the quality of SE-derived seedlings and to develop new in vitro culture strategies for plant propagation and manipulation.

5.3 Secretome Analysis of SE

Plant secretomics is a newly emerging area of plant proteomics (reviewed by Agrawal et al. 2010; Alexandersson et al. 2013). Agrawal et al. (2010) described it as "the global study of secreted proteins into the ECS by a cell, tissue, organ or organism at any given time and conditions through known and unknown secretory mechanisms involving constitutive and regulated secretory organelles." The plant cell wall, also called the extracellular matrix (ECM), is an extremely dynamic entity whose structure and composition change dramatically during cellular differentiation; it serves as the first mediator in cell-to-cell communications and protects cells from biotic and abiotic stresses. Extracellular proteins participate in cell morphology, cell division, proliferation, plant defense reactions, responses to stress, and cell-to-cell adhesion (Tian et al. 2009; Rose and Lee 2010). Despite the crucial role played by extracellular proteins in these diverse and significant processes, the extracellular proteome (secretome) has been less well characterized than other subcellular compartments.

SE in cell suspension cultures provides a good model system for investigating early plant development. The culture medium of plant cell cultures may be regarded as a large extension of the intercellular space; soluble secreted molecules that inhabit the apoplast *in planta* accumulate in the medium and exert their effect on cell growth and development. Indeed, conditioned medium harbors a complex array of molecules which play a significant role in SE by their ability to either promote or inhibit embryo development (for reviews, see Matthys-Rochon 2005; Ruiz-May et al. 2010; Fehér 2014).

We investigated the extracellular proteins in *Dactylis glomerata* L. embryogenic and nonembryogenic suspension cultures searching for potential early embryogenic markers.

Leaf explants from a highly embryogenic *D. glomerata* L. genotype were grown on induction SH30 medium containing the hormone dicamba. After 4 weeks, the younger basal tissue gave rise to callus which upon subculturing on the same medium segregated into embryogenic callus (EC) and nonembryogenic callus (NEC) (Fig. 5.1). Both calluses were morphologically distinct. The EC was more compact and of light yellow color, while the NEC was translucent showing needle-like structures. Somatic embryos from *D. glomerata* L. callus-derived suspension cultures fully developed on a hormone-containing medium only from the embryogenic callus. Two to three days after inoculation of induction medium, competent single cells in embryogenic suspension cultures started to divide intensively and formed microclusters (Fig. 5.1, panel a). Proembryogenic masses (PEMs) containing centers of embryonic growth were formed a week later (Fig. 5.1, panel b), and after 2 weeks, globular embryos began to differentiate from the cell masses (Fig. 5.1, panel c). Single cells from the nonembryogenic suspension culture were divided to form microclusters with blocked further development (Fig. 5.1, panel d).

In an attempt to identify SE-specific extracellular proteins, we analyzed the secretome of microclusters from *D. glomerata* L. embryogenic and nonembryogenic suspension cultures using 2-DE and LC-MS/MS. Fractions of microclusters were collected as described (Tchorbadjieva and Odjakova 2001), and after seven days in culture, the culture medium was used as a source for extracellular proteins. Representative 2-DE gels of extracellular proteins secreted from microclusters of both suspension cultures are shown in Fig. 5.2. Visual inspection of the gels revealed distinct protein patterns. Microclusters from the embryogenic suspension culture secreted approximately 40 proteins as compared to ca. 15 proteins secreted from the microclusters of the nonembryogenic line over

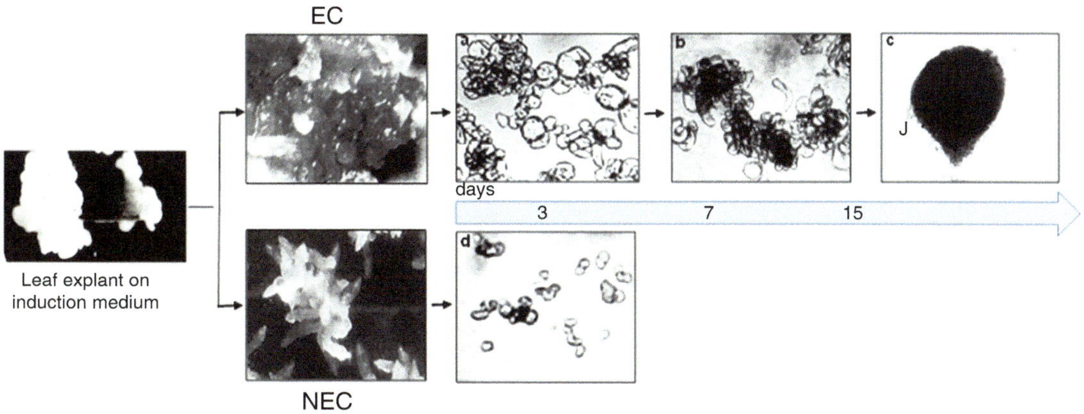

Fig. 5.1 Somatic embryogenesis of *Dactylis glomerata* L. Leaf explants cultured on induction (SH30) medium produce both embryogenic callus (EC) and nonembryogenic callus (NEC). Only EC is able to produce somatic embryos which fully develop in hormone-containing liquid medium before being transferred to a hormone-free medium for regeneration. Single cells divide to form microclusters (**a**), microclusters develop to form proembryogenic masses (PEMs) (**b**), embryos differentiate from PEMs (**c**). Single cells from NEC form microclusters only whose further development is blocked (**d**)

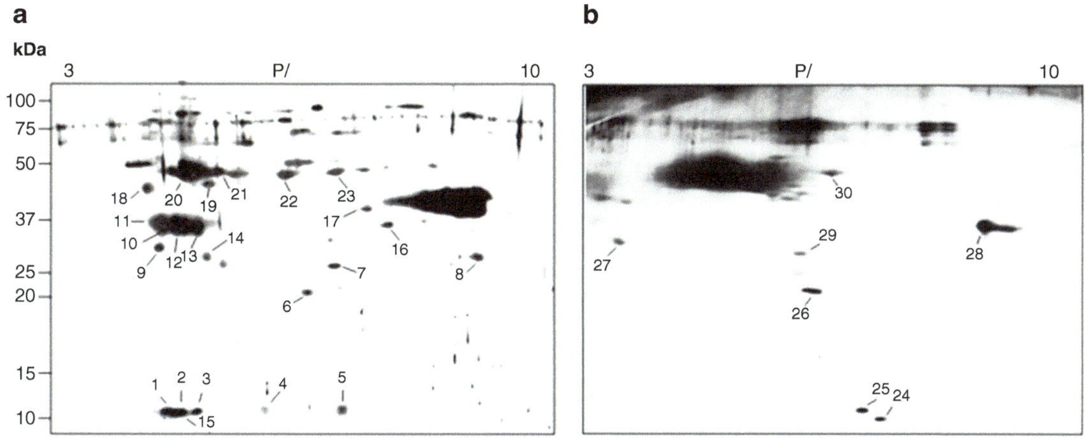

Fig. 5.2 2-DE analysis of proteins secreted into the culture medium of microclusters from *D. glomerata* L. embryogenic (**a**) and nonembryogenic (**b**) suspension cultures after 7 days in culture. Proteins were separated in the first dimension on an immobilized linear pH 3–10 gradient and in the second dimension on a 13 % acrylamide-SDS gel. Protein spots indicated in the gels were selected for protein identification. The positions of pI 3–10 and molecular mass markers are marked. Proteins were visualized by silver staining

a pH range of 3–10 and a size range of 10–100 kDa. In all, 30 differentially expressed proteins were excised, digested in-gel with trypsin, and subjected to LC-MS/MS identification. The proteins were identified by search against nonredundant protein database at the NCBI. Of the 30 candidate spots, only ten proteins (33 %) were successfully identified. This lower percentage of identification is typical of orphan organisms that are almost absent in public databases. In addition, because of the particular interest towards spots numbered 1, 2, 3, 4, and 5 provoked from previous studies and the unsuccessful attempt to identify them by MS/MS, we used N-terminal sequencing and identified five isoforms of an acidic cysteine proteinase inhibitor – cystatin (Table 5.2). For the remaining spots, a low score or no hits were observed. The spot ID corresponds to protein spots shown in

Table 5.2 Extracellular proteins secreted by microclusters from *D. glomerata* L. embryogenic suspension cultures identified by N-terminal sequencing and LC-MS/MS

Spot no	Protein name	Accession	Species	Theor. pI/MM	Exper. pI/MM	Peptides	Seq. coverage
			N-terminal sequencing				
1	Cystatin Hv-CP18	CAG38129.1	*Hordeum vulgare*	8.4/10.7	4.3/11.0	1	20
2	Cystatin Hv-CP18	CAG38129.1	*Hordeum vulgare*	8.4/10.7	4.5/11.0	1	20
3	Cystatin Hv-CP18	CAG38129.1	*Hordeum vulgare*	8.4/10.7	4.8/11.0	1	20
4	Cystatin Hv-CP18	CAG38129.1	*Hordeum vulgare*	8.4/10.7	5.2/11.0	1	20
5	Cystatin Hv-CP18	CAG38129.1	*Hordeum vulgare*	8.4/10.7	6.4/11.0	1	20
			MS/MS analysis				
8	Desiccation-related protein PCC13-62-like	XP_006653362.1	*Oryza brachyantha*	9.2/28.8	6.0/28.0	2	12
9	Chitinase	BAA22968.1	*Chenopodium amaranticolor*	8.6/26.1	4.2/32.0	4	16
10	Putative class III chitinase	BAC10165.1	*Oryza sativa*	9.3/28.7	4.2/34.0	3	14
11	Cathepsin B	CAC83720.1	*Hordeum vulgare*	6.0/34.9	4.3/36.0	7	22
12	Cysteine endopeptidase precursor	AAD20453.1	*Oryza sativa*	5.6/37.2	4.6/36.0	6	25
13	Cysteine protease Cp precursor	ABS12459	*Citrus sinensis*	6.0/36.9	4.8/36.0	7	34
14	Class III peroxidase	ACF08095.1	*Triticum aestivum*	9.0/31.3	4.8/30.0	2	7
15	Pathogenesis-related protein 4	AAT67050.1	*Triticum monococcum*	4.5/13.0	4.5/11.2	1	16
20	Barley α-amylase 2	1AVA_A	*Hordeum vulgare*	5.6/45.0	4.5/45.0	16	41
21	Barley α-amylase 2	1AVA_A	*Hordeum vulgare*	5.6/45.0	5.1/45.0	16	41

The table shows, spot no, spot number in Fig. 5.2; accession, NCBI accession number; peptides, number of identified tryptic peptides; theor. pI/MM, theoretical pI values and molecular mass; exper. pI/MM, the experimental pI values and molecular mass were determined according to the spot position in the 2-DE gel; seq. coverage, the percentage of sequence coverage of the match

Fig. 5.2. In most cases, the theoretical MMs agreed well with experimental values, still for some proteins (spots 9, 10, 12, 13, and 14) a discrepancy was observed. Additionally, two proteins (spots 20 and 21) matched the same gene sequence but had different pIs. Interestingly, for all spots identified, the experimental pI values were much lower and largely deviated from that of the corresponding theoretical ones. These phenomena are commonly found on 2-DE gels and are invariably due to post-translational modifications or protein degradation.

Studies of SE show that asymmetric cell division and controlled cell expansion are important mechanisms for the generation of embryogenic plant cells indicating a significant role for the plant cell wall in these processes (Fehér et al. 2003). A set of proteins was identified that take part in the remodeling of the cell wall architecture and stress defense, as well as in cell signaling. These included proteins acting on polysaccharides such as endochitinases and α-amylases, proteases and a protease inhibitor, an oxidoreductase, as well as pathogenesis-related proteins whose putative role in SE is discussed below.

In a previous study using a monoclonal antibody, we found an extracellular acidic 48-kDa glycoprotein (gp48) secreted into the medium of microclusters from embryogenic *D. glomerata* L. suspension cultures and proposed its use as an early marker for embryogenic potential (Tchorbadjieva et al. 2005). Further, we identified the gp48 as α-amylase (Fig. 5.2, spot 20), cloned the full-length cDNA of the α-amylase gene (designated *Dg*Amy1), and expressed the protein in *E. coli* (Rakleova et al. 2012). *Dg*Amy1 is a plant α-amylase that belongs to the glycoside hydrolase family 13 (GH13). Considerable differences in the structure of *Dg*Amy1 and other cereal α-amylases make unlikely its participation in starch degradation. Its transient expression in dividing microclusters during their developmental transition to proembryogenic masses and predominant localization at the regions of cell-to-cell adhesion make it more likely that the secreted *Dg*Amy1 may act on some unknown carbohydrate in the cell wall liberating signals necessary for the development of microclusters into PEMs or it may locally modify the cell wall and thus ensure close cell-to-cell contact between the developing embryogenic structures (Rakleova et al. 2012).

Previously, the protein corresponding to spot 9 (Fig. 5.2) was immunologically identified as an acidic endochitinase (Tchorbadjieva and Pantchev 2006). This enzyme was expressed in embryogenic suspension cultures only. Subsequently, using the sequence information from the MS/MS analysis of spots 9 and 10, the full-length cDNAs for both proteins were cloned and showed that one of the proteins (designated *Dg*ChiIV) is a class IV endochitinase (JN191351) and the other (designated *Dg*ChiIII) is a class III endochitinase (JN191350) (Tchorbadjieva et al., manuscript in preparation). Typically, chitinases are involved in plant defense, but a nondefensive role of these enzymes in plant development has been suggested as well (Kasprzewska 2003). Plant cells from embryogenic suspension cultures secrete into the medium arabinogalactan proteins (AGPs) believed to have important roles in cell–cell interaction and signaling (Quiroz-Figueroa et al. 2006). Interestingly, AGPs contain GlcNAc and Glc residues, sensitive to cleavage by chitinases, and it has been suggested that chitinase-modified AGPs are extracellular matrix signaling molecules able to control plant cell fate (van Hengel et al. 2001). The presence of *Dg*ChiIV in embryogenic suspension cultures of *D. glomerata* L. only (Tchorbadjieva and Pantchev 2006) and the secretion of AGPs in the latter (unpublished results) suggest a similar role in the process of SE.

Cell wall proteome and secretome analysis of *Arabidopsis*, *O. sativa*, *N. tabacum*, *M. sativa*, and *Vitis* revealed the presence of unexpected large number of proteases in the cell wall and the culture medium with putative signaling function (Albenne et al. 2013; Krause et al. 2013). In a previous study, we detected an extracellular cysteine proteinase activity band specific for embryogenic suspension cultures of *D. glomerata* L. (Rakleova et al. 2010). Mass spectrometry analysis showed that it contained two cysteine proteinases (Fig. 5.2, spot 11 and spot 13). Both cysteine proteinases were cloned and identified as cathepsin B and cathepsin L (GU067465.1 and GU067466.2, respectively). In an independent study, using N-terminal sequencing, we identified five isoforms of an acidic cysteine proteinase inhibitor – cystatin. The cystatin (designated *Dg*ECPI) was also cloned (GU065373.1) and expressed in *E. coli* (manuscript in preparation). It is well known that during SE, cystatins play a fundamental role in providing a suitable microenvironment of the embryogenic cell which is necessary for its further development into an embryo. At the same time, the surrounding somatic cells are prevented to dedifferentiate into competitive embryogenic cells (Rose and Lee 2010; Wolf et al. 2012). A putative role of the cystatin *Dg*ECPI might be the control of cell proliferation activated by cysteine proteinases or it might take part in the remodeling of the plant cell wall during embryo development. A cysteine proteinase inhibitor and cysteine proteinase were more abundantly expressed in embryogenic suspension cultures of *C. sativus* (Sharifi et al. 2012) and EC of cowpea (Nogueira et al. 2007). The opposite action of a cysteine proteinase (OsCP) and a cystatin (OC-1) was found to control cell proliferation in rice suspension cultures (Tian et al. 2009). Kusumawati et al. (2008) identified three secreted proteases only in an embryogenic cell line of *Medicago truncatula*. Redifferentiation processes during SE are involved in the general reprogramming of gene expression (chromatin remodeling, transcription machinery), and they require complex changes in the protein pattern and proteolysis.

Thus, a fine control of the protease activity during embryogenesis, organogenesis, programmed cell death, and tolerance to abiotic and biotic stresses mediated by protease inhibitors is of utmost importance (Benchabane et al. 2010; Martínez et al. 2012).

5.4 Concluding Remarks

Within the last 10 years, transcriptomics and proteomics studies have significantly contributed to an improved understanding of plant SE (Elhiti et al. 2013; Rocha and Dornelas 2013; Fehér 2014). Proteomics of SE has allowed the precise identification and quantification of differentially expressed proteins in embryogenic and nonembryogenic cell lines, in the different stages of SE, and in somatic and zygotic embryos. The results indicate that a complex molecular system is turned on during SE to control processes such as ROS detoxification, energy metabolism, protein synthesis and processing, and cell division among others. An interesting outcome of these studies is the potential use of certain proteins as molecular markers for specific developmental stages or to differentiate embryogenic from nonembryogenic genotypes.

Stress-induced in vitro somatic embryogenesis is an extreme example of plant developmental plasticity (Fehér 2014). The expression of totipotency in cultured somatic cells is part of a general stress-adaptive process that involves a fine regulation of hormone and stress signaling resulting in the restart of cell division and embryogenic competence acquisition. There is probably no single model applicable to all plant species for differentiating into an embryo. However, the observation that embryogenic tissues of different origins and obtained with different hormones display similar protein profiles indicates a general behavior of cellular metabolism that can give valuable insights into the mechanisms of SE (Correia et al. 2012).

The protein expression data complemented with that of transcriptomics and metabolomics will certainly help understand the molecular basis of plant SE and allow the development of more effective in vitro regeneration protocols for many plant species.

References

Agrawal G, Jwa N-S, Lebrun M-H, Job D, Rakwal R (2010) Plant secretome: unlocking secrets of the secreted proteins. Proteomics 10(4):799–827

Albenne C, Canut H, Jamet E (2013) Plant cell wall proteomics: the leadership of *Arabidopsis thaliana*. Front Plant Sci 4:111. doi:10.3389/fpls.2013.00111

Alexandersson E, Ali A, Resjö S, Andreasson E (2013) Plant secretome proteomics. Front Plant Sci 4:9. doi:10.3389/fpls.2013.00009

Almeida AM, Parreira JR, Santos R, Duque AS, Francisco R, Tome DF, Ricardo CP, Coelho AV, Fevereiro P (2012) A proteomics study of the induction of somatic embryogenesis in *Medicago truncatula* using 2DE and MALDI-TOF/TOF. Physiol Plant 146:236–249

Apel K, Hirt H (2004) Reactive oxygen species: metabolism, oxidative stress, and signal transduction. Annu Rev Plant Biol 55:373–399

Aslam J, Ilah A, Mujib A, Abdin MZ (2013) Proteomics during somatic embryogenesis. In: Aslam J, Srivastava PS, Sharma MP (eds) Somatic embryogenesis and gene expression. Narosa Publishing House, New Delhi, pp 245–258

Baba AI, Nogueira FCS, Pinheiro CB, Brasil JN, Jereissati ES, Juca TL, Soares AA, Santos MF, Domont GB, Campos FAP (2008) Proteome analysis of secondary somatic embryogenesis in cassava (*Manihot esculenta*). Plant Sci 175:717–723

Balbuena TS, Silveira V, Junqueira M, Dias LLC, Santa-Catarina C, Shevchenko A, Floh EIS (2009) Changes in the 2-DE protein profile during zygotic embryogenesis in the Brazilian Pine (*Araucaria angustifo*lia). J Proteomics 72:337–352

Benchabane M, Schlüter U, Vorster J, Goulet M-C, Michaud D (2010) Plant cystatins. Biochimie 92(11):1657–1666

Bian F, Zheng C, Qu F, Gong X, You C (2010) Proteomic analysis of somatic embryogenesis in *Cyclamen persicum* Mill. Plant Mol Biol Rep 28:22–31

Bozhkov PV, Filonova LH, Suarez MF (2005) Programmed cell death in plant embryogenesis. Curr Top Dev Biol 67:135–179

Businge E, Bygdell J, Wingsle G, Moritz T, Egertsdotter U (2013) The effect of carbohydrates and osmoticum on storage reserve accumulation and germination of Norway spruce somatic embryos. Physiol Plant. doi:10.1111/ppl.12039

Cangahuala-Inocente G, Villarino A, Seixas D, Dumas-Gaudot E, Terenzi H, Guerra MP (2009) Differential proteomic analysis of developmental stages of *Acca sellowiana* somatic embryos. Acta Physiol Plant 31:501–514

Chen F, Li Q, Sun L, He Z (2006) The rice 14–3-3 gene family and its involvement in responses to biotic and abiotic stress. DNA Res 13:53–63

Chen K, Wu HJ, Chen J, Cheng XF, Jing X, Wang XY (2012) Somatic embryogenesis and mass spectrometric identification of proteins related to somatic embryogenesis in *Eruca sativa*. Plant Biotechnol Rep 6:113–22

Correia S, Vinhas R, Manadas B, Lourenço AS, Veríssimo P, Canhoto JM (2012) Comparative proteomic analysis of auxin-induced embryogenic and nonembryogenic tissues of the solanaceous tree *Cyphomandra betacea* (Tamarillo). J Proteome Res 11(3):1666–1675

de Jong F, Mathesius U, Imin N, Rolfe BG (2007) A proteome study of the proliferation of cultured *Medicago truncatula* protoplasts. Proteomics 7:722–736

Denison FC, Paul AL, Zupanska AK, Ferl RJ (2011) 14-3-3 proteins in plant physiology. Semin Cell Dev Biol 22:720–727

Dixon DP, Lapthorn A, Edwards R (2002) Plant glutathione transferases. Genome Biol 3(3):reviews3004.1–3004.10

Dong JZ, Dunstan DI (1997) Endochitinase and beta-1,3-glucanase genes are developmentally regulated during somatic embryogenesis in *Picea glauca*. Planta 201:189–194

Dudits D, Bögre L, Györgyey J (1991) Molecular and cellular approaches to the analysis of plant embryo development from somatic cells in vitro. J Cell Sci 484:475–484

Dudits D, Gyorgyey J, Bogre L, Bako L (1995) Molecular biology of somatic embryogenesis. In: Thorpe TA (ed) In vitro embryogenesis in plants. Kluwer Academic Publisher, Dordrecht, pp 267–308

Elhiti M, Stasolla C, Wang A (2013) Molecular regulation of plant somatic embryogenesis. In Vitro Cell Dev Biol Plant 49:631–642

Fehér A (2005) Why somatic plant cells start to form embryos? In: Mujib A, Samaj J (eds) Somatic embryogenesis, vol 2, Plant cell monographs. Springer, Berlin/Heidelberg, pp 85–101

Fehér A (2008) The initiation phase of somatic embryogenesis: what we know and what we don't. Acta Biol Szeged 52(1):53–56

Fehér A (2014) Somatic embryogenesis: stress-induced remodeling of plant cell fate. Biochim Biophys Acta. doi:10.1016/j. bbagrm.2014.07.005

Fehér A, Pasternak TP, Dudits D (2003) Transition of somatic plant cells to an embryogenic state. Plant Cell Tiss Org Cult 74:201–228

Fraga HP, Agapito-Tenfen SZ, Caprestano CA, Nodari RO, Guerra MP (2013) Comparative proteomic analysis of off-type and normal phenotype somatic plantlets derived from somatic embryos of Feijoa (*Acca sellowiana* (O. Berg) Burret). Plant Sci 210:224–31

Ge X, Zhang C, Wang Q, Yang Z, Wang Y, Zhang X, Wu Z, Hou Y, Wu J, Li F (2015) iTRAQ Protein profile differential analysis between somatic globular and cotyledonary embryos reveals stress, hormone, and respiration involved in increasing plantlet regeneration of *Gossypium hirsutum* L. J Proteome Res 14(1):268–278. doi:10.1021/pr500688g

Glickman MH, Ciechanover A (2002) The ubiquitin-proteasome proteolytic pathway: destruction for the sake of construction. Physiol Rev 82:373–428

Gomez-Garay A, Lopez JA, Camafeita E, Bueno MA, Pintos B (2013) Proteomic perspective of *Quercus suber* somatic embryogenesis. J Proteomics 93:314–25

Guilloteau M, Laloi M, Michaux S, Bucheli P, McCarthy J (2005) Identification and characterization of the major aspartic proteinase activity in *Theobroma cacao* seeds. J Sci Food Agric 85:549–562

Guzmán-García E, Sánchez-Romero C, Panis B, Carpentier SC (2013) The use of 2D DIGE to understand the regeneration of somatic embryos in avocado. Proteomics 13:34983–507

Ikeda-Iwai M, Umehara M, Satoh S, Kamada H (2003) Stress-induced somatic embryogenesis in vegetative tissues of *Arabidopsis thaliana*. Plant J 34:107–114

Imin N, De Jong F, Mathesius U, van Noorden G, Saeed NA, Wang XD, Rose RJ, Rolfe BG (2004) Proteome reference maps of *Medicago truncatula* embryogenic cell cultures generated from single protoplasts. Proteomics 4:1883–1896

Imin N, Nizamidin M, Daniher D, Nolan KE, Rose RJ, Rolfe BG (2005) Proteomic analysis of somatic embryogenesis in *Medicago truncatula*. Explant cultures grown under 6-benzylaminopurine and 1-naphthaleneacetic acid treatments. Plant Physiol 137:1250–1260

Ito H, Iwabuchi M, Ogawa K (2003) The sugar-metabolic enzymes aldolase and triose-phosphate isomerase are targets of glutathionylation in *Arabidopsis thaliana*: detection using biotinylated glutathione. Plant Cell Physiol 44:655–660

Jo L, Dos Santos ALW, Bueno CA, Barbosa HR, Floh EI (2014) Proteomic analysis and polyamines, ethylene and reactive oxygen species levels of *Araucaria angustifolia* (Brazilian pine) embryogenic cultures with different embryogenic potential. Tree Physiol 34:94–104

Karami O, Aghavaisi B, Mahmoudi Pour A (2009) Molecular aspects of somatic-to-embryogenic transition in plants. J Chem Biol 2:177–190

Kasprzewska A (2003) Plant chitinases-regulation and function. Cell Mol Biol Lett 8:809–824

Krause C, Richter S, Knöll C, Jürgens G (2013) Plant secretome – from cellular process to biological activity. BBA 1834(11):2429–2441

Kumari S, Roy S, Singh P, Singla-Pareek S, Pareek A (2012) Cyclophilins: proteins in search of function. Plant Signal Behav 8, e22734. doi:10.4161/psb.22734

Kusumawati L, Imin N, Djordjevic MA (2008) Characterization of the secretome of suspension cultures of *Medicago* species reveals proteins important for defense and development. J Proteome Res 7:4508–4520

Laohavisit A, Davies JM (2011) Annexins. New Phytol 189:40–53

Li K, Zhu W, Zeng K, Zhang Z, Ye J, Ou W, Rehman S, Heuer B, Chen S (2010) Proteome characterization of cassava (*Manihot esculenta* Crantz) somatic embryos, plantlets and tuberous roots. Proteome Sci 8:10. doi:10.1186/1477-5956-8-10

Li Q, Zhang S, Wang J (2015) Transcriptomic and proteomic analyses of embryogenic tissues in *Picea balfouriana* treated with 6-benzylaminopurine. Physiol Plant. doi:10.1111/ppl.12276

Lippert D, Jun Z, Ralph S, Ellis DE, Gilbert M, Olafson R, Ritland K, Ellis B, Douglas CJ, Bohlmann J (2005) Proteome analysis of early somatic embryogenesis in *Picea glauca*. Proteomics 5:461–473

Liu JJ, Sturrock R, Ekramoddoullah AK (2010) The superfamily of thaumatin-like proteins: its origin, evolution, and expression towards biological function. Plant Cell Rep 29:419–436

Lyngved R, Renaut J, Hausman J-F, Iversen T-H, Hvoslef-Eide AK (2008) Embryo-specific proteins in *Cyclamen persicum* analyzed with 2-D DIGE. J Plant Growth Regul 27:353–369

Maga G, Hubscher U (2003) Proliferating cell nuclear antigen (PCNA): a dancer with many partners. J Cell Sci 116:3051–3060

Malerba M, Contran N, Tonelli M, Crosti P, Cerana R (2008) Role of nitric oxide in actin depolymerization and programmed cell death induced by fusicoccin in sycamore (*Acer pseudoplatanus*) cultured cells. Physiol Plant 133:449–457

Maraschin SF, de Priester W, Spaink HP, Wang M (2005) Androgenic switch: an example of plant embryogenesis from the male gametophyte perspective. J Exp Bot 56:1711–1726

Marsoni M, Bracale M, Espen L, Prinsi B, Negri AS, Vannini C (2008) Proteomic analysis of somatic embryogenesis in *Vitis vinifera*. Plant Cell Rep 27:347–356

Martínez M, Cambra I, González-Melendi P, Santamaría ME, Díaz I (2012) C1A cysteine-proteases and their inhibitors in plants. Physiol Plant 145(1):85–94

Matthys-Rochon E (2005) Secreted molecules and their role in embryo formation in plants: a mini review. Acta Biol Cracov 47:23–29

Morel A, Trontin JF, Corbineau F, Lomenech A-M, Beaufour M, Reymond I, Le Metté C, Ader K, Harvengt L, Cadene M, Label P, Teyssier C, Lelu-Walter M-A (2014) Cotyledonary somatic embryos of *Pinus pinaster* Ait. most closely resemble fresh, maturing cotyledonary zygotic embryos: biological, carbohydrate and proteomic analyses. Planta 240:1075–1095

Noah AM, Niemenak N, Sunderhaus S, Haase C, Omokolo DN, Winkelmann T, Braun H-P (2013) Comparative proteomic analysis of early somatic and zygotic embryogenesis in *Theobroma cacao* L. J Proteomics 78:123–33

Nogueira FCS, Gonçalves EF, Jereissati ES, Santos M, Costa JH, Oliveira-Neto OB, Soares AA, Domont GB, Campos FAP (2007) Proteome analysis of embryogenic cell suspensions of cowpea (*Vigna unguiculata*). Plant Cell Rep 26:1333–1343

Novinec M, Lenarčič B (2013) Papain-like peptidases: structure, function, and evolution. Biomol Concept 4(3):287–308

Pais SM, Tellez-Inon MT, Capiati DA (2009) Serine/threonine protein phosphatases type 2A and their roles in stress signaling. Plant Signal Behav 4:1013–1015

Pan Z, Guan R, Zhu S, Deng X (2009) Proteomic analysis of somatic embryogenesis in Valencia sweet orange (*Citrus sinensis* Osbeck). Plant Cell Rep 28:281–289

Pasternak T, Prinsen E, Ayaydin F, Miskolczi P, Potters G, Asard H, Van Onckelen H, Dudits D, Fehér A (2002) The role of auxin, pH and stress in the activation of embryogenic cell division in leaf protoplast-derived cells of alfalfa (*Medicago sativa* L.). Plant Physiol 129:1807–1819

Petrussa E, Bertolini A, Casolo V, Krajňáková J, Macrí F, Vianello A (2009) Mitochondrial bioenergetics linked to the manifestation of programmed cell death during somatic embryogenesis of *Abies alba*. Planta 231:93–107

Quiroz-Figueroa F, Rojas-Herrera R, Galaz-Avalos R, Loyola-Vargas VC (2006) Embryo production through somatic embryogenesis can be used to study cell differentiation in plants. Plant Cell Tiss Org Cult 86:285–301

Rakleova G, Keightley A, Panchev I, Tsacheva I, Tchorbadjieva M (2010) Cysteine proteinases and somatic embryogenesis in suspension cultures of orchardgrass (*Dactylis glomerata* L.). Gen Appl Plant Physiol 36(1–2):100–109

Rakleova G, Keightley A, Pantchev I, Tsacheva I, Tchorbadjieva M (2012) Identification, molecular cloning, and recombinant gene expression of an extracellular α-amylase from *Dactylis glomerata* L. embryogenic suspension cultures. Biotechnol Biotec Eq 26(4):3192–3200

Rocha DI, Dornelas MC (2013) Molecular overview on plant somatic embryogenesis. CAB Rev 8(022):1–17

Rode C, Gallien S, Heintz D, Van Dorsselaer A, Braun H-P, Winkelmann T (2011) Enolases: storage compounds in seeds? Evidence from a proteomic comparison of zygotic and somatic embryos of *Cyclamen persicum* Mill. Plant Mol Biol 75:305–319

Rode C, Lindhorst K, Braun HP, Winkelmann T (2012) From callus to embryo: a proteomic view on the development and maturation of somatic embryos in *Cyclamen persicum*. Planta 235:995–1011

Rose JKC, Lee SJ (2010) Straying off the highway: trafficking of secreted plant proteins and complexity in the plant cell wall proteome. Plant Physiol 153:433–436

Rose RJ, Nolan KE (2006) Genetic regulation of somatic embryogenesis with particular reference to *Arabidopsis thaliana* and *Medicago truncatula*. In Vitro Cell Dev Biol Plant 42:473–481

Rose JK, Bashir S, Giovannoni JJ, Jahn MM, Saravanan RS (2004) Tackling the plant proteome: practical approaches, hurdles and experimental tools. Plant J 39:715–733

Ruan SL, Ma HS, Wang SH, Fu Y-P, Xin Y, Liu W-Z, Wang F, Tong J-X, Wang S-Z, Chen H-Z (2011) Proteomic identification of OsCYP2, a rice cyclophilin that confers salt tolerance in rice (*Oryza sativa* L.) seedlings when overexpressed. BMC Plant Biol 11:34

Ruiz-May E, De la Pena C, Ayil-Gutierrez B, Nic-Can G, Myukul-Lopez H, Galaz-Avalos R, Loyola-Vargas V (2010) Protein secretion by cell cultures: an essential biological issue. Acta Hortic 849:213–222

Schwarzerova K, Vondrakova Z, Fischer L, Borikova P, Bellinvia E, Eliasova K, Havelkova L, Fiserova J, Vagner M, Opatrny Z (2010) The role of actin isoforms in somatic embryogenesis in Norway spruce. BMC Plant Biol 10:89. doi:10.1186/1471-2229-10-89

Sghaier B, Kriaa W, Bahloul M, Jorrin-Novo JV, Drira N (2009) Effect of ABA, arginine and sucrose on protein content of date palm somatic embryos. Sci Hortic 120:379–85

Sghaier-Hammami B, Drira N, Jorrin-Novo JV (2009) Comparative 2-DE proteomic analysis of date palm (*Phoenix dactylifera* L.) somatic and zygotic embryos. J Proteomic 73:161–177

Sharifi G, Ebrahimzadeh H, Ghareyazie B, Gharechahi J, Vatankhah E (2012) Identification of differentially accumulated proteins associated with embryogenic and non-embryogenic calli in saffron (*Crocus sativus* L.). Proteome Sci 10:3. doi:10.1186/1477-5956-10-3

Silva Rde C, Carmo LS, Luis ZG, Silva LP, Scherwinski-Pereira JE, Mehta A (2014) Proteomic identification of differentially expressed proteins during the acquisition of somatic embryogenesis in oil palm (*Elaeis guineensis* Jacq.). J Proteomics 104:112–27

Silveira V, Santa-Catarina C, Balbuena TS, Moraes FMS, Ricart CAO, Souza MV, Guerra MP, Handro W, Floh EIS (2008) Endogenous abscisic acid levels and comparative proteome during seed development of *Araucaria angustifolia* (Bert.) O. Ktze. Biol Plant 52:101–104

Smith DL, Krikorian AD (1990) Somatic proembryo production from excised, wounded zygotic carrot embryos on hormone-free medium: evaluation of the effects of pH, ethylene and activated charcoal. Plant Cell Rep 9:34–37

Sun L, Wu Y, Zou HD, Su S, Li S, Shan X, Xi J, Yuan Y (2013) Comparative proteomic analysis of the H99 inbred maize (*Zea mays* L.) line in embryogenic and non-embryogenic callus during somatic embryogenesis. Plant Cell Tiss Org Cult 113(1):103–119

Takáč T, Pechan T, Šamaj J (2011) Differential proteomics of plant development. J Proteomic 74:577–588

Tan BC, Chin CF, Liddell S, Alderson P (2013) Proteomic analysis of callus development in *Vanilla planifolia* Andrews. Plant Mol Biol Rep 31:1220–1229. doi:10.1007/s11105-013-0590-3

Tchorbadjieva M (2005) Protein markers for somatic embryogenesis. In: Mujib A, Samaj J (eds) Somatic embryogenesis, vol 2, Plant cell monographs. Springer, Heidelberg, pp 215–33

Tchorbadjieva M, Odjakova M (2001) An acidic esterase as a biochemical marker for somatic embryogenesis in orchardgrass (*Dactylis glomerata* L.) suspension cultures. Plant Cell Rep 20:28–33

Tchorbadjieva M, Pantchev I (2006) Secretion of a chitinase-like protein in embryogenic suspension cultures of *Dactylis glomerata* L. Biol Plant 50:142–145

Tchorbadjieva M, Kalmukova R, Pantchev I, Kyurkchiev S (2005) Monoclonal antibody against a cell wall marker protein for embryogenic potential of *Dactylis glomerata* L suspension cultures. Planta 222:811–819

Teyssier C, Maury S, Beaufour M, Grondin C, Delaunay A, Le Mette C, Ader K, Cadene M, Label P, Lelu-Walter MA (2014) In search of markers for somatic embryo maturation in hybrid larch (*Larix × eurolepis*): global DNA methylation and proteomic analyses. Physiol Plant 150(2):271–291

Tian L, Zhang L, Zhang J, Song Y, Guo Y (2009) Differential proteomic analysis of soluble extracellular proteins reveals the cysteine protease and cystatin involved in suspension-cultured cell proliferation in rice. Biochim Biophys Acta 1794:459–467

Tonietto Â, Sato JH, Teixeira JB, de Souza EM, Pedrosa FO, Franco OL, Mehta A (2012) Proteomic analysis of developing somatic embryos of *Coffea arabica*. Plant Mol Biol Rep 30(6):1393–1399

Turk V, Stoka V, Vasiljeva O, Renko M, Sun T, Turk B, Turk D (2012) Cysteine cathepsins: from structure, function and regulation to new frontiers. BBA 1824:68–88

Vale Ede M, Heringer AS, Barroso T, da Silva Ferreira AT, Nunes da Costa M, Perales JA, Santa-Catarina C, Silveira V (2014) Comparative proteomic analysis of somatic embryo maturation in *Carica papaya* L. Proteome Sci 12:37. doi:10.1186/1477-5956-12-37

Valledor L, Jorrin J (2011) Back to the basics: maximizing the information obtained by quantitative two dimensional gel electrophoresis analyses by an appropriate experimental design and statistical analyses. J Proteomic 74(1):1–18

van der Hoorn RA (2008) Plant proteases: from phenotypes to molecular mechanisms. Annu Rev Plant Biol 59:191–223

van Hengel AJ, Tadesse Z, Immerzeel P, Schols H, van Kammen A, de Vries SC (2001) N-Acetylglucosamine and glucosamine-containing arabinogalactan proteins control somatic embryogenesis. Plant Physiol 125:1880–1890

van Loon LC, Rep M, Pieterse CM (2006) Significance of inducible defense-related proteins in infected plants. Annu Rev Phytopathol 44:135–162

Varhaníková M, Uvackova L, Skultety L, Pretova A, Obert B, Hajduch MC (2014) Comparative quantitative proteomic analysis of embryogenic and non-embryogenic calli in maize suggests the role of oxylipins in plant totipotency. J Proteomic 104:57–65

Vianello A, Zancani M, Peresson C, Petrussa E, Casolo V, Krajňáková J, Patui S, Braidot E, Macrì F (2007) Plant mitochondrial pathway leading to programmed cell death. Physiol Plant 129:242–252

Vierstra RD (2009) The ubiquitin-26S proteasome system at the nexus of plant biology. Nat Rev Mol Cell Biol 10:385–397

Vogel G (2005) How does a single somatic cell become a whole plant? Science 309(5731):86. doi:10.1126/science.309.5731.86

Wang W, Vinocur B, Shoseyov O, Altman A (2004) Role of plant heat-shock proteins and molecular chaperones in the abiotic stress response. Trends Plant Sci 9:244–252

Waters ER (2011) The evolution, function, structure, and expression of the plant sHSPs. J Exp Bot 64:391–403

Winkelmann T, Heintz D, Van Dorsselaer A, Serek M, Braun HP (2006) Proteomic analyses of somatic and zygotic embryos of *Cyclamen persicum* Mill. reveal new insights into seed and germination physiology. Planta 224:508–519

Wolf S, Hematy K, Hofte H (2012) Growth control and cell wall signaling in plants. Annu Rev Plant Biol 63:381–407

Yang X, Zhang X (2010) Regulation of somatic embryogenesis in higher plants. Crit Rev Plant Sci 29:36–57

Zavattieri MA, Frederico AM, Lima M, Sabino R, Arnholdt-Schmitt B (2010) Induction of somatic embryogenesis as an example of stress-related plant reactions. Electron J Biotechnol 13:1–9

Zhang J, Ma H, Chen S, Ji M, Perl A, Kovacs L, Chen S (2009) Stress response proteins' differential expression in embryogenic and non-embryogenic callus of *Vitis vinifera* L. cv. Cabernet Sauvignon-a proteomic approach. Plant Sci 177:103–113

Zhao J, Wang B, Wang X, Zhang Y, Dong M, Zhang J (2015) iTRAQ-based comparative proteomic analysis of embryogenic and non-embryogenic tissues of Prince Rupprecht's larch (*Larix principis*-rupprechtii Mayr). Plant Cell Tiss Org Cult 120:655–669. doi:10.1007/s11240-014-0633-y

Somatic Embryogenesis in *Camellia japonica* L.: Challenges and Future Prospects

6

M.C. San José, J.L. Couselo, M.T. Martínez, P. Mansilla, and Elena Corredoira

Abstract

Camellia japonica L. is a popular ornamental plant, widely grown in parks and private gardens throughout the world. This review provides a summary of the advances made in the application of somatic embryogenesis in the propagation of *C. japonica*. The factors affecting the induction (type of explant, growth conditions, mineral media, plant growth regulators), maintenance and multiplication of the embryogenic cultures by repetitive embryogenesis and the maturation and conversion into plants of somatic embryos are explained. The results achieved on the in vitro storage of *C. japonica* embryogenic lines are also mentioned.

Keywords

Artificial seed • Cold storage • Ornamental plant • Plant conversion • Somatic embryos

6.1 Introduction

Camellia is the most economically important genus in the family *Theaceae*, which comprises numerous tropical and subtropical trees and shrubs. The genus was first described by Linnaeus in 1735 in the earliest edition of *Systema Naturae*. Since then, the number of species identified as belonging to the genus *Camellia* has continued to grow, and at present, the genus is believed to comprise more than 300 species (Izco 2014). Among these, *C. sinensis* L., the common tea plant, is the most economically important species. The three most popular species used as ornamentals are *C. japonica* L., *C. sasanqua* Thunb. and *C. reticulata* Lindl. By far the majority of existing ornamental camellia cultivars belong to these three species and their hybrids (Vieitez and Barciela 1990).

It is thought that around 90 % of *Camellia* species originated in South and Southeast China. However, *C. japonica* (the Japanese or common

M.C.S. José • M.T. Martínez • E. Corredoira (✉)
Instituto de Investigaciones Agrobiológicas de Galicia, CSIC, Apartado 122, 15705 Santiago de Compostela, Spain
e-mail: elenac@iiag.csic.es

J.L. Couselo • P. Mansilla
Estación Fitopatolóxica do Areeiro, EFA, Subida a la Robleda, 36153 Pontevedra, Spain

A. Mujib (ed.), *Somatic Embryogenesis in Ornamentals and Its Applications*,
DOI 10.1007/978-81-322-2683-3_6

camellia) originated in Southern Japan (Mondal 2011) where it is called "rose of winter" and is a national symbol. *Camellia japonica* arrived in Europe from the Orient during the seventeenth century. The East India Company introduced *C. japonica* to England by mistake as the plant was thought to be a tea variety. Despite this mistake, camellias quickly became prized for their beauty. *Camellia japonica* was introduced to the USA at the beginning of the eighteenth century, and the plants became so popular in the southern states that the flower of this species was designated the state flower of Alabama (Gao 2005). The camellia was also introduced from England to Australia during the nineteenth century. The plant was unknown in this continent before this time, despite the proximity to the natural area of distribution of the genus.

6.1.1 General Features

Camellia japonica is a small and slow-growing evergreen tree that can reach a height of 8 m. The leaves are generally oval, dark green and leathery. The flowers are terminal, non-fragrant and with overlapping sepals (Fig. 6.1a); the type species has 5–7 petals surrounding a mass of stamens. Geneticists and gardeners have transformed many of the stamens into petals, so that numerous varieties with flowers of different colours (ranging from pure white to intense red and including all shades of red and pink) and shapes are now available. The dry fruit only appears in varieties with simple or semi-double flowers with fertile carpels. The fruit comprises bi- or tri-lobulate spherical capsules and each compartment contains one or two seeds.

6.1.2 Diseases

Several fungi have been reported as causal agents of *Camellia* diseases. *Phytophthora ramorum*, *Phytophthora cinnamomi*, *Armillaria mellea*, *Rosellinia necatrix* or *Glomerella cingulata* are not specific pathogens of *C. japonica* but can cause severe diseases and even death of the camellias. *C. japonica* varieties differ in their susceptibility to these diseases, and plants weakened by transplanting shock, poor soil drainage, excessive nitrogen fertilization, drought or other stress factors are more susceptible to infection. There are fungi that attack specifically camellias such as *Pestalotiopsis camelliae* causing grey blight disease on leaves of *C. japonica* (Zhang et al. 2012) or *Exobasidium camelliae* causing the camellia leaf gall disease (Shirai 1896). The damages caused from these two fungi are not severe, and there are also effective phytosanitary treatments against them.

The ascomycete *Ciborinia camelliae* is the causal agent of a disease known as *Camellia* flower blight (CFB) or *Camellia* petal blight. *C. camelliae* is a hemibiotrophic and host-specific fungal pathogen of all camellia flowers, and it is regarded as the main phytosanitary problem of *C. japonica* since it damages their ornamental value. In addition, the incidence of CFB is high in areas where the fungus is present, and it is considered as plant quarantine pathogen (EPPO 1997, 2003). Despite the efforts, effective control of CFB under field conditions by using chemicals, biological control agents (BCAs) or mulching so far has not proved to be effective (van Toor 2002; Couselo et al. 2014), and only plant breeding using biotechnological tools seems a real alternative for disease control.

6.1.3 Economic Importance and Uses

Camellia japonica is an evergreen perennial with more than 32,000 registered cultivars (Savige 1993) of great ornamental value mainly due to its beautiful flowers. The possibility of growing *C. japonica* as a houseplant and for gardening and landscaping has strengthened its economic value as an ornamental plant in many European countries, the USA, Japan and Australia (Vela et al. 2013). *C. japonica* is also cultivated for oil production due to its seed that can contain up to 30 % oil of which the 80 % is oleic acid (Salinero et al. 2012; De Ron et al. 2014). This oil can be used for human consumption after refining process (Lee et al. 2014). It is also considered a moisturizing agent and therefore a possible anti-

Fig. 6.1 Somatic embryogenesis in *Camellia japonica*. (**a**) Flower of the common type of *C. japonica*. (**b**, **c**) Differentiation of somatic embryos on a cotyledonary explant (**b**) and a root explant (**c**) of *C. japonica*. (**d**) Plantlet regeneration from a somatic embryo of *C. japonica* after 2 months in cold storage and 8 weeks culture on MS medium supplemented with 5 mg/l GA_3 and 1 mg/l AIA. (**e**) Somatic embryos of *C. japonica* encapsulated in alginate beads

ageing and wrinkle-reducer candidate for cosmetics production (Jung et al. 2007; Kim et al. 2007). Uses for human health have also been investigated as inhibiting metastasis of melanoma BL6 cells (Miuraa et al. 2007), anti-inflammatory activity (Kim et al. 2012) or antiviral activity (Akihisa et al. 2004). Further *C. japonica* oil has been evaluated for the production of biofuels (Chung 2010).

Leaves, bark or flowers have been used traditionally in oriental ethnomedicine for health purposes (Yoshikawa et al. 2007). Modern science has made it possible the research of several potential medical properties of leaf, bark or flower extracts as antimicrobial (Kim et al. 2001), antioxidant (Onodera et al. 2006; Mizutani and Masaki 2014), antitumoural (Thao et al. 2010), neuronal cell protective (Jeong et al. 2010), antiviral (Park et al. 2002), antihistaminic and antiallergic (Kuba et al. 2008; Lee et al. 2008; Onodera et al. 2010) agents.

6.1.4 Propagation

Increasing interest in ornamental camellias has led to continuous efforts by researchers, growers and breeders to produce new varieties and hybrids with improved floral and growth characteristics and to develop more efficient methods for their propagation. In *C. japonica*, few fruits are produced per shrub and small numbers of viable seeds are produced by each fruit. Like many other ornamental woody species, this species has traditionally been propagated by means of cuttings, or by stooling or grafting, although not always with satisfactory results (Samartín et al. 1986). This has led to the search for other methods of propagation, and, since the 1990s, attention has turned to biotechnological methods, specifically in vitro or tissue culture techniques. In vitro tools can provide backup collections and provide alternative means of propagating and conserving species (Reed et al. 2011). Bennett (1977) was one of the first researchers to suggest the use of in vitro tissue culture for rapid clonal propagation of *Camellia* species of economic importance. Use of tissue culture technology may provide a supplementary method for propagating and breeding *Camellia* species. This approach has several potential advantages over conventional methods: clonal propagation enables rapid multiplication of select mother plants and is independent of the season; in vitro culture of very small meristems or shoot tips should rid several highly prized existing camellia clones of the viruses with which they are now naturally infected; and in the not-too-distant future it should be possible to improve existing clones to produce new types of camellia. The improvement of plant quality would probably further increase the popularity of camellias, and the possibility of all-year-round production would make it easier for growers to respond to any increase in demand (Vieitez et al. 1992).

The first attempts to regenerate *C. japonica* by in vitro culture techniques were probably made by Beauchesne (1978). Since then, several protocols have been developed for the micropropagation of this species by axillary shoot culture from juvenile and/or adult material (Vieitez et al. 1992) and for the initiation of somatic embryogenesis (Vieitez 1995; Pedroso and Pais 1999). Somatic embryogenesis (SE) offers several advantages over other methods of micropropagation. The use of somatic embryos in clonal propagation, artificial seed production (with the use of bioreactors) and cryopreservation for long-term storage would enable establishment of gene banks of ornamental camellias. In particular, induction of somatic embryos on zygotic embryos produced by interspecific cross-pollination leads to propagation of hybrids whose seeds fail to develop or have poor viability (Vieitez 1995). Somatic embryogenesis is a very effective method for regeneration of material from genetically transformed cells. Moreover, study of the processes of induction and development of somatic embryos may help us to understand the genetic regulation of embryo development (Correia et al. 2011).

In this chapter, the state of the art of somatic embryogenesis in *C. japonica*, an important ornamental species of the genus *Camellia*, is summarized.

6.2 Culture Initiation

Induction of somatic embryogenesis in *C. japonica* mainly depends on the explant type, media composition and lighting conditions used during culture. Successful establishment of a SE system depends on the initial explant used. In almost all reports concerning SE in *C. japonica*, embryogenic cultures were initiated from juvenile tissue explants. Somatic embryos have been obtained from seeds and zygotic embryos (or parts of these) explanted at various developmental stages (Kato 1986; Nakamura 1988; Vieitez and Barciela 1990; Vieitez 1995), and they have also been obtained using roots (Vieitez et al. 1991) and leaves (San José and Vieitez 1993; Pedroso and Pais 1993a) derived from axillary shoot cultures established from juvenile material (Table 6.1). Of all the known reports concerning SE in *C. japonica*, only one has reported induction of somatic embryos on leaf explants derived from mature material (Pedroso and Pais 1993a).

With the exception of Bennett and Scheibert (1982), who used mineral Knop's medium (Tabachnik and Kester 1977) to initiate somatic embryos, other researchers have initiated somatic embryogenic cultures of *C. japonica* by using Murashige and Skoog medium (MS; 1962) as the basal medium (Table 6.1). The presence of cytokinins appears to be more important than the presence of auxin, and induction media without plant growth regulators (PGRs) or only including 6-benzylaminopurine (BA) can be used to induce embryo formation. Among auxins, naphthaleneacetic acid (NAA) and indole-3-butyric acid (IBA) are the most used for the initiation of somatic embryos.

The effects of the light conditions during induction of somatic embryos of *C. japonica* have also been considered, although contradic-

Table 6.1 Various explants used for induction and germination of somatic embryos in various PGRs added medium in *Camellia japonica*

Origin (explant)[a]	Induction medium (mg/l)[b]	Germination medium (mg/l)[b]	Growth response	References
ZE (mature cotyledon)	Knop+BA (0.5-5)	–	Indirect somatic embryo formation and plantlet development	Bennett and Scheibert (1982)
ZE (mature cotyledon)	MS+BA (0–5)+IBA (0–2)	MS+GA_3 (1)	Direct somatic embryo formation and plantlet development	Kato (1986)
ZE (mature cotyledon)	MS+BA (3)	–	Somatic embryo formation	Nakamura (1988)
ZE (embryonic axes and immature and mature cotyledons)	MS+BA (0, 0.5, 1, 2)+IBA (0, 0.5, 1, 2)	MS+GA_3 (5)+IAA (1)	Direct somatic embryo formation and plantlet development	Vieitez and Barciela (1990)
S (roots from in vitro plantlets)	MS+Z (1) or MS+BA (0–2)+IBA (0, 0.1, 1)	MS+GA_3 (5)+IAA (1)	Direct somatic embryo formation and plantlet development	Vieitez et al. (1991)
S (leaves from shoot cultures from seedlings)	MS+BA (8)+IBA (0.5) or WPM+BA (8)+IBA (0.5)	MS+GA_3 (3)+IAA (1)	Indirect somatic embryo formation and adventitious root, plantlet development	San José and Vieitez (1993)
SG (leaves from shoot cultures from in vitro germinated seeds)	1/2MS liquid+IBA (1000) for 20 min	MS+BA (1) or MS+BA (1)+0.1 (IBA or IAA)	Direct somatic embryo formation and plantlet development	Pedroso and Pais (1993a)
MT (anthers and anther-petaloids)	MS+BA (0.5)	1/2MS	Direct somatic embryo formation and plantlet development	Pedroso and Pais (1993b, 1997)
MT (microspores)	MS+2,4-D (1.1)+Kinetin (0.1)	–	Proembryo formation	Pedroso and Pais (1994b)
SG (stems from shoot cultures from in vitro germinated seeds)	MS+BA (1)+IBA (0.1)	–	Direct somatic embryo formation	Pedroso and Pais (1999)
SG (leaves from shoot cultures from in vitro germinated seeds)	MS+2,4-D (1)+kinetin (0.1) or MS liquid+2,4-D (0.5)+BA (1)	MS+BA (1)+0.1 (IBA or IAA)	Direct somatic embryo formation and plantlet development	Pedroso and Pais (1999)

[a]*S* 3–4-month seedlings, *SG* in vitro germinated seeds, *MT* mature trees, *ZE* zygotic embryos

[b]Basal media abbreviations: MS Murashige and Skoog (1962), 1/2MS MS with half-strength macronutrients, WPM Woody Plant Medium (Lloyd and McCown 1980)

– Data not available

tory effects have been reported. Thus, San José and Vieitez (1993) found that light inhibited formation of both callus and somatic embryos in leaf explants, whereas Pedroso and Pais (1999) found that somatic embryos were not produced in the dark, or the numbers produced were significantly lower than in cultures maintained under photoperiodic conditions. These authors recommended use of glass culture vessels (300 cm^3) with opaque plastic covers that reduce light intensity to 30 % for embryo formation.

6.2.1 Somatic Embryogenesis from Zygotic Embryos

As in most woody species, zygotic embryos are the most suitable explants for successful induction of SE in *C. japonica* (Table 6.1). In this species, induction of somatic embryogenesis in cotyledon explants was first reported by Bennett and Scheibert (1982). Kato (1986) induced SE in cotyledons explants by using MS medium supplemented with different combinations of BA, NAA and IBA. Subsequently, Nakamura (1988) reported an SE induction rate of 48–58 % in cotyledon explants cultured on MS medium with 3 mg/l BA.

A considerable progress was made by Vieitez and Barciela (1990), who defined the conditions necessary for successful induction in *C. japonica* of SE on embryonic axes and cotyledons of both immature and mature zygotic embryos. Immature seeds of two different sizes (2–5 and 6–8 mm), collected in July, and zygotic embryos (12–14 mm), divided into five types of explants (two distal semi-cotyledons, two proximal semi-cotyledons and the embryonic axis) and collected in September and October, were used as starting material. Explants were cultured on MS medium alone or supplemented with BA (0.5, 1, 2, 4 mg/l) and/or IBA (0.5, 1, 2 mg/l). As has been observed in many woody species, the embryogenic response was clearly affected by the developmental stage of the zygotic embryos used as initial explants (Canhoto et al. 1999; Kong et al. 2012; Corredoira et al. 2013). Although an embryogenic response was observed in seeds collected in July (25 % for the 2–5 mm explants and 40 % for the 6–8 mm explants), the best results were achieved with immature embryogenic axes collected in September (94 %). This was also higher than the response obtained with embryogenic axes from mature seeds collected in October (20 %). Cotyledonary explants also produced somatic embryos, but the rate of embryogenesis was lower than for embryogenic axes (Fig. 6.1b). Somatic embryos are generally formed by direct embryogenesis, but embryogenic callus has also been obtained on cotyledonary explants. Unlike most woody species, in which the presence of 2,4-dichlorophenoxyacetic acid (2,4-D), either alone or in combination with a cytokinin, is required for initiation of embryogenic cultures, in *C. japonica*, somatic embryos were formed on MS medium without any growth regulators or with different combinations of IBA and BA.

Subsequently, Barciela and Vieitez (1993) carried out a detailed comprehensive study of the origin and anatomical development of somatic embryos differentiated on cotyledon explants. Only the abaxial surface of the cotyledons was morphogenetically competent; somatic embryos only formed on this side regardless of whether the explant was placed abaxial or adaxial side down on the culture medium. Embryos developed on abaxial parenchymatic protuberances or nodules arising by dedifferentiation and active cell division in the epidermis and subepidermis. After 12–15 days of culture, successive divisions at the surface of the nodules led to the formation of embryogenic precursor cells, which dedifferentiated into embryogenic cells. Most somatic embryos were apparently derived from multicellular proembryogenic complexes, although a number of few-celled proembryos within a common wall seemed to have originated unicellularly. Between days 24 and 27, somatic embryos at the heart, torpedo and cotyledonary stages were evident. By day 30, many embryos at cotyledonary stage with shoot and root meristems and differentiation of procambial tissue were observed. Vieitez (1995) indicated that the nodules associated with the formation of somatic embryogenesis can be considered as localized callus tissue, which is necessary for redetermination of embryogenic cells. However, these nodules cannot be considered as true callus, which can be subcultured.

6.2.2 Somatic Embryogenesis from Non-embryogenic Tissues

In *C. japonica*, somatic embryogenesis from non-embryogenic tissues has been reported by two research groups: the Spanish group led by Dr Ana María Vieitez and the Portuguese group led by Dr Salomé Pais. Somatic embryos were initiated from leaf (San José and Vieitez 1993; Pedroso and Pais 1993a, 1999), stem (Pedroso and Pais 1999) and roots explants (Vieitez et al. 1991).

6.2.2.1 Somatic Embryogenesis from Leaf and Stem Explants

In all reports, somatic embryos have been initiated from leaf explants (Table 6.1) derived from stock axillary shoot cultures (San José and Vieitez 1993; Pedroso and Pais 1993a, 1999). The use of explants excised from shoot cultures to initiate embryogenic systems offers advantages over the use of zygotic embryo tissues, and clonal material may be a suitable source of explants for induction of somatic embryogenesis from selected mature genotypes (Corredoira et al. 2006). The use of leaves or roots isolated from in vitro cultures provides advantages such as the lack of need to use sterilization procedures and the fact that experiments can be programmed all year around. Somatic embryos derived from zygotic embryos are only slightly influenced by growth regulators, even in basal medium; however, in leaf explants the presence of cytokinins and auxins is necessary for differentiation of somatic embryos.

San José and Vieitez (1993) described the induction of somatic embryogenesis in leaf explants excised from axillary shoot proliferation cultures originally derived from a 3-month-old plant. Three to four fully expanded leaves from the distal ends of 6-week-old shoots were used. The leaves were cut perpendicular to the mid vein, and both apical and basal halves were cultured, abaxial side down, on MS or Woody Plant Medium (Lloyd and McCown 1980; WPM), supplemented with BA (1–8 mg/l), IBA (0.1–2 mg/l) or NAA (0.5–1 mg/l). The best results were obtained with apical explants cultured in the dark for at least 14 weeks with MS medium supplemented with 8 mg/l BA and 0.5 mg/l IBA. Embryo formation occurred predominantly in the distal region of the basal explants and the proximal region of the apical explants. Embryo differentiation was indirect, and six to eight embryos become visible after 7–10 weeks in the callus developed on the cut surface of the explant. In two of the seven treatments evaluated, somatic embryogenesis was induced and adventitious buds were also formed.

Similarly, Pedroso and Pais (1993a, 1999) reported the induction of somatic embryos from leaves derived from axillary shoot cultures established from in vitro germinated seeds. These authors achieved the induction by using a large variety of combinations of growth regulators and conditions. In early studies, entire leaves were immersed in an IBA solution at 1 g/l for 20 min, transferred to half-strength MS solid or liquid medium with 25 g/l D-glucose and incubated in darkness for 11 days. The cultures were then transferred to light, and the embryogenic response was observed 4 weeks after treatment with auxin. Under these conditions, the embryogenesis response ranged from 11.1 to 100 % (Pedroso and Pais 1993a). Subsequently, these authors successfully induced SE in leaves cultured on solid MS supplemented with 20 g/l sucrose or 25 g/l D-glucose, 1 mg/l 2,4-D and 0.1 mg/l kinetin or on MS liquid medium with 20 g/l sucrose or 25 g/l D-glucose, 0.5 mg/l 2,4-D and 1 mg/l BA. In the liquid treatment, the rate of embryogenesis was 31 %, whereas in solid medium the rate of embryogenesis ranged between 35 and 79 % (Pedroso and Pais 1999).

The physiological state of shoot cultures has also been found to influence the embryogenic response, and the best results were observed with 1–3 uppermost entire leaves excised from shoot subcultured every 12 weeks (Pedroso and Pais 1993a). These authors suggested that the differences in morphogenic competence depended on the leaf region considered. Direct somatic embryo formation only occurred in the marginal leaf section, whereas direct root formation was only observed in a well-defined region of the leaf. The Portuguese group also report the induction of somatic embryos from leaves taken from shoot cultures established from an adult *C. japonica* tree (more 50 years); however, they did not differentiate the rate of embryogenesis in this type of material and juvenile material (Pedroso and Pais 1993a, 1999).

Anatomical and ultrastructural studies and studies of the distribution of chemical elements during the induction of direct embryogenesis in leaf explants showed that initiation of embryogenic cells may be associated with the increase in Na, Ca, K, P, Fe and S in the induced cells (Pedroso and Pais 1994a) and/or changes in the cell wall composition (Pedroso and Pais 1992). Pedroso and Pais (1995) pointed out that the appearance of a callus and secondary deposition of cutin are necessary for the development of somatic embryo in leaf explants of *C. japonica*.

Pedroso and Pais (1999) also reported induction of somatic embryos from stem explants cultured on MS supplemented with 25 g/l D-glucose, 1 mg/l BA and 0.1 mg/l IBA. Although initiation of embryogenesis was observed in 78 % of stem explants, very few somatic embryos were produced per stem explant.

6.2.2.2 Somatic Embryogenesis from Adventitious Roots

Somatic embryos have also been initiated on the roots of *C. japonica* plantlets from two in vitro clones derived from 3- to 4-month-old seedlings (Vieitez et al. 1991). Rooting was induced by dipping the basal ends of 2- to 3-cm-long shoots in IBA at 1 g/l for 15 min, after which the shoots were placed in jars containing MS basal medium with half-strength macronutrients and no growth regulators. After 1 month in rooting medium, rooted shoots were transferred to basal MS medium without PGRs or supplemented with 1 mg/l zeatin or 1–2 mg/l BA, alone or in combination with 0.1 or 1 mg/l IBA. Cultures were incubated under standard conditions (16 h/8 h photoperiod and 25/20 °C day/night), and the percentage of plantlets exhibiting somatic embryos was recorded after 8 weeks. Somatic embryogenesis occurred in all treatments, including basal medium without PGRs, and was apparently not related to treatment conditions. The overall rates of embryogenesis were 10 % for one clone and 26 % for another. Yellow-white somatic embryos differentiated directly on plantlet roots either singly or in groups (Fig. 6.1c). Most of these embryos developed in the proximal region of a single root. These somatic embryos appear to have originated from epidermic cells and were connected with the root via few parenchymatous cells.

6.2.3 Androgenesis

The use of microspore or anther culture to generate haploid and doubled haploids is an important tool for plant breeding. The production of homozygous plants is essential for genetic mapping and the first step allowing gene identification. As explained above, pathogens such as *C. camelliae* cause serious damage to the flower of *C. japonica*. The identification of genes involved in the fungus-plant interaction is one of the key steps which will allow obtaining *Camellia* cultivars with low susceptibility to the fungus. Pedroso and Pais (1993b, 1994b) claim to have obtained gametic embryos by anther and microspore culture (Table 6.1) in *C. japonica* (cv. Vila de Nantes and cv. Elegans). The tetrad and early uninucleate microspore stages were found the best stages for androgenesis. MS supplemented with 2,4-D and kinetin were reported to be the best for embryogenic induction in microspore culture. MS supplemented with BA were reported to be the best for androgenesis from anthers both under light and dark conditions (Pedroso and Pais 1997). However, the ploidy of the regenerated plants was not subsequently verified, and no further work has been reported so far.

6.3 Characteristics of Somatic Embryos

One particular feature of somatic embryogenesis in *C. japonica* is the ability of somatic embryos to differentiate and mature in the induction medium, without any need for transfer to a second medium (Nakamura 1988; Vieitez and

Barciela 1990; Vieitez et al. 1991; Vieitez 1995; Pedroso and Pais 1999). Embryo development is not synchronized, and different stages of embryo ontogeny are observed in the same explant.

In most reports, and independently of the type of explant used, two morphological types of somatic embryos were observed in embryogenic cultures of *C. japonica*: "seedlike" and "budlike" embryos (Vieitez 1995; Pedroso and Pais 1999). Formation of these two morphological types appears to be influenced by the presence of a relatively high concentration of BA in the medium (Vieitez 1995). The "seedlike" embryos were yellowish white with large cotyledons as in mature camellia zygotic embryos, whereas "budlike" embryos showed cotyledons resembling true leaves. Both morphological types were clearly bipolar embryos with shoot and root poles. Similar morphology to that described for somatic embryogenesis in *C. japonica* has also been reported for embryogenic cultures of *C. sinensis* (Jha et al. 1992), although these authors do not relate embryo formation to the presence of cytokinins or other types of regulators in the medium. Other morphological abnormalities observed included the presence of more than two cotyledons and cotyledon hypertrophy.

6.4 Maintenance of Embryogenic Cultures

The multiplication and maintenance of embryogenic capacity in *C. japonica*, independently of the initial explant used (zygotic embryos, leaves or roots), were mainly achieved by secondary or repetitive embryogenesis (Kato 1986; Vieitez and Barciela 1990; Vieitez et al. 1991; San José and Vieitez 1993; Vieitez 1995; Pedroso and Pais 1999). Multiplication of somatic embryos based on subculture of embryogenic callus has not been described in this species, probably because the somatic embryos are usually produced directly. Medium-term maintenance of various *C. japonica* culture lines on semi-solid medium was comprehensively described by Vieitez (1995) and later updated by Pedroso and Pais (1999).

The Spanish research group isolated somatic embryos from the initial explants and cultured them on MS medium with or without plant growth regulators, observing secondary embryos after 3–4 weeks. Secondary embryos mainly appeared in the "crown-like" part of the hypocotyl zone of the primary embryo (Vieitez 1995). Clusters of small somatic embryos (3–4 mm long) or isolated cotyledonary embryos (7–9 mm) showed the highest embryogenic capacity. The embryogenic capacity of the two morphological types of somatic embryos (seedlike or budlike) was evaluated in MS supplemented with different combinations of BA and IBA (Vieitez et al. 1991). After 8 weeks of culture, budlike embryos produced more secondary embryos than yielded by seedlike embryos, and the former type displayed a higher capacity for germination. A multiplication medium containing 2 mg/l BA plus 0.1 mg/l IBA was selected for maintenance of embryogenic cultures as this combination favours production of budlike somatic embryos.

The Portuguese group reported the direct transfer of embryos to germination medium for plant development or maintenance of somatic embryos by secondary embryogenesis, although they did not provide details of the type of cultures used (Pedroso and Pais 1999).

6.5 Germination and Plantlet Recovery

The low rate of germination achieved in most embryogenic systems in woody plants is mainly due to the nonsynchronous embryo development, the poor development of shoot meristem and the difficulty in halting secondary embryogenesis. The same is also true for *C. japonica*, and although most reports of somatic embryogenesis of *C. japonica* have described plantlet regeneration (Table 6.1), scant details on culture conditions and conversion frequencies are usually provided (Vieitez 1995). Thus, although Kato (1986) outlines the use of gibberellic acid (GA_3) 1 mg/l to induce germination of somatic embryos of *C. japonica* obtained from cotyledon cultures, the authors do not report the germination rates obtained. Vieitez et al. (1991) obtained a conversion frequency of 62 % on cultivating budlike embryos (Fig. 6.1d) on MS medium supplemented with 5 mg/l GA_3 and 1 mg/l indole-3-acetic acid (IAA). In a later study, Vieitez (1995) improved the conversion efficiency by storing the somatic embryos in the cold for 2 months prior to germination (86 % with cold storage, relative to 54 % without cold storage). Cold storage not only improved the conversion rate but also improved the quality of the shoots. These results were confirmed by Janeiro et al. (1995) who observed that chilling somatic embryos for 2 months significantly improved the plantlet conversion capacity of secondary embryos derived from different embryogenic lines, with rates ranging from 76 to 100 % relative to 32–65 % in the absence of cold treatment. Finally, Pedroso and Pais (1999) obtained conversion frequencies of 18–43.9 % by transferring cotyledonary somatic embryos derived from leaf explants directly to germination medium consisting of MS medium supplemented with 25 mg/l D-glucose, 1 mg/l BA and 0.1 mg/l IBA or IAA.

6.6 In Vitro Storage of Embryogenic Cultures

In vitro culture technology for germplasm storage is carried out in two different ways: slow-growth storage and cryoconservation (storage in liquid nitrogen). Minimal growth storage is a suitable option for medium-term conservation of species. The main advantage of this method is that the collection is maintained under sterile conditions in a controlled environment, thus reducing the need for subculture; the cultures can be rapidly recovered and propagated as required (Grout 1995). The aim of this method is to reduce the plant's metabolism without affecting its physiology, viability or capacity to regenerate under normal conditions. Cryoconservation is the long-term storage of biological material in liquid nitrogen at −196 °C (Reed 2004). The method is based on the arrest of cell division and metabolic processes as a result of storage at ultralow temperatures (Niino and Sakai 1992). Both of these techniques are routinely used for the storage of numerous species, in both temperate and tropical zones (Reed et al. 2004; Engelmann 2011).

The potential use of in vitro systems for conservation and multiplication of germplasm collections has been widely discussed in several reviews and feature articles (Engelmann 1997, 2011; Reed et al. 2004). The application of in vitro techniques to germplasm storage is of particular interest for the conservation of plants such as *Camellia* spp., which are usually propagated vegetatively and/or have recalcitrant seeds. However, to date, few studies have considered methods of storing somatic embryos of *C. japonica*, and currently used methods are generally only successful in the short to medium term. The only study concerning the cryoconservation of somatic embryos of *C. japonica* that we have found in the relevant literature states that somatic embryos did not survive immersion in liquid nitrogen in any of the treatments tested (Janeiro et al. 1996).

6.6.1 Short- to Medium-Term Storage

In vitro slow-growth techniques are routinely used for medium-term conservation of numerous species (Ashmore 1997; Engelmann 2011). The goal of such techniques is to modify the physical environment of the culture, the medium composition or both, to slow down plant growth and thus increase the interval between subcultures. The most widely applied strategy is to maintain cultures at low temperature, either in complete darkness or under low irradiance (Marco-Medina and Casas 2012). These techniques are clearly useful because of their flexibility, simplicity and practicality, and they are the most direct way of restricting the growth and development of explants in vitro (Engelmann 1997; Turner et al. 2001). However, as with most in vitro techniques, the successful implementation of minimal growth technology requires the establishment of specific protocols for each type of explant and species under consideration.

The effects of short- to medium-term cold storage on the maintenance of embryogenic capacity and the germination and conversion of somatic embryos of *C. japonica* have been investigated by Janeiro et al. (1995) in three embryogenic lines. Newly formed cotyledonary secondary embryos removed from 8-week-old proliferated embryo clusters were used in these experiments. Samples were stored for 3, 6, 9 or 12 months at 2–4 °C. Following this treatment, the embryo clusters were transferred to fresh medium and maintained under standard conditions (16 h/8 h photoperiod and 25/20 °C day/night) for another 2 weeks before being examined. The authors reported that increasing the duration of cold storage (from 3 to 12 months) decreased the frequency of surviving embryo clusters and the mean number of surviving embryos per cluster. However, as already mentioned, cold storage improved the plantlet conversion ability. The rate of secondary embryogenesis during germination also decreased after cold storage. Taking into account these results, Janeiro et al. (1995) concluded that cold storage of somatic embryos of *C. japonica* reduced their competence for secondary embryogenesis, but increased plant conversion capacity. These effects appear to reflect a general cold-induced increase in the maturity of the embryos (Vieitez and Ballester 1995).

The capacity for secondary embryogenesis in plants is inversely proportional to the maturation state of the somatic embryo (Williams and Maheswaran 1986; Ammirato 1987; Wann 1988), which is also an important factor in embryo germination. These findings are consistent with other reports stating that germination of somatic embryos of certain woody species requires or is enhanced by cold storage (Corredoira et al. 2006, 2012). Germination may be enhanced by chilling due to interruption of embryo dormancy and may involve changes in the balance between endogenous levels of abscisic acid and gibberellic acid. The increased maturity and germination capacity may also be related to increased levels of storage proteins in chilled embryos (Lecouteux et al. 1993; Jutsuyama et al. 2002).

6.6.2 Cold Storage of Encapsulated Somatic Embryos

The potential uses for artificial seeds are numerous and include the storage, handling and delivery of elite germplasm, as well as the production of hand-pollinated hybrids with reduced seed fertility and genetically engineered plants with sterile or unstable genotypes. The possible use of cold storage to preserve synthetic camellia seeds was investigated by Janeiro et al. (1997). In this study, the effects of cold storage on the maintenance of embryogenic competence and germination of encapsulated embryos of *C. japonica* were determined. The encapsulated embryos were stored for 1 or 2 months at 4 °C in darkness (Fig. 6.1e). After 1 month of cold storage, the encapsulated embryos exhibited a significant reduction in both survival rate and competence for secondary embryogenesis, although cold storage for 2 months had little further effect. The rates of survival and secondary embryogenesis after storage of somatic embryos for 2 months at 4 °C were acceptable (68 %). However, the number of secondary embryos produced was greatly reduced.

Janeiro et al. (1997) also compared the survival and germination capacity of naked and encapsulated embryos after cold storage for 1 or 2 months. Cold storage significantly reduced the survival and germination of both encapsulated and nonencapsulated embryos and increased the percentage of embryos showing only radicle elongation. This change was particularly evident in naked embryos. In general, encapsulated embryos displayed better resistance to storage at 4 °C than naked embryos, although chilling of naked embryos for 2 months reduced the germination rate to half that of non-chilled encapsulated embryos. According to the authors, the fact that higher rate of incomplete germination was associated with lower plant recovery rates suggested that cold storage was detrimental to shoot meristem development.

The reduced capacity for secondary embryogenesis of cold-stored encapsulated embryos of camellia appears to reflect increased maturity, since their capacity for germination is higher than their embryogenic competence. In this respect, short- to medium-term cold storage of synthetic camellia seeds destined for germination appears to be feasible, if a 30–50 % fall in plant recovery rate is considered acceptable. However, cold storage alone cannot be used to maintain embryogenic competence, since the productivity of cold-stored encapsulated embryos is greatly reduced.

6.7 Concluding Remarks

Somatic embryogenesis technology shows a great potential for improvement of woody plants and has applications in rapid large-scale clonal propagation, genetic transformation and cryopreservation of desirable select lines (Park 2002). Previous studies have shown that *C. japonica* has a considerable inherent capacity for direct somatic embryogenesis, at least in the juvenile phase. This ability is particularly well illustrated by the production of somatic embryos from zygotic embryos in basal medium without the addition of growth regulators. Induction of somatic embryogenesis without exogenous auxins is relatively rare in woody species. The ease with which somatic embryos form directly on explants may reflect an inherent capacity of *Camellia* species to produce "pre-embryogenic determined cells" from which somatic embryos develop. Indirect embryogenesis, which requires the presence of auxin and/or cytokinin, may involve the redetermination of callus cells as "induced embryogenic determined cells". Somatic embryos have also been initiated from non-embryogenic tissues such as leaf, root and stem explants excised from axillary shoot cultures established from germinated seeds. Initial explants derived from adult trees have scarcely been used for somatic embryo induction in *C. japonica*.

The long-term maintenance of the embryogenic capacity, especially by repetitive embryogenesis, enables the continuous supply of somatic embryos at an adequate multiplication rate. Moreover, it should be noted that in *C. japonica*, most embryogenic systems involve direct embryogenesis, which theoretically guarantees the genetic stability that is important for preservation of genetic fidelity.

Although there are several potential advantages associated with the use of synthetic seeds (e.g. the possibility of long-term storage and for automation and commercialization purposes), information regarding the feasibility of encapsulation in *C. japonica* is scant. The use of encapsulation techniques to produce artificial seeds would open up new possibilities for handing, transportation and delivery of in vitro tissue cultures of this species. Cold storage of somatic embryo clusters (2–6 months) and synthetic seeds (2 months) of camellia destined for germination appears feasible. However, cold treatment is not by itself sufficient to maintain embryogenic capacity, as the productivity of cold-stored embryos is severely reduced. Further research should aim to improve cold storage and germination rates.

Cryopreservation ensures safe, long-term conservation of genetic resources of plant species with recalcitrant seeds, of vegetatively propagated species and of biotechnology products such as somatic embryos, cell lines and genetically transformed material. In *C. japonica*, none of the cryogenic procedures applied to somatic embryo cultures were successful. Further research on the cryopreservation of camellia embryogenic cultures is required to enable long-term storage of clonal somatic embryos. It is our hope that the experience gained with other woody species will help in the development of protocols for cryopreservation of *C. japonica* somatic embryos.

Although transgenic technology is of enormous potential application in *Camellia*, transgenic plants have not yet been developed. As the species can multiply by cross-pollination and by outbreeding with wild relatives, there is a high level of diversity in the cultivated gene pool, some of which could be exploited in future breeding programmes. Future efforts should be aimed toward transformation of this material using genetic constructions that would confer resistance to the diseases currently threatening the species. Finally, research involving transgenic camellia must be carried out to address some of the problems faced by the *Camellia* nursery growers and to exploit this highly regenerative system.

Acknowledgements The members of the Biotechnology and Forest Improvement research group thank Dr Ana Mª Vieitez for sharing her enthusiasm about the topic of somatic embryogenesis.

References

Akihisa T, Tokuda H, Ukiya M, Suzuki T, Enjo F, Koike K, NikaidoT,NishinoH(2004)3-epicabraleahydroxylactone and other triterpenoids from *Camellia* oil and their inhibitory effects on Epstein-Barr virus activation. Chem Pharm Bull 52:153–156

Ammirato PV (1987) Organizational events during somatic embryogenesis. In: Green CE, Somers DA, Hackett WP, Biesboer DD (eds) Plant tissue and cell culture. Alan R. Liss Inc, New York, pp 57–81

Ashmore SE (1997) Status report on the development and application of in vitro techniques for the conservation and use of plant genetic resources. International Plant Genetic Resources Institute, Rome

Barciela J, Vieitez AM (1993) Anatomical sequence and morphometric analysis during somatic embryogenesis on cultured cotyledon explants of *Camellia japonica* L. Ann Bot 71:395–404

Beauchesne M (1978) Riproduzione meristematica della *Camellia*. Notiziario della Societa Italiana de la *Camellia*, Cannero Riviera, anno XIV, March 1978

Bennett WY (1977) Tissue culture for *Camellia* I. Am Camellia Yearb 1997:188–190

Bennett WY, Scheibert P (1982) In vitro regeneration of callus and plantlets from cotyledons of *Camellia japonica*. Camellia J 37:12–15

Canhoto JM, Lopes ML, Cruz GS (1999) Somatic embryogenesis in myrtaceous plants. In: Jain SM, Gupta PK, Newton RJ (eds) Somatic embryogenesis in woody plants, vol 4. Kluwer Academic Publishers, Dordrecht, pp 293–340

Chung KH (2010) Transesterification of *Camellia japonica* and *Vemicia fordii* seed oils on alkali catalysts for biodiesel production. J Ind Eng Chem 16:506–509

Corredoira E, Ballester A, Vieitez FJ, Vieitez AM (2006) Somatic embryogenesis in chestnut. In: Mujib A, Samaj J (eds) Plant cell monogr (2): somatic embryogenesis. Springer, Berlin/Heidelberg, pp 177–199

Corredoira E, San José MC, Vieitez AM (2012) Induction of somatic embryogenesis from different explants of shoot cultures derived from young *Quercus alba* trees. Trees 26:881–891

Corredoira E, Valladares S, Martínez MT, Vieitez AM, San José MC (2013) Somatic embryogenesis in *Alnus glutinosa* (L.) Gaertn. Trees 27:1597–1608

Correia S, Lopes ML, Canhoto JM (2011) Somatic embryogenesis induction system for cloning an adult *Cyphomandra betacea* (Cav.) Sendt. (tamarillo). Trees 25:1009–1020

Couselo JL, Vela P, Salinero C, Mansilla P (2014) Limiting factors for the biological and chemical control of Camellia Flower Blight. In: Proceedings of the International *Camellia* Congress, Pontevedra, Spain, pp 266–279

De Ron AM, Salinero MC, Vela P (2014) Variation in oil content in *Camellia japonica* seeds. In: Proceedings of the International *Camellia* Congress, Pontevedra, Spain, pp 286–291

Engelmann F (1997) In vitro conservation methods. In: Ford-Lloyd BV, Newburry HJ, Callow JA (eds) Biotechnology and plant genetic resources: conservation and use. CABI, Wallingford, pp 119–162

Engelmann F (2011) Use of biotechnologies for the conservation of plant biodiversity. In Vitro Cell Dev Biol Plant 47:5–16

EPPO (1997) *Ciborinia camelliae*. In: Quarantine pest for Europe, 2nd edn. CAB International, Wallingford, pp 684–687

EPPO (2003) *Ciborinia camelliae*. EPPO Bull 33: 257–264

Gao JY (2005) Collected species of the genus *Camellia*-an illustrated outline. Zhejiang Science and Technology Publishing House, Hangzhou

Grout BW (1995) Minimal growth storage. In: Grout BW (ed) Genetic preservation of plant cells in vitro. Springer, Berlin, pp 21–28

Izco J (2014) Camellias: diversity and globalization processes. In: Proceedings of the International *Camellia* Congress, Pontevedra, Spain, pp 9–16

Janeiro LV, Ballester A, Vieitez AM (1995) Effect of cold storage on somatic embryogenesis systems of *Camellia*. J Hortic Sci 70:665–672

Janeiro LV, Vieitez AM, Ballester A (1996) Cryopreservation of somatic embryos and embryonic axes of *Camellia japonica* L. Plant Cell Rep 15:699–703

Janeiro LV, Ballester A, Vieitez AM (1997) In vitro response of encapsulated somatic embryos of camellia. Plant Cell Tissue Organ Cult 51:119–125

Jeong C, Kim JH, Choi GN, Kwak JH, Kim D, Heo HJ (2010) Protective effects of extract with phenolics from Camellia (*Camellia japonica*) leaf against oxidative stress-induced neurotoxicity. Food Sci Biotechnol 19:1347–1353

Jha TB, Jha S, Sen SK (1992) Somatic embryogenesis from immature cotyledons of an elite Darjeeling tea clone. Plant Sci 84:209–213

Jung E, Lee J, Baek J, Jung K, Lee J, Huh S, Kim S, Koh J, Park D (2007) Effect of *Camellia japonica* oil on human type I procollagen production and skin barrier function. J Ethnopharmacol 112:127–131

Jutsuyama Y, Sususki T, Harada T, Fujikawa S (2002) Sucrose incubation increases freezing tolerance of asparagus (*Asparagus officinalis* L.) embryogenic cell suspensions. CryoLetters 29:145–156

Kato M (1986) Micropropagation through cotyledon culture in *Camellia japonica* L. and *Camellia sinensis* L. Jpn J Breed 36:31–38

Kim KY, Davidson PM, Chung HJ (2001) Antibacterial activity in extracts of *Camellia japonica* L. petals and its application to a model food system. J Food Protect 64:1255–1260

Kim YH, Kim KS, Han CS, Yang HC, Park SH, Ko KI, Lee SH, Kim KH, Lee NH, Kim JM, Son KH (2007) Inhibitory effects of natural plants of Jeju Island on elastase and MMP-1 expression. J Cosmet Sci 58:19–33

Kim S, Jung E, Shin S, Kim M, Kim YS, Lee J, Park D (2012) Anti-inflammatory activity of *Camellia japonica* oil. BMB Rep 45:177–182

Kong D-M, Preece JE, Shen H-L (2012) Somatic embryogenesis in immature cotyledons of Manchurian ash (*Fraxinus mandshurica* Rupr.). Plant Cell Tissue Organ Cult 108:485–492

Kuba M, Tsuha K, Tsuha K, Matsuzaki G, Yasumoto T (2008) In vitro analysis of the anti-allergic activities of *Camellia japonica* extract and okicamelliaside, a degranulation inhibitor. J Health Sci 54:584–588

Lecouteux CG, Lai FM, Mckersie BD (1993) Maturation of alfalfa (*Medicago sativa* L.) somatic embryos by abscisic acid, sucrose and chilling stress. Plant Sci 94:207–213

Lee JH, Kim JW, Ko NY, Mun SH, Kim DK, Kim JD, Kim HS, Lee KR, Kim YK, Radinger M, Her E, Choi WS (2008) *Camellia japonica* suppresses immunoglobulin E-mediated allergic response by the inhibition of Syk kinase activation in mast cells. Clin Exp Allergy 3:794–804

Lee SY, Jung MY, Yoon SH (2014) Optimization of the refining process of camellia seed oil for edible purposes. Food Sci Biotechnol 23:65–73

Lloyd G, McCown B (1980) Commercially-feasible micropropagation of mountain laurel, *Kalmia latifolia*, by use of shoot tips culture. Proc Int Plant Prop Soc 30:420–427

Marco-Medina A, Casas JL (2012) Polyamine content during minimal growth storage of *Thymus moroderi* explants. Biol Plant 56:590–594

Miuraa D, Kidab Y, Nojimac H (2007) *Camellia* oil and its distillate fractions effectively inhibit the spontaneous metastasis of mouse melanoma BL6 cells. FEBS Lett 581:2541–2548

Mizutani T, Masaki H (2014) Anti-photoaging capability of antioxidant extract from *Camellia japonica* leaf. Exp Dermatol 23:23–26

Mondal TK (2011) Camellia. In: Kole C (ed) Wild crop relatives: genomic and breeding resources. Springer, Berlin/Heidelberg, pp 15–39

Murashige T, Skoog F (1962) A revised medium for rapid growth and bioassays with tobacco tissue cultures. Physiol Plant 15:473–497

Nakamura Y (1988) Efficient differentiation of adventitious embryos from cotyledon culture of *Camellia sinensis* and other *Camellia* species. Tea Res J 67:1–12

Niino T, Sakai A (1992) Cryopreservation of alginate-coated in vitro-grown shoot tips of apple, pear and mulberry. Plant Sci 87:199–206

Onodera K, Hanashiro K, Yasumoto T (2006) Camellianoside, a novel antioxidant glycoside from the leaves of *Camellia japonica*. Biosci Biotechnol Biochem 70:1995–1998

Onodera K, Tsuha K, Yasumoto-Hirose M, Tsuha K, Hanashiro K, Naoki H, Yasumoto T (2010) Okicamelliaside, an extraordinarily potent anti-degranulation glucoside isolated from leaves of *Camellia japonica*. Biosci Biotechnol Biochem 74:2532–2534

Park Y-S (2002) Implementation of conifer somatic embryogenesis in clonal forestry: technical requirements and deployment considerations. Ann For Sci 59:651–656

Park JC, Hur JM, Park JG, Hatano T, Yoshida T, Miyashiro H, Min BS, Hattori M (2002) Inhibitory effects of Korean medicinal plants and camelliatannin H from *Camellia japonica* on human immunodeficiency virus type 1 protease. Phytother Res 16:422–426

Pedroso MC, Pais MS (1992) A scanning electron microscope and x-ray microanalysis study during induction of morphogenesis in *Camellia japonica* L. Plant Sci 87:99–108

Pedroso MC, Pais MS (1993a) Direct embryo formation in leaves of *Camellia japonica* L. Plant Cell Rep 12:639–643

Pedroso MC, Pais S (1993b) Regeneration from anthers of adult *Camellia japonica*. In Vitro Cell Dev Biol Plant 29:155–159

Pedroso MC, Pais MS (1994a) Early detection of embryogenic competence and of polarity in *Camellia japonica* L. by electron probe x-ray microanalysis. Plant Sci 96:189–201

Pedroso MC, Pais MS (1994b) Induction of microspore embryogenesis in *Camellia japonica* cv. Elegans. Plant Cell Tissue Organ Cult 37:129–136

Pedroso MC, Pais MS (1995) Factors controlling somatic embryogenesis. Plant Cell Tissue Organ Cult 43:147–154

Pedroso MC, Pais MS (1997) Anther and microspore culture in *Camellia japonica*. In: Jain SM, Sopory SK, Veilleux RE (eds) In vitro haploid production in higher plants vol 5: Oil, ornamental and miscellaneous plants, Current plant science and biotechnology in agriculture. Kluwer Academic Publishers, Dordrecht, pp 89–107

Pedroso MC, Pais MS (1999) Direct somatic embryogenesis from leaves of *Camellia japonica*. In: Jain SM, Gupta PK, Newton RJ (eds) Somatic embryogenesis in woody plants, vol 5. Kluwer Academic Publishers, London, pp 163–178

IPGRI Handbooks for Genebanks N°7. International Plant Genetic Resources Institute, Rome, Italy.

Reed BM, Engelmann F, Dulloo ME, Engels JMM (2004) IPGRI handbooks for genebanks No 7. International Plant Genetic Resources Institute, Rome

Reed BM, Sarasan V, Kane M, Bunna E, Pence VC (2011) Biodiversity conservation and conservation biotechnology tools. In Vitro Cell Dev Bio Plant 47:1–4

Salinero C, Feás X, Mansilla JP, Seijas JA, Vázquez-Tato MP, Vela P, Sainz MJ (2012) ^{1}H-Nuclear Magnetic Resonance Analysis of the triacylglyceride composition of cold-pressed oil from *Camellia japonica*. Molecules 17:6716–6727

Samartín A, Vieitez AM, Vieitez E (1986) Rooting of tissue cultured camellias. J Hortic Sci 61:113–120

San José MC, Vieitez AM (1993) Regeneration of *Camellia* plantlets from leaf explants cultures by embryogenesis and caulogenesis. Sci Hortic 54:303–315

Savige TJ (1993) The international *Camellia* register. The International Camellia Society, Wirlinga

Shirai M (1896) Descriptions of some new Japanese species of *Exobasidium*. Bot Mag 10:51–54

Tabachnik L, Kester DE (1977) Shoot cultures for almond and almond-peach hybrid clones in vitro. HortSci 12:545–547

Thao NTP, Hung TM, Lee MK, Kim JC, Min BS, Bae K (2010) Triterpenoids from *Camellia japonica* and their cytotoxic activity. Chem Pharm Bull 58:121–124

Turner S, Krauss SL, Bunn E, Senaratna T, Dixon K, Tan B, Touchell D (2001) Genetic fidelity and viability of *Anigozanthos viridis* following tissue culture, cold storage and cryopreservation. Plant Sci 161:1099–1106

van Toor RF (2002) Development of biocontrol methods for *Camellia* flower blight caused by *Ciborinia camelliae* (Kohn). Ph D thesis, Lincoln University, New Zealand

Vela P, Salinero C, Sainz MJ (2013) Phenological growth stages of *Camellia japonica*. Ann Appl Biol 162:182–190

Vieitez AM (1995) Somatic embryogenesis in *Camellia* spp. In: Jain S, Gupta P, Newton R (eds) Somatic embryogenesis in woody plants, vol 2. Kluwer Academic Publishers, Dordrecht, pp 235–276

Vieitez AM, Ballester A (1995) Cold storage and cryopreservation of *Camellia* embryogenic cultures. In: Jain SM, Gupta PK, Newton RJ (eds) Somatic embryogenesis in woody plants. Kluwer Academic Publishers, Dordrecht, pp 209–238

Vieitez AM, Barciela J (1990) Somatic embryogenesis and plant regeneration from embryonic tissues of *Camellia japonica* L. Plant Cell Tissue Organ Cult 21:267–274

Vieitez AM, San José MC, Vieitez FJ, Ballester A (1991) Somatic embryogenesis from roots of *Camellia japonica* plantlets cultured in vitro. J Am Soc Hortic Sci 116:753–757

Vieitez AM, Vieitez ML, Ballester A, Vieitez E (1992) Micropropagation of *Camellia* spp. In: Bajaj YPS (ed) Biotechnology in agriculture and forestry, vol 19: High-tech and micropropagation III. Springer, Berlin/Heidelberg, pp 361–387

Wann SR (1988) Somatic embryogenesis in woody species. Hortic Rev 10:153–181

Williams EG, Maheswaran G (1986) Somatic embryogenesis factors influencing coordinated behavior of cells as an embryogenic group. Ann Bot 57:443–462

Yoshikawa M, Morikawa T, Asao Y, Fujiwara E, Nakamura S, Matsuda H (2007) Medicinal flowers. XV. The structures of noroleanane- and oleanane-type triterpene oligoglycosides with gastroprotective and platelet aggregation activities from flower buds of Camellia japonica. Chem Pharm Bull 55:606–612

Zhang YM, Maharachchikumbura SSN, Wei JG, McKenzie EHC, Hyde KD (2012) *Pestalotiopsis camelliae*, a new species associated with grey blight of *Camellia japonica* in China. Sydowia 64:335–344

Somatic Embryogenesis and Genetic Transformation of Carnation (*Dianthus caryophyllus* L.)

7

Anelia Iantcheva

Abstract

The chapter summarizes the progress of somatic embryogenesis and genetic transformation of carnation obtained during the last 40 years. Factors that determine the processes of indirect and direct somatic embryogenesis of this ornamental plant are described and discussed. The present chapter outlines the current results in the field and focuses on the primary guidelines of genetic transformation of carnation. Future applications and intentions identified by Bulgarian research group in this direction have also been highlighted.

Keywords

Carnation • *in vitro* regeneration • Somatic embryogenesis • Genetic transformation

7.1 Introduction

The genus *Dianthus* comprises around 300 species including *D. caryophyllus*, *D. barbatus*, and *D. chinensis*, which possess important commercial value for cut flower production. Among those, *Dianthus caryophyllus* L. is the most important cut flower plant. Carnation is one of the most valuable commercial cut flowers around the globe. Due to its excellent storage quality, wide range of forms, ability to withstand long-distance transportation, and remarkable ability to rehydrate after continuous shipping, it is preferred by growers to roses and chrysanthemums in several flower-exporting countries. In many countries, it ranks next to rose in popularity and it is widely cultivated on a large scale in Italy, Spain, Columbia, Kenya, Sri Lanka, the Canary Islands, France, Holland, Germany and the United States. In Bulgaria, 75 % of the greenhouses designated for ornamentals are occupied by this crop. Commercial carnation cultivars are distributed in two main groups – standard SIM

A. Iantcheva (✉)
Agrobioinstitute,
Blvd. Dragan Tzankov 8, 1164 Sofia, Bulgaria
e-mail: aneliaiancheva@abi.bg

A. Mujib (ed.), *Somatic Embryogenesis in Ornamentals and Its Applications*,
DOI 10.1007/978-81-322-2683-3_7

group and spray carnation (Hughes 1993). Standard or single flower SIM is characterized by a big single flower with a high and solid flower stalk. Breeding of this group started in France, North Italy, and later on in the United States. In 1938 American breeder W. Sim created a cultivar White Sim, which gives rise to many cultivars with worldwide distribution. Cultivars belonging to this group are characterized by flower color richness, but are very sensitive to the fungal pathogen *Fusarium oxysporum*. Mediterranean carnation cultivars created in France and Italy are more resistant to *Fusarium*. Worldwide existing single flower carnation cultivars are the product of SIM and Mediterranean cultivars (Holley and Baker 1991). The spray carnation was created by Thompson in 1956 in the United States and later on distributed in Europe since 1964. In spray carnation, the axillary flower bud is removed and adventitious flower buds are preserved, which leads to the spray flower architecture.

The global cut flower market is maintained by the introduction of new improved cultivars. Classical breeding has long been the main route for generation of new traits into a wide range of commercial cultivars. The genetic variability within carnation is relatively poor; therefore, the breeding potential for new flower colors and patterns as well as resistance to biotic and abiotic stresses is also very limited. Carnation is a vegetatively propagated plant, which further reduces its genetic pool availability.

The modern cell and molecular techniques could be regarded as an alternative and an additional complementation tool to the classical improvement methods of carnation. In the last 20 years, genetic engineering opened up the possibility for creation of novelty (Zuker et al. 1998). The transgenic carnations of the "Moon Series" composing eight shades varying from soft lavender pearl to violet color (Moonaqua, Moonlite, Moonshade, Moonvista, Moonpearl, Moonberry, Moonique, Moonvelvet) are the first genetically engineered commercial flowers in the world (distributed from 1992 to 2013 around the globe) and demonstrated the success of genetic manipulation of flower color (Tanaka et al. 2005) through plant transformation. Since last 40 years plant tissue culture technique focusing on ornamental plants; in carnation, tissue culture method begins with the development of micropropagation scheme (Kozak and Hampel 1979; Davis et al. 1977) and production of virus-free carnation plant material (Hollings and Stone 1972). In spite of the richness of carnation cultivars on the global market, the efforts are directed to create new cultivars with novel ornamental value, petal shape, and increased resistance against biotic and abiotic stress.

The routine application of gene transfer methods requires existence of efficient regeneration procedure. The existing cultivars are characterized by high heterozygosity and genotypic variation is observed within even cultivars, all determine genotypes' role in plant regeneration. In carnation, adventitious shoot formation has been described from different plant explants, e.g., shoot tips (Earle and Langhans 1975), petals (Kakehi 1979; Gimelli et al. 1984; Frey and Janick 1991), stems (Nugent et al. 1991), ovules (Demmink et al. 1987), anthers (Villalobos 1981), and leaf bases (Van Altvorst et al. 1994; Yantcheva et al. 1997a). Regeneration via indirect somatic embryogenesis was also reported. Protocols for indirect somatic embryo formation were reported in different varieties of carnation (Frey et al. 1992; Iantcheva et al. 2005; Kanwar and Kumar 2009); similarly, direct origin of somatic embryos was observed in a liquid medium of group SIM and spray carnation (Yantcheva et al. 1998; Pareek and Kothari 2003).

7.2 Induction of Somatic Embryogenesis

7.2.1 Type of Somatic Embryogenesis

Somatic embryogenesis is a process whereby a cell or group of cells originated from somatic tissue form embryos. Developmental stages of somatic embryogenesis resemble the process of zygotic embryo formation. Somatic embryogenesis is the preferred pathway, because it offers large-scale potential for multiplication. It occurs either in dedifferentiated cells or directly in predetermined embryogenic cells (Ammirato 1985; De Jong et al. 1993). Somatic embryogenesis in *Dianthus caryophyllus* is a very rarely described event.

7.2.2 Indirect Somatic Embryogenesis of Carnation

The indirect somatic embryogenesis systems are characterized by a sequence of events that includes the stimulation of cell proliferation, dedifferentiation, acquisition of embryogenic competence, and the induction of embryos. Application of an auxin, usually 2,4-dichlorophenoxyacetic acid (2,4-D), is an important treatment for the early stages, but subsequent embryo development requires removal of exogenous auxin. One feature of indirect systems is that the initial activation of cell proliferation is temporary and physically separated from the induction of embryo-specific cell division.

Indirect somatic embryogenesis in carnation is characterized by the formation of highly dedifferentiated callus tissue and was reported for the first time in the research of Frey et al. (1992). Internodal callus was induced from *cvs.* Scania, White Sim, and Sandra. A medium for induction of callus tissue was augmented with the auxin 2,4-D. Later on, embryo development continued on medium lacking the auxin but supplemented with cytokinin benzyladenine (BA) or kinetin, which supported this process. Performed histological assay confirmed that the somatic embryos originated from single cells and early embryo development proceeded via globular, heart-shaped, and torpedo stages. The reported frequency of embryo conversion to plantlets was low and authors linked this low conversion rate to the lack of well-formed apical and root meristem of somatic embryos. This study (Frey et al. 1992) confirmed that somatic embryos could be induced from internodal callus. Most of the obtained somatic embryos were abnormal and lacked either apical or root meristems, but a few embryos were developed successfully into valuable plants and acclimatized to greenhouse condition. Optimum embryo initiation was obtained on a liquid callus induction medium supplemented with 2.4-D for 4–6 weeks, with embryo development occurring on 2.4-D-free medium after an additional month.

Iantcheva et al. (2005) induced callus tissue in dark and light conditions from leaf explants in two-spray Bulgarian carnation *cvs.* Fea and Rossitza. A solid callus induction medium was supplemented with auxin 2,4-D in combination with cytokinin 6-benzylaminopurine (BAP) and casein hydrolysate. This combination supported the induction of massive embryogenic callus tissue for a period of 30–60 days. Unorganized tissue became visible on the surface of the explants and on the wounded edges 15–20 days of induction. Dark condition led to increased callus production for both cultivars tested. The calli obtained in *cv.* Fea were white and friable developed on the surface of the explant. In *cv.* Rossitza, the obtained tissue was more structured and yellowish and was produced mostly on the edges of the explant. Around 85–90 % of the explants reacted and no difference was observed for the tested cultivars. At light condition, the intensity of obtained callus was significantly less than at dark. During the first 30 days, the yield of callus tissue was very low, but later on from day 30 to 60, the unorganized callus tissue increased profusely at dark condition for both cultivars. The color of tissue was also changed and varied from yellow to green and the structure was more compact for both cultivars. The transfer of embryogenic callus to a medium amended with low cytokinin 0.05 mg/l BAP and 250 mg/l casein hydrolysate promoted embryo formation process in *cvs.* Fea and Rossitza for a period of 15–20 days. In the case of *cv.* Fea, embryogenic callus induced at dark condition was able to form embryos. On the contrary, *cv.* Rossitza callus, induced at light, was more responsive to somatic embryo formation. The frequency of obtained embryos was 1–3 per explant for *cv.* Rossitza and 1–2 embryos per explant in *cv.* Fea formed bipolar structures possessing fully developed cotyledonary leaves and converted to plants.

Regenerated plants obtained from somatic embryos had normal phenotype and rooted easily in in vitro condition.

In an indirect embryogenesis carnation system (Frey et al. 1992; Iantcheva et al. 2005; Karami et al. 2006, 2007, 2008), induction of embryogenic callus tissue was achieved on a medium supplemented with synthetic auxin 2,4-D alone or in combination with cytokinin. Karami and Kordestani (2007) obtained embryogenic callus from two *cvs*. Impulse and Sagre in the presence of auxin picloram at different concentrations. Besides the combination of auxin 2,4-D and cytokinin BAP, the effect of different sucrose concentrations on induction of embryogenic callus was investigated from petal explants in four carnation *cvs*. Nelson, Sagre, Spirit, and Impulse by Karami et al. (2010). The authors found that high sucrose concentration promoted embryogenic callus induction and positively affected somatic embryo development with formation of high percentage of embryo. Karami and Kordestani (2007) also investigated the effect of sugar alcohol sorbitol on the process of embryo formation from callus induced with picloram. The authors observed positive osmotic effect of sorbitol on production of embryo numbers.

7.2.3 Direct Somatic Embryogenesis

Direct somatic embryogenesis is characterized by the formation of embryos directly from differentiated tissue without the apparent requirement of dedifferentiation stage involving disorganized cell proliferation. For example, the somatic embryogenesis system in carnation (Yantcheva et al. 1998) involved direct formation of embryos from young leaf explants in response to an induction treatment. There are two different models to explain this phenomenon. The first model proposes that there are cells within the tissue that are already embryogenically competent and require the inductive signal to trigger direct embryo formation (Maheswaran and Williams 1984; Williams and Maheswaran 1986; Carman 1990). It has also been argued that in the direct system, embryogenesis does not differ significantly from the indirect procedure at the molecular level and both proceed through similar stages of genetic reprogramming at different rates (De Jong et al. 1993). These models have different explanations for cell division activation in the process of direct somatic embryogenesis. In the first model, the inductive signal acts as a mitotic trigger and reactivates cell division in cells that are already competent to switch from somatic to embryogenic type and proceed into asymmetric cell division to form embryos. In the second model, the induction of cell proliferation is required for dedifferentiation, which then permits the acquisition of embryogenic competence in certain cells, just as in the indirect system.

Direct somatic embryogenesis in *Dianthus caryophyllus* is a very rarely described event. Protocols for direct somatic embryo formation were reported for a limited number of cultivars. The development of somatic embryos both through the culture of small white callus and directly on the leaf surface was observed by Nakano and Mii (1993). However, in the mentioned case, the regeneration efficiency was quite low. Yantcheva et al. (1998) described for the first time a system for direct formation of somatic embryos in liquid medium for SIM *cv*. Lena and four spray Bulgarian carnation *cvs*. Nasslada, Yanita, Regina, and Line 84. Later on, a very similar procedure was developed for the cultivars of economically important species of *D. caryophyllus*, *D. barbatus*, and *D. chinensis* (Pareek and Kothari 2003).

In direct somatic embryogenesis system described by Yantcheva et al. (1998), 35-day-old leaf explants were used as starting explants. The preparation of starting leaf tissue was also found to be very important. The leaves were cut into small 2–4 mm^2 pieces for all tested cultivars. The appearance of somatic embryos was recorded after 20–25 days of cultivation in a liquid medium supplemented with 1 mg/L 2.4-D, 0.2 mg/L BAP and 2 g/L casein hydrolysate. About 65 % of the explants produced globular embryos directly on explant surface (Fig. 7.1a). In this system of direct embryo induction, the importance of age and size of explants for initiation of somatic

Fig. 7.1 Direct somatic embryogenesis of carnation (*Dianthus caryophyllus* L.). (**a**): direct somatic embryo formation on explant in a liquid medium in *cv*. Lena; (**b**): torpedo stage embryos after PEG treatment in a liquid medium; (**c**): conversion of embryos and polyembryos from *cv*. Nasslada; and (**d**): carnation plants regenerated from somatic embryos of *cv*. Yanita

embryos were noted. If explants were detached from older donor plants originating from greenhouse, the induction of somatic embryos was delayed and extended up to 45–50 days. When the size of explants was too small (~1 mm^2), these explants became brownish and failed to produce embryo structures. The number of induced somatic embryos was remarkably low when suspension culture was initiated from big leaf explants (half or one third of the leaf). In the optimal embryo initiation condition, the number of primary embryos ranged from 5–15 per explant and most of the embryos appeared on the cut edges of plant material. Most of the small explants (4 mm^2) resembled clusters. Some of the induced embryos detached from the explants continued to develop separately or remain attached to the starting tissues, which were defined as polyembryos. No deviation in embryo morphology was detected at this stage of development.

No significant difference in the numbers of produced somatic embryos was also observed in the studied cultivars.

In the presence of higher 2.4-D concentration (4 mg/L), total somatic embryo induction period (35 days) was reduced by 5 days. Morphology of embryos induced in the presence of higher concentration of 2,4-D were different, however the embryos possessed a more friable surface and disposed to recallus. The development of induced somatic embryos from globular up to torpedo stage was the most critical moment in this regeneration procedure. Development up to torpedo stage was observed on medium supplemented with polyethylene glycol (Fig. 7.1b). The obtained data clearly indicated the positive effect of PEG at low concentrations. The addition of 1 % (PEG) as an osmoticum and desiccative factor plays an important role in embryo development in carnation species. In the absence of PEG treatment (when the suspension culture was transferred directly to a maturation medium), the induced globular embryos could not reach to the next "torpedo" stage. The embryos usually transformed into hard green callus. In 1 % PEG added medium, somatic embryo transformation rate to torpedo stage varied from 40.7 % for *cv*. Yanita to 60.65 % observed in *cv*. Lena. In higher level of PEG (2.5–5.0 %), the differentiated embryo structures turned brown and further development was stalled. Some morphological differences (color, shape) of "torpedo" somatic embryos between tested cultivars were observed. The positive effect of PEG for somatic embryo development was later confirmed on other observations (Pareek and Kothari 2003). The authors observed PEG concentration of 2.5 % was optimal for embryo development of tested cultivars. The embryo maturation from torpedo to cotyledonary stage was found to be genotype specific, Yantcheva et al. (1998) observed. Some of the cultivars like Nasslada, Yanita, and Lena required hormone free liquid medium amended with casein hydrolysate for successful conversion of somatic embryos to vigor plantlets. Conversion rate was found to be highest in *cv*. Nasslada (180 plantlets were obtained from 20 plated embryos and polyembryos, Fig. 7.1c), followed by *cvs*. Yanita (45 plantlets) and Lena (34 plantlets). In the case of *cv*. Regina, the addition of BAP alone was necessary for successful embryo maturation and conversion to plantlets. Formation of secondary embryos was observed if maturation of embryos was performed on liquid medium added with cytokinin and auxin inhibitor. Plants regenerated from somatic embryos, for all the tested cultivars possess normal phenotype and were adapted successfully to greenhouse conditions. Performed biometric analyses of ten traits related to ornamental value of regenerants confirmed normal flower morphology without any aberration (Fig. 7.1d). The study of Yantcheva et al. (1998) demonstrated direct somatic embryogenesis of carnation in liquid media, which has several benefits and advantages:

- Regeneration via direct somatic embryogenesis was developed for economically important carnation cultivars.
- Somatic embryo-regenerated plants had normal phenotype and were adapted and grown in green house condition successfully.
- The established culture conditions for inducing secondary embryogenesis may offer the opportunity to exploit different gene transfer methods for obtaining nonchimeric transgenic plants.

7.3 Genetic Transformation of Carnation

The flower market demands flowers with improved traits, such as new colors, new shapes, better fragrance, and longer vase life. Every year, new and more attractive varieties are created by breeders. Unfortunately high heterozygosity and limited genome information restrict breeding programs in carnation. Genetic engineering is a valuable tool, currently used in classical breeding methods for the production of novel carnation varieties, and these technologies have been exploited to introduce new traits into commercial cultivars. In the last 20 years, *Agrobacterium tumefaciens*-mediated transformation systems have been developed using different plant explants like leaves, petals, and stems (Lu et al. 1991; Van Altvorst et al. 1995, 1996; Yantcheva et al. 1997b; Nontaswatsri et al. 2004). Direct gene transfer by using microprojectile bombardment or with *Agrobacterium* was also successfully conducted (Zuker et al. 1995, 1999) in creating transgenic carnations with novel traits. The Moon Series composed of eight carnation varieties possessing soft to dark violet flower color was produced by Florigene. These newly obtained carnation varieties with modified anthocyanin biosynthesis were achieved by introduction of flavonoid 3′,5′-hydroxylase (F3′,5′ is a key enzyme in the synthesis of delphinidin) in carnation. Long vase life, which is a very important characteristic of cut flowers, was achieved by downregulation of ethylene production by antisense ACC synthase (Kiss et al. 2000; Veres et al. 2005) and antisense ACC oxidase (Savin et al. 1995). A lot of efforts were concentrated on the establishment of *Fusarium oxysporum f.* sp. *dianthi*-resistant transgenic carnation cultivars, and different genes like osmotin, PR-1, and chitinase were introduced into susceptible varieties (Zuker et al. 2001). The same research group successfully downregulated the flavanone 3-hydroxylase gene, which modified carnation flower color and fragrance simultaneously (Zuker et al. 2002).

Together with modern biotechnological methods, in the last 5 years, researchers focused on structural analysis of the whole genome of carnation in order to accelerate the process of molecular breeding. Genetic linkage maps of carnation have been constructed and were used to identify quantitative trait loci (QTL) responsible for resistance to bacterial wilt (Yagi et al. 2006, 2012). Application of next-generation sequencing (NGS) technology and large-scale transcriptome analysis (RNA-seq) have also been conducted (Tanase et al. 2012). Recently, a reference genetic linkage map for carnation using simple sequence repeat (SSR) markers derived from RNA-seq analysis was constructed (Yagi et al. 2013). Thus, mapping QTLs with various traits will improve breeding programs of elite carnation cultivars (Table 7.1).

During the last 20 years, a lot of reports for transgenic carnation establishment procedure were published; the efficiency of successful transformation events is, however, low and in most of the cases is genotype specific. In published protocols, several factors that have been identified as affecting transformation frequency were discussed. In most cases, bacterial strain and density, period of cocultivation, selective pressure, and regeneration method were identified to be very crucial for the efficiency of transformation. Different *Agrobacterium* strains like AGLO, LBA 4404, and EHA 105 were used and bacterial density level varies widely (0.5–1 OD_{600}) in tested transformation experiments (Zuker et al. 1999; Nontaswatsri et al. 2004; Szoke et al. 2006; Arici and Koc 2009). The positive effect of acetosyringone in different concentrations was reported in different research studies (Nontaswatsri and Fukai 2006; Arici and Koc 2009). Nontaswatsri and Fukai (2006) were able to obtain fairly good transformation with efficiency varying from 7 to 95 % depending on the genotype. By using different ways like efficient regeneration protocol, bacterial strain, cocultivation, and application of acetosyringone, the authors were able to establish highly effective transgenic protocol. In early studies of Yantcheva et al. (1997b), transgenic plants of carnation were obtained by *Agrobacterium*-mediated transformation. A regeneration/transformation procedure was developed based on leaf base and petal as initial explants (Yantcheva et al. 1997a). Transformation of SIM *cv.* Lena and spray *cvs.* Line 84 and Yanita was made with *Agrobacterium tumefaciens* C58C1 carrying a plasmid (p2260 I 170), which contains marker genes – *npt* II and *hpt* encoding neomycin phosphotransferase and hygromicin phosphotransferase, respectively. For SIM *cvs.* Scania and Lena and spray *cv.* Nasslada, the *Agrobacterium tumefaciens* LBA 4404 with plasmid vip 104-carrying antisense of chalcone-synthase-A gene (CHS-A) from petunia was used. Cultivar Regina was transformed with plasmid vip 165, in which marker GUS gene was under control of endogenous chalconesynthase-J (*chs-J*) promoter. Transformation efficiency varied from 3.9 to 7.5 % among tested cultivars.

Table 7.1 Explant, PGRs, and somatic embryogenesis (indirect and direct) in carnation

Carnation (*Dianthus caryophyllus* L.)			
Species/cultivar	Explant	PGRs for induction of SE	References
Indirect somatic embryogenesis			
SIM			
cv. Scania	IN	2.4-D	Frey et al. (1992)
cv. White Sim			
cv. Sandra			
Spray	L	2.4-D+BAP+CH	Iantcheva et al. (2005)
cv. Fea			
cv. Rossitza			
SIM	P	2.4-D+BAP	Karami et al. (2006, 2008, 2010)
cv. Nelson			
cv. Sagre			
cv. Impulse			
cv. Spirit			
SIM	PB	Picloram	Karami and Kordestani (2007)
cv. Impulse			
cv. Sagre			
cv. Nelson			
Direct somatic embryogenesis			
Dianthus sp.	L	Antibiotics	Nakano and Mii (1993)
		Cefotaxime	
		Penicillin G	
		Carbenicillin	
SIM	L	2.4-D+BAP+CH	Yantcheva et al. (1998)
cv. Lena			
Spray			
cv. Nasslada			
cv. Yanita,			
cv. Regina			
Line 84			
D. caryophyllus	L	2.4-D	Pareek and Kothari (2003)
D. barbatus			
D. chinensis			

Abbreviations: *P* petiole, *PB* petiole base, *L* leaf, *IN* internode, *2.4-D* dichlorophenoxyacetic acid, *BAP* 6-benzylaminopurine, *CH* casein hydrolysate

The presence of *npt* II gene in plants was confirmed by PCR and Southern blot analysis. Downregulation of CHS-A was estimated by Northern blot. The main features of selected protocols for gene transfer developed for different carnation cultivars from SIM and spray type are presented in Table 7.2.

Histochemical GUS activity was detected in the flower buds – flower bed and calyx, from regenerated plants of *cv.* Regina. Vegetatively propagated clones of transgenic plants from *cvs.* Scania, Nasslada, Yanita, and Line 84 were evaluated for plant and flower morphology. Ten morphological traits were selected for biometrical analyses in order to estimate ornamental value of obtained transgenic carnation plants. Downregulation of CHS-A gene in flowers of *cvs.* Scania and Nasslada was low and could be characterized with thin white or brighter strips across petals (Fig. 7.2a). Slight visual detection

Table. 7.2 Explant, regeneration mode, bacterial strain, and markers used in genetic transformation of carnation

Agrobacterium-mediated gene transfer						
Species/cultivar	Explant	Regeneration system	Bacterial strain	Selectable marker gene	Specific feature	References
D. caryophyllus	Leaf base	Direct organogenesis	C58C1	*npt II*	–	van Altvorst et al. (1995)
30 cvs.			AGLO			
D. caryophyllus-SIM	Leaf base	Direct organogenesis	C58C1	*npt II*	–	Yantcheva et al. (1997b)
cv. Lena			LBA 4404			
cv. Scania						
D. caryophyllus-spray						
cv. Regina						
cv. Nasslada						
cv. Yanita						
Line 84						
D. caryophyllus-SIM	Stem	Direct organogenesis	AGLO	*npt II*	Wounding by bombardment	Zuker et al. (1999)
cv. WS			EHA 105			
cv. Visa					100 μM AS	
cv. Desio					Two-cycle selection	
D. caryophyllus-spray						
cv. Eilat						
cv. Darling						
cv. Lior						
D. caryophyllus	Petal	Direct organogenesis	CBE 21	*npt II*	–	Miroshnichenko and Dolgov (2000)
cv. Dia 12	Leaf base		EHA 105	*hpt*		
D. caryophyllus	Node	Direct organogenesis	AGLO	*npt II*	50 μM AS	Nontaswatsri and Fukai (2006)
cv. Tanga						
cv. Killer						
cv. Laurella						
cv. Master						
cv. Otone						

(continued)

Table. 7.2 (continued)

Species/cultivar	Explant	Regeneration system	Bacterial strain	Selectable marker gene	Specific feature	References
Agrobacterium-mediated gene transfer						
D. chinensis	Leaf base	Direct organogenesis	LBA 4404	*npt II*	–	Szoke et al. (2006)
cv. Simon						
D. caryophyllus						
cv. Bibor						
cv. IWS						
D. caryophyllus	Leaf segments	Direct organogenesis	EHA 105	*npt II*	200 μM AS	Arici and Koc (2009)
cv. Turbo						
D. caryophyllus	Leaf segments	Indirect and direct organogenesis	EHA 105	*npt II*	Compare transformation efficiency by using direct and indirect organogenesis	Kanwar and Kumar (2011)
cv. Tempo						
Direct gene transfer						
D. caryophyllus-SIM	Stem	Direct organogenesis	Plasmid DNA	Bar	Two-cycle selection	Zuker et al. (1995)
cv. WS						
cv. Ondina						
cv. Visa						
D. caryophyllus-spray						
cv. Lior						

Abbreviations: *AS* acetosyringone, *npt II* neomycin phosphotransferase, *hpt* hygromycin phosphotransferase, *bar* phosphinothricin acetyltransferase

Fig. 7.2 Genetically modified carnation (*Dianthus caryophyllus* L.). (**a**): downregulation of CHS-A gene in petals of *cv*. Scania with white or brighter strips; (**b**): flower of transgenic Line 84 with white to light pink edge on the end of the petals; and (**c**): transgenic carnation plant from *cv*. Yanita

of downregulated gene from petunia could be due to expression of gene in heterologous plant species under existing sequence homology between genes involved in biosynthetic pathways of flavonoids of two ornamental species: petunia and carnation. Some deviations were observed in leaf length and width of genetically modified plants in comparison with controls. Transgenic plants from *cvs*. Yanita and Line 84 also possessed morphological differences from controls. Transgenic flowers of Line 84 acquired white to light pink edges, 3–4 mm at the end of the petals (Fig. 7.2b). Besides changes in flower color, decreased flower bud numbers and flower stalks were also observed. Differences in plant morphology, i.e., increased leaf length, width, and number of flower buds and flower stalk, were identified in Yanita cultivar (Fig. 7.2c).

Recently, the same research group initiated a project titled "Integrated functional and comparative genomics studies on the model legumes *Medicago truncatula* and *Lotus japonicus*." In this project, *M. truncatula* and *L. japonicus* are used for functional genomic studies for the establishment of a platform, which will facilitate the genetics and breeding of important crops (pea, faba bean, alfalfa, and clover) in Bulgaria. The project aims were to identify and understand functions of genes involved in plant organogenesis/ embryogenesis. Nine genes involved in basic processes of plant growth and development were selected for experimental work (Revalska et al. 2011; Iantcheva et al. 2014; Boycheva et al. 2014). One of the cloned genes encoding 1-amin ocyclopropane-1-carboxylate oxidase contains InterPro domain(s) IPR005123 oxoglutarate and iron-dependent oxygenase MT5G085330 (PLAZA2,5). The systematic name of enzyme coding by this gene is 1-aminocyclopropane-1-carboxylate oxygenase (ethylene forming). Enzymes with Fe (2+) and 2-oxoglutarate (2OG)-dependent oxygenase domain typically catalyze the oxidation of an organic substrate using a dioxygen molecule, mostly by using ferrous iron as the active site cofactor and 2OG as a cosubstrate, which is decarboxylated to succinate and CO_2. In plants, Fe(II) 2OG oxygenase domain enzymes catalyze the formation of plant hormones, such as ethylene, gibberellins, and pigments anthocyanidins and flavones. Using GATEWAY technology, the abovementioned gene was cloned in overexpression vector and using RNAi methodology in vector for downregulation of gene expression. The cloned complementary DNA of the investigated *M. truncatula* gene 1-aminocyclopropane-1-carboxylate oxidase (MT5G085330) showed 81 % similarity with Dca55566.1 1-aminocyclopropane-1-carboxylate oxidase *Dianthus caryophyllus* clover pink. In order to create 2OG-Fe(II) oxygenase RNAi construct, the 132 bp fragment from the 2OG-Fe(II) oxygenase mRNA was used, part of which correspond to positions 397–437 bp of the Dca55566.1 (http://carnation.kazusa.or.jp). Based on the above-described genetic transformation system, this gene was heterologously expressed in carnation (Yantcheva et al. 1997b), and overexpressed (OE) and downregulated (RNAi) transgenic plants from *cv.* Scania and downregulated transgenic plants from spray carnation *cv.* Red Barbara were obtained. Preliminary biometric measurements, i.e., plant height, leaf length, leaf width, root, and secondary root numbers of in vitro plant material, were evaluated and compared. The obtained results of root length and root number indicated that OE and RNAi transgenic plants from *cv.* Scania had longer root length with more numbers of roots compared to control carnation plants. In the case of Red Barbara cultivar, root length and root numbers of RNAi plants were more compared to control plants.

A gene, *DMR6*, was identified from *Arabidopsis thaliana* (At5G24530) encoding 2OG-Fe(II) oxygenase with unknown function. Induction of DMR6 transcript was detected during *Hyaloperonospora parasitica* infection, which causes downy mildew disease to plants of family Brassicaceae and the accumulated transcript confirmed the participation of this gene in protective responses of plant (van Damme et al. 2008). In model *L. japonicas* plant, overexpression of the *MT5G085330* gene caused long and well-developed root with more numbers of nodules on the upper part of the root. Downregulated *Lotus* plants on the other hand had long and well-developed roots with induced nodules distributed along the whole length of the root, which also confirmed the involvement of genes in protective response against infection.

The experimental works of upregulated and downregulated 2OG-Fe (II) oxygenase transgenic plants of carnation are in progress. The analyses will be concentrated on the response of transgenic plants to infection with different carnation pathogens. Further analyses will also be conducted with plant morphology and vase life of transgenic plants.

7.4 Conclusion

Although it is a rare event, direct somatic embryogenesis is an important way to regenerate plants from single cells, or a small group of somatic cells, in carnation. One important application of somatic embryogenesis is rapid mass propagation of commercially valuable genotypes especially in liquid media, which is the most attractive use of this morphogenic pathway. Single-cell origin also permits synchronized, homogeneous, and stable plant material; thus, somatic embryogenesis is the preferred method of regeneration rather than organogenesis. Induction and development of large-scale somatic embryos is easy and is possible in different cultivars of carnation groups – SIM and spray. Another important use is in the generation of transgenics by using direct somatic embryos or secondary somatic embryos. The development of new carnation cultivars with novel plant morphology, physiology, and flower characteristics will depend on successful use of plant tissue culture tool, molecular biology, genomics, and transcriptomics established in plants including model species.

References

Ammirato PV (1985) Patterns of development in culture. In: Henke RR, Hughes KW, Constantin MP, Hollander A (eds) Tissue culture and forestry and agriculture. Plenum, New York, pp 9–29

Arici SE, Koc NK (2009) Regeneration and *Agrobacterium* mediated transformation studies in carnation (*Dianthus caryophyllus* L. cv. Turbo). Afr J Biotechnol 8:6094–6100

Boycheva I, Vassileva V, Iantcheva A (2014) Histone acetyltransferases in plant development and plasticity. Curr Genomics 15:28–37

Carman JG (1990) Embryogenic cell in plant tissue cultures: occurrence and behaviour. In Vitro Cell Dev Biol Plant 26:746–753

Davis MJ, Baker R, Hanan JJ (1977) Clonal multiplication of carnation by micropropagation. J Am Soc Hortic Sci 102:48–53

De Jong AJ, Schmidt EDL, De Vries SC (1993) Early events in higher plant embryogenesis. Plant Mol Biol 22:367–377

Demmink JF, Custers JBM, Bergervoet JHW (1987) Gynogenesis to bypass crossing barriers between diploid and tetraploid *Dianthus* species. Acta Horticult 216:343–344

Earle ED, Langhans RW (1975) Carnation propagation from shoot tips cultured in liquid medium. HortSci 10:608–610

Florigene http://www.florigene.com

Frey L, Janick J (1991) Organogenesis in carnation. J Am Soc Hortic Sci 116:1108–1112

Frey L, Saranga Y, Janick J (1992) Somatic embryogenesis in carnation. HortSci 27:63–65

Gimelli F, Ginatta G, Venturo R, Positan S, Bulatti M (1984) Plantlet regeneration from petals and floral induction in vitro in the Mediterranean carnation (*Dianthus caryophyllus* L.). Riv Ortoflorofruttic Ital 68:107–121

Holley DW, Baker R (1991) Carnation production II. Kendall Hunt Publishing Company, Dubuque, 156 pp

Hollings M, Stone OM (1972) Productivity of virus tested carnation clones and the rate of reinfection with virus. J Hortic Sci 47:141–149

Hughes S (1993) Carnations and pinks. The complete guide. The Crowood Press, Ramsbury, 221 pp

Iantcheva A, Vlahova M, Atanassova B, Atanassov A (2005) Plant regeneration via direct organogenesis and somatic embryogenesis of two new Bulgarian spray carnation cultivars. Biotechnol Biotechnol Equip 19:15–19

Iantcheva A, Revalska M, Zehirov G, Vassileva V (2014) *Agrobacterium* mediated transformation of *Medicago truncatula* cell suspension culture provides a system for functional analysis. In Vitro Cell Dev Biol Plant 50:149–157

Kakehi M (1979) Studies on the tissue culture of carnation. V. Induction of redifferentiated plants from the petal tissue. Bull Hiroshima Agric Coll 6:159–166

Kanwar JK, Kumar S (2009) Influence of growth regulators and explants on shoot regeneration in carnation. Hortic Sci 36(4):140–146

Kanwar JK, Kumar S (2011) Recovery of transgenic plants by *Agrobacterium*-mediated genetic transformation in *Dianthus caryophyllus* L. (carnation). Adv Appl Sci Res 2:357–366

Karami O, Kordestani GK (2007) Proliferation, shoot organogenesis and somatic embryogenesis in embryogenic callus of carnation. J Fruit Ornam Plant Res 15:167–175

Karami O, Deljou A, Esna-Ashari M, Ostad-Ahmadi P (2006) Effect of sucrose concentrations on somatic embryogenesis in carnation (*Dianthus caryophyllus* L.). Sci Hortic 110:340–344

Karami O, Deljou A, Mahmodi Pour A (2007) Repetitive somatic embryogenesis in carnation on picloram supplemented media. Plant Growth Regul 50:33–39

Karami O, Deljou A, Kordestani GK (2008) Secondary somatic embryogenesis of carnation (Dianthus caryophyllus L.). Plant Cell Tissue Org Cult 92:273–280

Karami O (2010) Saidi A. The molecular basis for stress-induced acquisition of somatic embryogenesis 37:2493–2507

Kiss E, Veres A, Zs G, Nagy N, Tyth E, Varga A, Hrazdina G, Heszky L (2000) Production of transgenic carnation with antisense ACS (1-aminocyclopropane-1-carboxylate synthase) gene. Int J Hortic Sci 6:104–107

Kozak D, Hampel M (1979) Studies of in vitro multiplication of carnations III. The optimization of multiplantlet formation. Acta Horticult 91:333–337

Lu CY, Nugent G, Wardley-Richardson T, Chandler SF, Young R, Dalling MJ (1991) *Agrobacterium*-mediated transformation of carnation (*Dianthus caryophyllus* L.). Biotechnology 9:864–868

Maheswaran G, Williams EG (1984) Direct somatic embryoid formation on immature embryos of *Trifolium repens*, *T. pratense* and *Medicago sativa*, and rapid clonal propagation of *T. repens*. Ann Bot 54:201–211

Miroshnichenko DN, Dolgov SV (2000) Production of transgenic hygromicin resistant carnation (Dianthus caryophyllus L.) plants after cocultivation with Agrobacterium tumefaciens. In: Cadic A(ed) Prot 19 Int'l symposium improvement ornamental plants. Acta Hort 508, ISHS

Nakano M, Mii M (1993) Antibiotics stimulate somatic embryogenesis without plant growth regulators in several *Dianthus* cultivars. J Plant Physiol 141:721–725

Nontaswatsri C, Fukai S (2006) Carnation (*Dianthus caryophyllus* L.). In: Wang K (ed) Agrobacterium protocols, vol 2. Humana Press, Totowa, pp 311–320

Nontaswatsri C, Fukai S, Goi M (2004) Revised cocultivation conditions produce effective *Agrobacterium* mediated genetic transformation of carnation (*Dianthus caryophyllus* L.). Plant Sci 166:59–68

Nugent G, Wardley RT, Lu CY (1991) Plant regeneration from stem and petal of carnation (*Dianthus caryophyllus* L.). Plant Cell Rep 10:477–480

Pareek A, Kothari SL (2003) Direct somatic embryogenesis and plant regeneration from leaf cultures of ornamental species of *Dianthus*. Sci Hortic 98:449–459

Revalska M, Vassileva V, Goormachtig S, Van Hautegem T, Ratet P, Iantcheva A (2011) Recent progress in development of *Tnt1* functional genomics platform for *Medicago truncatula* and *Lotus japonicus* in Bulgaria. Curr Genomics 12:147–152

Savin KW, Baudinette SC, Graham MW, Michael MZ, Nugent GD, Lu CY, Chandler SF, Cornish EC (1995) Antisense ACC oxidase RNA delays carnation senescence. Hortic Sci 30:970–972

Szoke A, Kiss E, Toldi O, Heszky L (2006) Production of transgenic carnation with a heterologous 6-phosphofructo-2-kinase/fructose 2,6-bisphosphatase bifunctional enzyme cDNA. Int J Hortic Sci 12(4):75–79

Tanaka Y, Kastumoto Y, Brugliera F, Mason J (2005) Genetic engineering in floriculture. Plant Cell Tiss Org Cult 80:1–24

Tanase K, Nishitani C, Hirakawa H, Isobe S, Tabata S, Ohmiya A, Onozaki T (2012) Transcriptome analysis of carnation (*Dianthus caryophyllus* L.) based on next-generation sequencing technology. BMC Genomics 13:292

Van Altvorst AC, Koehorst H, Bruinsma T, Dons HJM (1994) Improvement of adventitious shoot formation from carnation leaf explants. Plant Cell Tiss Org Cult 37:87–90

Van Altvorst AC, Rikesen T, Koehorst H, Dons JJM (1995) Transgenic carnations obtained by *Agrobacterium tumefaciens* mediated of leaf explants. Transgenic Res 4:105–113

Van Altvorst AC, Koehorst HJJ, Dons JJM (1996) Transgenic carnation plants obtained by Agrobacterium tumefaciens mediated transformation of petal explants. Plant Cell Tissue Org Cult 169:169–173

Van Damme M, Huibers RP, Elberse J, Van den Ackerveken G (2008) Arabidopsis DMR6 encodes a putative 2OG-Fe(II) oxygenase that is defence-associated but required for susceptibility to downy mildew. Plant J 54:785–793

Veres A, Kiss E, Tyth E, Tyth A, Heszky L (2005) Downregulation of ethylene production in carnation (*Dianthus caryophyllus* L.) by an apple derived ACC-cDNA. Int J Hortic Sci 11:101–104

Villalobos V (1981) Floral differentiation in carnation (*Dianthus caryophyllus* L.) from anthers cultured *in vitro*. Phyton 41:71–75

Williams E, Maheswaran G (1986) Somatic embryogenesis: factors influencing coordinated behaviour of cells as an embryogenic group. Ann Bot 57:443–462

Yagi M, Onozaki T, Taneya M, Watanabe H, Yoshimura T, Yoshinari T, Ochiai Y, Shibata M (2006) Construction of a genetic linkage map for the carnation by using RAPD and SSR markers and mapping quantitative trait loci (QTL) for resistance to bacterial wilt caused by *Burkholderia caryophylli*. J Jpn Soc Hortic Sci 75:166–172

Yagi M, Kimura T, Yamamoto T, Isobe S, Tabata S, Onozaki T (2012) QTL analysis for resistance to bacterial wilt (*Burkholderia caryophylli*) in carnation (*Dianthus caryophyllus*) using an SSR-based genetic linkage map. Mol Breed 30:495–509

Yagi M, Yamamoto T, Isobe S, Hirakawa H, Tabata S, Tanase K, Yamaguchi H, Onozaki T (2013) Construction of a reference genetic linkage map for carnation (*Dianthus caryophyllus* L.). BMC Genomics 14:734

Yantcheva A, Vlahova M, Atanassova B, Atanassov A (1997a) Direct organogenesis and plant regeneration of carnation (*Dianthus caryophyllus* L.). Biotechnol Biotechnol Equip 11:60–65

Yantcheva A, Vlahova M, Todorovska E, Atanassov A (1997b) Genetic transformation of carnation (*Dianthus caryophyllus* L.). Biotechnol Biotechnol Equip 11:21–25

Yantcheva A, Vlahova M, Antanassov A (1998) Direct somatic embryogenesis and plant regeneration of carnation (*Dianthus caryophyllus* L.). Plant Cell Rep 18:148–153

Zuker A, Chang P-FL, Ahroni A, Cheah K, Woodson WR, Bressan RA, Watad AA, Hasegawa PM, Vainstein A (1995) Transformation of carnation by microprojectile bombardment. Sci Hortic 64:177–185

Zuker A, Tzfira T, Vainstein A (1998) Genetic engineering for cut flower improvement. Biotechol Adv 16:33–79

Zuker A, Ahroni A, Tzfira T, Ben-Meir H, Vainstein A (1999) Wounding by bombardment yields highly efficient *Agrobacterium*-mediated transformation of carnation (*Dianthus caryophyllus* L.). Mol Breed 5:367–375

Zuker A, Shklarman E, Scovel G, Ben-Meir H, Ovadis M, Neta-Sharir I, Ben-Yephet Y, Weiss D, Watad A, Vainstein A (2001) Genetic engineering of agronomic and ornamental traits in carnation. Acta Horticult 560:91–94

Zuker A, Tzfira T, Ben-Meir H, Ovadis M, Shklarman E, Itzhaki I, Forkmann G, Martens S, Neta-Sharir I, Weiss D, Vainstein A (2002) Modification of flower color and fragrance by antisense suppression of the flavanone 3-hydroxylase gene. Mol Breed 9:33–41

8 Cryopreservation of Somatic Embryos of Ornamental Plants

Emrah Kırdök, Hamit Ekinci,
and Yelda Özden Çiftçi

Abstract

Ornamental plants play a social and economic role in human society since antic ages, and its production consists about 78 % of total production. Thus, in situ and *ex situ* germplasm conservation techniques must be applied to preserve elite varieties. Since in situ strategies are more prone to environmental factors (e.g., biotic and abiotic stress) and *ex situ* approaches are open to cross-pollination or homologous recombination during gamete formation, currently in vitro strategies are the good complementary mechanism to avoid these problems. Among in vitro conservation techniques, cryopreservation seems to be the best candidate as it enables to preserve selected germplasm theoretically unlimited period of time with maintaining genetic stability, which is very important in ornamental plant cultivation. Besides, embryogenic cultures are used for in vitro propagation; the tissues are also utilized as target materials for gene transfer studies. These cultures have the potential to produce cisgenic or intragenic plants and cryogenic technology offers opportunities to conserve germplasm to introduce genes from crossable ornamental plants especially for cisgenesis. Thus cryopreservation also plays an important role in maintaining transgenic, cisgenic, or intergenic somatic embryos of ornamentals in a stable way. In this chapter, efficient cryopreservation technique including one-step and two-step freezing methods together with vitrification- and dehydration-based techniques for conservation of somatic embryos and related tissues of ornamentals is discussed.

Keywords

Dehydration • Freezing • Vitrification

E. Kırdök • H. Ekinci • Y.Ö. Çiftçi (✉)
Department of Molecular Biology and Genetics,
Gebze Technical University,
41400 Çayırova, Kocaeli, Turkey
e-mail: yelda75@yahoo.com

A. Mujib (ed.), *Somatic Embryogenesis in Ornamentals and Its Applications*,
DOI 10.1007/978-81-322-2683-3_8

8.1 Introduction

Today, about 157 million types of plant species are cultivated as ornamental plants which consists 78 % of total production, and there are a lot of studies dealing with in vitro culture, new variety production, and germplasm conservation (reviewed in Rout et al. 2006). Ornamental plants play an important role in social improvement in many countries around the world, increasing and improving the quality of life with psychological and social benefits for humans (Thomsen et al. 2011). Besides their aesthetic appeal, ornamental plants also serve some less obvious uses such as for fragrance, attracting wildlife, and cleaning the air. In addition, decorating offices with ornamental plants can reduce the levels of discomfort (like cough, fatigue) and improve health (Fjeld et al. 1998), probably due to their ability to remove volatile organic compounds from the air (Dela Cruz et al. 2014). Moreover, ornamental plants have great economic value; thus, the conservation of valuable ornamental genetic resources is very important in order to meet the market demands (Kulus and Zalewska 2014). Traditional *ex situ* conservation strategies in greenhouses or field conditions require intensive care of pot cultures (Reed et al. 2006); biotechnological storage techniques could be utilized to access large material variety for breeding and horticultural production (Halmagyi et al. 2004). Since the plants have those benefits, intensive care must be applied to maintain the genetic stability of cultures and to conserve important ornamental plants of endangered nature (Swarts and Dixon 2009). Up to now, cryopreservation is accepted to be the most promising and valuable biotechnological storage technique for long-term preservation (Kulus and Zalewska 2014).

The combination of somatic embryogenesis and cryopreservation in ornamental plants allows to preserve high-value clonal varieties by recovering quality characteristics (color, odor, etc.) by avoiding the risk of loss via contamination, human error, or somaclonal variation, associated with frequent culture transfer for maintenance by repetitive embryogenesis (George et al. 2008; Panis 2009). It also helps to store embryogenic lines from new hybrid species or transgenic plants (Wang et al. 2012). The selected genotypes then could be used for advanced breeding programs and commercial plantations. Interestingly, Mikula and coworkers (2011) demonstrated that cryopreserved embryogenic tissues of *Gentiana cruciata* had significantly higher morphogenetic potential than the nonfrozen controls. In this chapter, effective cryopreservation methods for conservation of somatic embryos of ornamental plants are discussed. It should also be noted that as protocorm-like bodies (PLBs) is reported as somatic embryos for orchids (Lee et al. 2013), cryopreservation of PLBs is also included.

8.1.1 Concepts of Cryopreservation

Conservation defines the effort for management and preservation of germplasm which are threatened (e.g., endemic species) or valuable (e.g., ornamental plants) via in situ, *ex situ*, or in vitro strategies for further utilization. While in situ strategies aim to conserve plant germplasm in their own ecosystem (e.g., national parks), *ex situ* approaches try to preserve plants in vegetative forms (e.g., separated field pots), seeds (mostly orthodox seeds according to their moisture content), or in vitro methods (e.g., plant tissue culture). Since in situ strategies are more prone to environmental factors (e.g., biotic and abiotic stress) and *ex situ* approaches are open to cross-pollination or homologous recombination during gamete production, in vitro strategies are employed to avoid these problems. In vitro strategies are based on plant tissue culture techniques, which are defined as the usage of plant organs, cells, or tissues for producing new plant cells, tissues, or metabolites in aseptic culture conditions. With this approach, it is possible to produce virus-free (e.g., meristem culture), genetically stable and clonally produced ornamental plants in high quantities (e.g., in vitro propagation) (Engelmann 2010).

The utilization of in vitro approaches for conservation of plant germplasm could be in two ways: medium- and long-term conservation. The former aims to decrease plant metabolism, thus incubating in limited growth conditions (e.g., storing cultures in darkness and low temperature) which is widely used to conserve different plant germplasm including ornamental plants (Akdemir et al. 2010), while the latter aims to stop all metabolic reactions in liquid nitrogen (LN), theoretically for unlimited time.

Nowadays, conservation efforts have arisen since plant diversity is getting lost at an alarming rate (100–1000-fold) higher than recent past. Also, in situ conservation strategies are constrained with economic issues. For example, it is figured out that conservation of Cape Floristic Region of South Africa, which contains 20 % of the continent flora, costs about 330 $ per plant species for annum. On the contrary, 1 year conservation expense takes approximately 13–20 € for *Solanum tuberosum* and related species per accession for 1 year with the usage of in vitro tools like cryopreservation (reviewed in Li and Pritchard 2009).

8.1.2 Cryopreservation

As mentioned before, cryopreservation aims to stop nearly all biologic reactions in vapor (−150 °C) or liquid phase (−196 °C) of nitrogen and thus enables to maintain genetic stability of wide range of plant explants (suspension cultures, in vitro buds, seeds, zygotic and somatic embryos) without somaclonal variation theoretically to *ad infinitum* (reviewed in Kulus and Zalewska 2014). As the concept of this book chapter is limited to cryopreservation of somatic embryos, the application of long-term conservation techniques for other plant explants of ornamental plants is excluded and not discussed.

Somatic and zygotic embryos have some common features, i.e., both are large, complex structures and formed of differentiated tissues; somatic embryos have generally relatively smaller size (several hundred μm to several mm) in comparison to zygotic embryos (up to 1 cm); somatic embryos are highly hydrated, while zygotic embryos have lower water content; somatic embryos often comprise heterogeneous structure in size and developmental stage, whereas zygotic embryos are homogenous populations. The steps of cryopreservation are, however, little different (introduction to in vitro conditions are before cryopreservation in somatic embryos, while zygotic embryos are introduced to in vitro conditions after cryostorage). Moreover, the preservation of proliferation capacities of somatic embryos is adequate for successful post-survival, but in the case of zygotic embryos, the whole structure should be conserved for the germination (Engelmann 2010).

In cryopreservation protocols, the ultimate goal is to vitrify the cells and tissues by decreasing their water content with cryoprotective solutions (including ethylene glycol, Dimethyl sulfoxide (DMSO), and sucrose) or dehydration-based protocols (e.g., using air desiccation) to prevent lethal ice formation. To achieve this context, two different cryopreservation approaches are employed based on the nature of the explants (e.g., dehydration sensitive): (I) two-step freezing and (II) one-step freezing (Engelmann 2010).

8.1.2.1 Two-Step Freezing

Also called as "traditional methods," "controlled-rate freezing," or "slow cooling," this practice includes pre-culture of the plant material, for instance, somatic embryos, with low concentrated cryoprotectants containing medium, and then application of slow cooling step with cryoprotectants before transferring to LN. The key points of this technique are the optimization of not only the time and concentration of the cryoprotectants but also cooling rate and terminal temperature prior to immersion to LN. For instance, cooling rate should not be too fast or too slow; thus, a pace of 0.1–1.0 °C min^{-1} to −40 °C is used in general (Kulus and Zalewska 2014). During slow freezing, crystallization takes place initially in the external medium, and water flows from out of the cells to the external ice that results in dehydration of the samples (Engelmann 2010).

Although this two-step process is employed to somatic embryos, it requires specific instruments and could only be employed for cold-tolerant species. However, in recent years, the requirement of specific freezing apparatus is overcome by the usage of freezing container, Nalgene® Mr. Frosty, which is a nonmechanical device that only requires 100 % isopropyl alcohol and a mechanical freezer.

8.1.2.2 One-Step Freezing

The process, also called "rapid freezing," depends on vitrification-based methods in which a transition of intracellular water to an amorphous glass state occurs without ice crystallization. Therefore, the samples have to be sufficiently dehydrated physically (under airflow of the laminar cabinet or over silica gel) or chemically (with the usage of vitrification solutions) before immersion to LN. Vitrification is a multistep procedure containing pre-culture treatment with loading solution (a mixture of diluted nontoxic cryoprotectants usually contains 2 M glycerol and 0.4 M sucrose in MS medium), dehydration with vitrification solution, and rapid immersion to LN. Thawing is followed by washing of samples in unloading solution (1.2 M sucrose) and culture on recovery media. However as naked PLBs are found to be very sensitive to dehydration, and vitrification solutions could be toxic; vitrification-based methods are broadened to include encapsulation-dehydration, encapsulation-vitrification, and droplet-vitrification. The encapsulation of samples in alginate beads prior to dehydration or vitrification has benefits as the beads provide enhanced protection of materials both from mechanical and oxidative damage and ease of sample handling during pre- and post-cryopreservation (Teixeira da Silva et al. 2014).

The droplet-vitrification technique is developed by Panis and coworkers (2005) which consists of application of vitrification solution in droplets on a strip of heat-sterilized aluminum foil. When the vitrification treatment is over, strips are rapidly placed into the cryovials (one strip/cryovial) filled with LN, and the lids of the cryovials are closed and cryovials are plunged to LN. For thawing, the strips are rinsed in 10 ml unloading solution in a Petri dish and kept there for 30 min followed by their transfer to recovery media. Vitrification-based protocols have mainly been used to cryostore shoot apices, but recently encouraging results have also been obtained with embryogenic tissues of several ornamental species (reviewed in Ozudogru et al. 2010).

8.2 Cryogenic Applications to Conserve Embryogenic Cultures of Ornamental Plants

Although cryopreservation was employed for *ad infinitum* conservation of many economically important plants, there are only a limited number of articles published for cryopreservation of somatic embryos of ornamental plants. A standard cryopreservation technique comprises successive steps including preconditioning, pre-culture, dehydration time or cryoprotective solution incubation period, cooling, thawing, and post-treatments (Fig. 8.1). As the relative survival rate after vitrification cryopreservation of embryogenic suspension cells of ornamental plants like *Anthurium andraeanum* has a certain relationship with subculture time, type and concentration of osmotic regulating agent, preculture time, loading solution type and pretreatment time, dehydration time of PVS2, and thawing temperature (Wang et al. 2010), each of these steps should be optimized every time for different plant species and/or different cultivars of the same species.

8.2.1 Pre-culture

To achieve maximum regrowth after cryopreservation, the water content of the tissue must be discarded to avoid ice crystal formation, which is lethal for the cells (Engelmann 2010). In literature, this pre-culture period generally employed via MS medium (or other optimized media) supplemented with various concentrations of sucrose, sorbitol, or other sugar alcohols (Table 8.1). Pre-culture period could be employed with a particular or in combination with increased concentration of sugar content. For instance, *Anthurium andraeanum* embryogenic cell suspensions were pre-cultured on ½ MS (Murashige and Skoog 1962) medium supplemented with 0.5 M sorbitol for 2 days (Wang et al. 2010).

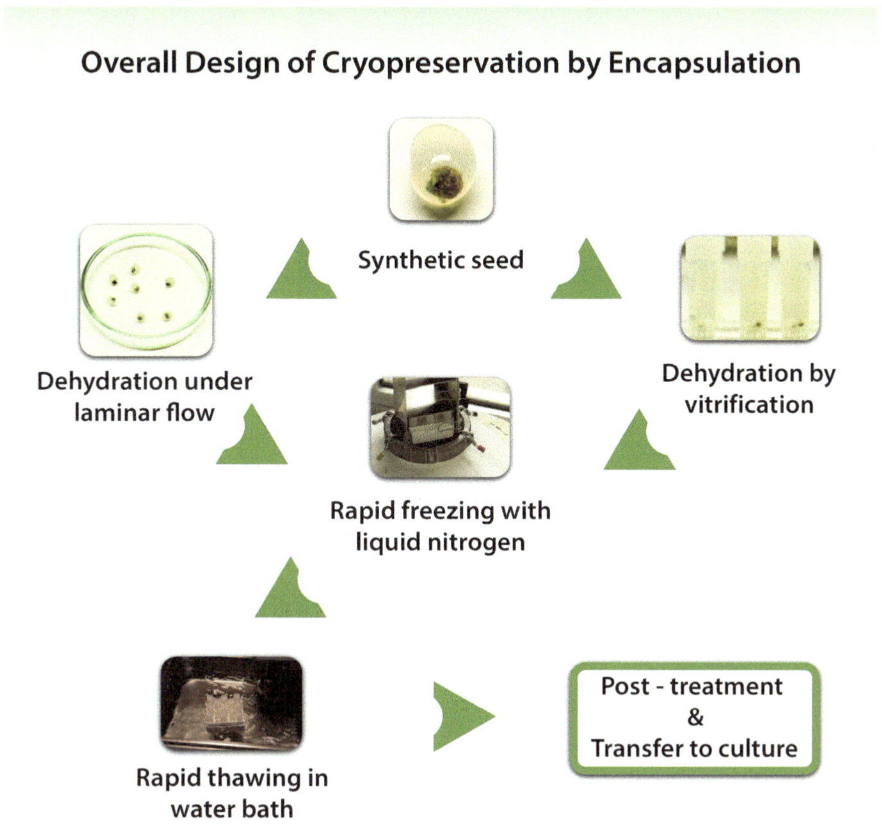

Fig. 8.1 Overall design of cryopreservation methods based on encapsulation of the plant material

Differently, *Doritaenopsis* (*Phalaenopsis X Doritis*) embryogenic cell suspensions were pre-cultured on New Dogashima (ND, Tokuhara and Mii 1993) medium with 0.1 M sucrose for 1 week (Tsukazaki et al. 2000), while *Quercus suber and Quercus robur* cultures were incubated on semi-solid MS medium supplemented with 0.3 M (Valladares et al. 2004 and Sánchez et al. 2008, respectively) or 0.7 M (Fernandes et al. 2008) sucrose. In *Dendrobium* 'Sonia-28', PLBs were cultured on different sucrose concentrations (0.25–1 M) by several researchers (Hwa et al. 2009; Zainuddin et al. 2011; Ching et al. 2012; Poobathy et al. 2013b). *Dendrobium nobile* PLBs were pre-cultured on 0.5 M sucrose for 2 days prior to encapsulation (Khoddamzadeh et al. 2011). In addition, somatic embryos of *Iris nigricans* were pre-cultured on 0.75 M sucrose for 3 days at 22 °C and 1 day on 30 °C (Shibli 2000), whereas *Clitoria ternatea* somatic embryos were pre-cultured on 0.75 M sucrose and 3 % DMSO for 1 day after encapsulation (Nair and Reghunath 2008).

The elevation of sugar content and duration of time could also be used if explants like somatic embryos or encapsulated materials are suitable since both are more resistant to dehydration stress. For instance *Picea omorika* embryogenic tissues pretreated with ½ LM medium with 0.25–1 M sucrose prior to desiccation (Hazubska-Przybył et al. 2012). In *Camellia japonica* somatic embryos, different sucrose concentrations (0.3–0.75) with different additives (DMSO, ethylene glycol, glycerol) could be used up to 2 h prior to encapsulation (Janeiro et al. 1996). For *Gentiana* species first 48 h on 0.3 and 0.5 M then 24 h on 1 M sucrose pre-culture was used (Mikula et al. 2008). *Melia azedarach* L. encapsulated somatic embryos were pre-cultured on MM medium with daily increasing sugar content 0.5,

Table 8.1 Cryopreservation methods optimized for embryogenic cell suspensions, somatic embryos, and PLBs of ornamental plants

Species	Method	Protocol	Recovery	References
Embryogenic cell suspensions				
Anthurium andraeanum	Vitrification	Pre-culture: ½ MS medium with 0.5 M sorbitol for 2 days, then 24 h at 4 °C, and pretreatment with 25 % PVS2 for 15 min at RT	32.1 % survival rate	Wang et al. (2010)
		Vitrification: 100 % PVS2 for 10 min at 0 °C. Thawing: 40 °C for 3 min and washing in 1.2 M sucrose for three times (10 min each)		
Doritaenopsis	Vitrification	Pre-culture: liquid ND medium with 0.1 M sucrose and 1.0 mg/l abscisic acid for 1 week at 25 °C	64 % viability by TTC test	Tsukazaki et al. (2000)
		Vitrification: PVS2 for 1–3 h on ice		
		Viability assessment: cell clumps stained with TTC		
Picea omorika	Vitrification, air-drying	2-year-old embryogenic tissue	99 % regrowth	Hazubska-Przybył et al. (2012)
		Pre-treatment: ½ LM medium 0.25–1 M sucrose		
		Desiccation: 2 h air-drying		
Quercus robur	Dehydration	Pre-treatment: 0.3 and 0.7 M sucrose	56 % regeneration	Martínez et al. (2003)
		Desiccation: airflow of a laminar flow cabinet 1–4 h		
		Thawing: 2 min immersion in a 40 °C and washing in MS with 1.2 M sucrose		
Quercus robur	Vitrification	Pre-culture: 0.3 M sucrose medium	70 % regeneration	Martínez et al. (2003)
		Vitrification: CM and PVS2 incubation for 30 min (25 °C) or 60 min (0 °C)		
		Thawing: 2 min immersion in a 40 °C and washing in MS with 1.2 M sucrose		
Quercus suber	Vitrification	Pre-culture: 0.3 M sucrose for 3 days	88–93 % recovery (around 60 % germination) with 60 min PVS2 incubation	Valladares et al. (2004)
		Vitrification: PVS2 at 0 °C for 0–12 min (30 min intervals). Thawing: 2 min 40 °C and washed with 1.2 M sucrose		

Somatic embryos				
Clitoria ternatea L.	Encapsulation-dehydration	Preconditioning: germinated somatic embryos inoculated ½ MS medium devoid of glycine and preconditioned on 30–50 μmol m^{-2} s^{-2} photon flux intensity	55 % regeneration with 5 h of desiccation	Nair and Reghunath (2008)
		Encapsulation: ½ MS with 2.5 % Na-Alg, 0.5 M sucrose and hardening in ½ MS with 0.1 M $CaCl_2$ and 10 % sucrose both devoid of glycine		
		Pre-culture: ½ MS supplemented with 0.75 M sucrose and 3 % DMSO in Erlenmeyer flasks with 25 ml medium and incubated at 4 °C in darkness		
		Dehydration: beads were air desiccated for 0–5 h		
		Thawing: 40 °C for 30–60 s then beads transferred to ½ MS supplemented with 30 g/l sucrose, 0.5 mg/l GA, 0.2 mg/l BA		
Gentiana spp.	Encapsulation-dehydration	Encapsulation: 3 % Na-Alg and 0.1 M $CaCl_2$ both contain 60 gr/l sucrose	5–6 h dehydration optimal for *G. tibetica* 68 % and *G. cruciate* 83 % viability	Mikula et al. (2008)
		Pre-culture: MM medium with 0.3, 0.5 and 0.75 M sucrose for 48 h then 1 M sucrose for 24 h		
		Dehydration: air-drying 1–7 h at room temperature		
		Thawing: 35 °C for 2–3 min and placed MM medium	Cryopreservation did not affect ploidy level. *G. tibetica* showed higher DNA content after cryopreservation	
		Viability assessment: TTC test at 490 nm. Fresh weight and somatic embryo regeneration rates calculated		
		Genetic stability: flow cytometry analyses		
Galanthus nivalis L.	Encapsulation-dehydration	Encapsulation: MS medium with 3 % Na-Alg and 0.75 M sucrose	No regeneration was observed	Bozena (2008)
		Pre-culture: MS medium with increasing concentration of sucrose for 7 consecutive days (day 1, 0.3 M; day 3, 0.5; day 5, 0.75 M; and day 7, 1 M) for tissue dehydration		
Galanthus elwesii Hook		Desiccation: sterile laminar air for 5 h		
		Thawing: in water bath at 35 °C for 2 min then cultured on ½ MS with 30 g/l of sucrose, 5 μM BA, and 0.1 μM NAA		
Iris nigricans	Encapsulation-dehydration	Pre-culture: 0.75 M sucrose for 3 days at 22 °C and additional 30 °C for 1 day	60 %	Shibli (2000)

(continued)

Table 8.1 (continued)

Species	Method	Protocol	Recovery	References
Melia azedarach L.	Encapsulation-dehydration	Encapsulation: somatic embryo MS medium with 3 % Na-Alg, 4.54 μM TDZ, and 0.1 M $CaCl_2$	Highest recovery with encapsulation – dehydration for 36 % and 30 % with pregrowth-dehydration	Scocchi et al. (2007)
		Pre-culture: MS medium with 0.5, 0.75, and 1 M sucrose every 24 h		
		Dehydration: silica gel up to 4 h		
		Pregrowth-dehydration: non-encapsulated somatic embryos dehydrated with silica gel up to 5 h		
		Desiccation: non-pretreated somatic embryos desiccated on silica gel up to 4 h		
		Cryopreservation: slow freezing (–30 °C with a rate of 1 °C /min) and rapid immersion techniques employed		
		Thawing: 30 °C for 2 min		
		Viability assessment: recovery measured after 3 weeks		
Paeonia lactiflora Pall.	Air desiccation	Pre-culture: MS medium with 0.3 mg/L GA_3 for 1 day	>66 % survival with 1.5 h desiccation	Kim et al. (2006)
		Dehydration: desiccation under laminar flow cabinet 0–2 h		
		Thawing: 40 °C water bath for 5 min and cultured on MS medium with 0.3 mg/ml GA_3		
		Viability assessment: a ratio of germinated embryos to planted embryos		
Quercus robur (seven genotypes)	Vitrification	Pre-culture: 0.3 M sucrose for 3 days	57–92 % embryo recovery	Sánchez et al. (2008)
		Vitrification: PVS2 solution for 60 min at 24 °C		
		Thawing: 2 min 40 °C and washing in 1.2 M sucrose		
Quercus suber	Encapsulation-dehydration	Encapsulation: MS medium with 3 % Na-Alg, 0.5 M sucrose, and 0.1 M $CaCl_2$	>90 % survival	Fernandes et al. (2008)
		Pre-culture: MS medium with 0.7 M sucrose for 3 days		
		Dehydration: air desiccation under laminar flow cabinet to 25 % and 35 % water content	No difference observed for genome content	
		Thawing: 38 °C for 2 min	No variation	
		Genetic stability: flow cytometry, AFLP (six primer), and SSR (four primer)		

PLBs				
Cymbidium eburneum L.	Encapsulation-vitrification	Encapsulation: 3 % Na-Alg and 0.1 M $CaCl_2$ solution	66 % regeneration	Gogoi et al. (2012)
		Pre-culture: in liquid MS medium supplemented with different concentrations of sucrose (0.1–0.8 M, higher regeneration at 0.7 M), on a rotary shaker set at 130 rpm, at 25 ± 2 °C for 24 h		
		Vitrification: the protocorms were transferred to cryovials and added 1 ml loading solution and 1 ml PVS2 supplemented with 0.4 M sucrose in liquid MS medium (20 min)		
		Thawing: in a water bath at 42 ± 2 °C for 2 min		
Cymbidium eburneum L.	Vitrification	Pre-culture: in liquid MS medium supplemented with different concentrations of sucrose (0.1–0.8 M, higher regeneration at 0.2 M), on a rotary shaker set at 130 rpm, at 25 ± 2 °C for 24 h	50 % regeneration	Gogoi et al. (2012)
		Vitrification: the protocorms were transferred to cryovials and added 1 ml loading solution and 1 ml PVS2 supplemented with 0.4 M sucrose in liquid MS medium (20 min)		
		Thawing: in a water bath at 42 ± 2 °C for 2 min		
Dendrobium candidum Wall ex Lindl.	Encapsulation-vitrification	Pre-culture: MS medium with 0.5 g/l BA, 0.2 mg/l NAA, and increased concentrations of sucrose (0, 0.25, 0.5, 0.75, and 1 M) used for 5 days	At 25 °C 120 min of dehydration gave 72.6 % survival	Yin and Hong (2009)
		Encapsulation: MS medium with 2 % Na-Alg and 0.5 M $CaCl_2$ for 15 min both contain 0.5 M sucrose		
		Osmoprotection: 2 M glycerol and 1 M sucrose for 0–120 min	At 0 °C 150 min dehydration resulted with 89.4 % survival	
		Vitrification: PVS2 at 0 and 25 °C for 0–240 min (30 min intervals)		
		Thawing: 40 °C for 3 min and washing with 1.2 M sucrose		
Dendrobium chrysanthum Wall ex Lindl.	Encapsulation-vitrification	Encapsulation: 3 % Na-Alg and 0.1 M $CaCl_2$ supplemented with 0.4 M sucrose and 2 M glycerol	Best regrowth %60 for 100 min PVS2 at 0 °C	Mohanty et al. (2013)
		Pre-culture: liquid MS with 0.06, 0.3, 0.5, and 0.7 M sucrose for 3 days in rotary shaker (23 °C)		
		Osmoprotection: loading solution for 0–120 min	60 % for 80 min PVS2 at 25 °C	
		Vitrification: PVS2 for 0–150 min at 0 or 25 °C		
		Thawing: 38 ± 2 °C for 2 min		

(continued)

Table 8.1 (continued)

Species	Method	Protocol	Recovery	References
Dendrobium nobile Lindl.	Encapsulation-dehydration	Encapsulation: 3 % Na-Alg and 0.1 M $CaCl_2$ both supplemented with 0.4 M sucrose and 2 M glycerol	50.2 % regrowth	Mohanty et al. (2012)
		Pre-culture: ½ MS liquid medium with 0.3, 0.5, and 0.7 M sucrose for 1/2/3 days in shaker at 98 rpm 25 °C		
		Dehydration: 1 h under laminar airflow hood under aseptic conditions		
		Thawing: in water bath at 38 ± 2 °C for 2 min		
Dendrobium nobile Lindl.	Encapsulation-vitrification	Pre-culture: ½ MS liquid medium with 0.5 M sucrose for 2 days in shaker at 98 rpm 25 °C	75.9 %	Mohanty et al. (2012)
		Encapsulation: 3 % Na-Alg and 0.1 M $CaCl_2$ supplemented with 0.4 M sucrose and 2 M glycerol		
		Vitrification: treatment for various durations 0–120 min at 25 °C and PVS2 for 0–150 min at 0 or 25 °C		
		Thawing: in water bath at 38 ± 2 °C for 2 min		
Dendrobium wardianum Warner	Vitrification	Osmoprotection: ½ MS with 2 mol/l glycerol and 0.4 M sucrose at room temp 40 min	20 %	Wu and Shen (2011)
		Dehydration: PVS2 for 40 min at 0 °C		
		Thawing: 40 °C for 1 min and washing in ½ MS with 1.2 M sucrose		
Dendrobium 'Bobby Messina'	Encapsulation-vitrification	Pre-culture: MS medium with sucrose (0–1.2 M; 0.2 M increments) and sorbitol (0–1.4 M; 0.2 M increments) for 24 h	Highest absorbance achieved with 0.6 M sucrose (0.2) and 1.2 M sorbitol (0.4) at TTC test	Antony et al. (2010, 2011, 2012)
		Osmoprotection: 1.5 ml LS for 20 min		
		Vitrification: PVS2 at 0 °C 20 min	In cryopreserved samples, denser cytoplasm, cell wall damage, plasmolysis, nuclear shrinkage, cell rupture, storage materials around nucleus were observed	
		Thawing: 40 °C for 90–120 s and washed in ½ MS supplemented with 1.2 M sucrose		
		Viability assessment: TTC test at 490 nm	Genetic fidelity maintained two polymorphic bands	
		Genetic fidelity: RAPD markers (ten primers)		

Dendrobium 'Bobby Messina'	Encapsulation-dehydration	Pre-culture: ½ MS semisolid medium with 0.2, 0.4, 0.6, 0.8, 1.0 M sucrose	Best results 1–2 mm PLBs with 1 M sucrose pre-culture and 2.5 % Na-Alg encapsulation	Zainuddin et al. (2011)
		Encapsulation: ½ MS with 2.5, 3, 3.5 % Na-Alg and 0.4 M sucrose. Hardening, MS medium with 0.1 M $CaCl_2$ and 0.4 M sucrose		
		Osmoprotection: MS medium with 0.75 M sucrose for 24 h		
		Dehydration: silica gel for 3 h		
		Thawing: 40 ± 2 °C for 90 s	Total chlorophyll content decreased 1.2–0.5	
		Viability assessment: Evans blue test at 600 nm		
		Chlorophyll content, total soluble protein, and peroxidase activity measured after cryopreservation		
Dendrobium 'Bobby Messina'	Encapsulation-dehydration	Pre-culture: ½ semisolid MS media supplemented with 0.4 M sucrose at 25 °C for 3 days	Desiccation up to 10 h shows cell shrinkage and denser cytoplasm	Antony et al. (2014)
		Encapsulation: 3 % Na-Alg and 0.1 M $CaCl_2$		
		Osmoprotection: ½ liquid MS medium supplemented with 0.75 M sucrose on an orbital shaker (110 rpm) at 25 °C for 24 h		
		Dehydration: 50 g of oven-sterilized (120 °C) silica gel in a laminar airflow cabinet for 0, 5, 9, and 10 h	SEM analyses shows stomatal damage and irregular cell orientation	
		Thawing: in water bath at 40 °C for 90 s then cultured for 3 months		
		Microscopy: scanning electron microscopy and histology		
Dendrobium 'Bobby Messina'	Vitrification	Pre-culture: MS medium with 0.2 M sucrose for 0–5 days at 25 or 0 °C	40 % regeneration with 0.25 M sucrose pre-culture for 1 day, LS for 20 min at 25 °C, and PVS2 incubation for 20 min at 0 °C	Antony et al. (2013)
		Osmoprotection: loading solution for 20 min at 25 °C		
		Vitrification: PVS2 for 0–100 min (20 min intervals) at 25 or 0 °C		
		Thawing: 40 °C for 90–120 min	Regeneration was 10 % using ascorbic acid	
		Viability assessment: TTC assay at 490 nm		
		Medium additives: 0.6 M ascorbic acid in every step		
		Microscopy: scanning electron microscopy		

(continued)

Table 8.1 (continued)

Species	Method	Protocol	Recovery	References
Dendrobium 'Sonia-17'	Encapsulation-dehydration	Encapsulation: 3 % Na-Alg and 0.1 M $CaCl_2$	Highest absorbance (1.032) achieved with 0.5 M sucrose pre-culture and 3 h dehydration	Subramaniam et al. (2011)
		Pre-culture: MS medium with 0.3, 0.5, and 0.7 M sucrose		
		Dehydration: 0, 1, 3, and 5 h on 50 g silica gel		
		Thawing: 38 °C water bath		
		Viability assessment: TTC test at 530 nm		
Dendrobium 'Sonia-28'	Droplet freezing	Pre-culture: ½ MS semisolid medium with 0.25, 0.5, 0.75, 1 M sucrose	20 min of PVS2 incubation had optimal TTC absorbance	Poobathy et al. (2013a)
		Osmoprotection: loading solution for 15–35 min (5 min intervals)		
		Dehydration: in PVS2 at 0 °C for 10, 20, 30, 40 min		
		Cryopreservation: rapid freezing with PVS2 droplets in LN		
		Thawing: foils removed and thawed in ½ MS medium with 1.2 M sucrose		
		Measurement: TTC assay at 490 nm		
Dendrobium 'Sonia-28'	Encapsulation-vitrification	Pre-culture: MS medium with 0.25, 0.5, 0.75, and 1 M sucrose at 25 °C for 0, 3, 6, and 9 days	0.3 absorbance after TTC test at 490 nm	Ching et al. (2012)
		Encapsulation: ½ MS medium with 3 % Na-Alg and 0.1 M CaCl2 both supplemented with 0.4 M sucrose and 1 M glycerol		
		Osmoprotection: with ½ MS medium supplemented with 0.4 M sucrose and 1 M glycerol	0.5 M sucrose for 6 days and 150 min PVS2	
		Vitrification: PVS2 with 0–180 min (30 min intervals)		
		Thawing: 40 ± 2 °C for 2 min and 1.2 M sucrose for unloading	Histology analyses confirmed disruption of cell integrity	
		Survival assessment: TTC spectrophotometric analysis at 490 nm		
		Histological analyses conducted		
Dendrobium 'Sonia-28'	Vitrification	Pre-culture: with sucrose 0.06 M, 0.10 M, 0.25 M, 0.50 M, and 0.75 M	The highest viability was observed in PLBs treated with 20 min PVS2 at 0 °C	Hwa et al. (2009)
		Vitrification: with PVS2 at 0 °C and 24 °C: six different periods 0.06 M, 0.10 M, 0.25 M, 0.50 M, and 0.75 M		
		Viability assessment: TTC assay		

Dendrobium 'Sonia-28'	Vitrification	Pre-culture: ½ MS medium with 0.4, 0.6, and 1 M of sucrose for 24, 48, and 72 h, respectively	4.4 % regeneration 0.4 M sucrose pre-culture for 48 h	Poobathy et al. (2013b)
		Osmoprotection: loading solution for 20 min		
		Vitrification: PVS2 at 0 °C for 0–120 (20 min intervals) min		
		Thawing: 40 °C for 90 s	With usage of 0.6 mM ascorbic acid and 50 min of PVS2 16 % regeneration was obtained	
		Medium additives: all solutions include 0.6 mM l-ascorbic acid, and recovery medium includes 2 g/l activated charcoal		
		Survival assessment: TTC at 490 nm		
Phalaenopsis bellina (Rchb.f.) christenson	Encapsulation-dehydration	Encapsulation: 4 % Na-Alg and 75 mM $CaCl_2$	The highest viability recorded based on TTC reduction after cryopreservation was 46.6 % (A_{530nm}0.2) after 5 h of dehydration	Khoddamzadeh et al. (2011)
		Pre-culture: ½ strength liquid MS medium supplemented with 0.75 M sucrose for 3 days with agitation (75 rpm)		
		Dehydration: using silica gel for 5 h resulting in a moisture content of 39 %		
		Thawing: 38 ± 2 °C for 2 min, water bath		
		Viability assessment: TTC assay		

0.75, and 1 M (Scocchi et al. 2007). *Galanthus nivalis* and *elwesii* somatic embryos were pre-cultured for 7 days on 0.3 M for 1 day, 0.5 M for 3 days, 0.75 M for 2 days, and 1 M for 2 days. For *Dendrobium* species, sucrose concentration could be used around 0.7 (Subramaniam et al. 2011; Mohanty et al. 2012, 2013) and 1 M (Yin and Hong 2009) or increased up to 1.2 M (Antony et al. 2010; Zainuddin et al. 2011; Antony et al. 2012; Antony et al. 2013) prior to encapsulation. In addition, plant growth regulators like 6-benzylaminopurine (BAP) and 1-naphthaleneacetic acid (NAA) could also be used in combination with sucrose pre-culture (Yin and Hong 2009).

8.2.2 Cryoprotective Solution-Based Vitrification

Structurally and functionally cryoprotection of cells during freezing and thawing stages of cryopreservation is usually induced artificially by treatment of permeating or non-permeating cryoprotectants that influence not only ice formation but also activity of the electrolytes present in the solution. DMSO (usually %5–10) and glycerol (%10–20) are known as permeating compounds, while polyethylene glycol (PEG), sugars, and sugar alcohols are regarded as non-permeating compounds. It should also be noted that treatment of cells with DMSO could have cytotoxic effect and should be used for short incubation periods and/or usual application is carried out at low temperature (nearly 0 °C). In general, the combination of permeating and non-permeating compounds is used together in cryoprotective solutions.

Up to now, plant vitrification solution 2 (PVS2), which is developed by Sakai and coworkers in 1990, is widely used in cryopreservation of various plant species including ornamentals. The solution contains 30 % glycerol, 15 % ethylene glycol, 15 % DMSO, and 0.4 M sucrose; however, it could be toxic to tissues due to the presence of DMSO (Suzuki et al. 2008). Thus, in order to reduce the toxic effect of PVS2 and hence increase recovery rates, several vitrification solutions including PVS3 (Nishizawa et al. 1993) and VSL (Suzuki et al. 2008) are developed. DMSO concentration is lowered in VSL (20 % glycerol, 30 % ethylene glycol, 5 % sucrose, %10 DMSO, 10 mM $CaCl_2$), while it is totally disused in PVS3 (50 % glycerol and %50 sucrose). However, PVS2 is the most frequently used cryoprotective solution for cryopreservation of embryogenic tissues of ornamental plants. It could be used either at 0 °C or at room temperature. Nevertheless, 30 min at 25 °C PVS2 vitrification in *Quercus robur* resulted in higher recovery rates (70 %) among other protocols like air desiccation and cryoprotectant mixture (CM, Jekkel et al. 1998) vitrification (Martínez et al. 2003). With seven different genotypes of *Q. robur*, up to 92 % viability was achieved with PVS2 vitrification about 60 min at room temperature (Sánchez et al. 2008). The most efficient application is 0 °C due to DMSO toxicity that was mentioned above. The viability of different *Quercus* species (*Q. suber*) was 93 %, and around 60 % germination was obtained with treatment of PVS2 at 0 °C for 60 min (Valladares et al. 2004). Similarly, incubation of PVS2 at 0 °C for 20 (Hwa et al. 2009) and 50 min (Poobathy et al. 2013b) was best in *Dendrobium* 'Sonia-28' PLBs. Moreover, incubation of PVS2 at 0 °C for 40 min was also suitable for *Dendrobium wardianum* Warner (Wu and Shen 2011). With droplet freezing method, the incubation time of PVS2 was decreased to 20 min in *Dendrobium* 'Sonia-28' (Poobathy et al. 2013a).

PVS2 vitrification procedures could be applied as naked or encapsulated explants. In the case of naked explants, 10 min incubation with PVS2 at 0 °C resulted with 32.1 % recovery in *Anthurium andraeanum* embryogenic cell suspensions (Wang et al. 2010) while PVS2 treatment was relatively longer (1–3 h) for *Doritaenopsis* (64 % regeneration, Tsukazaki et al. 2000).

Incubation time could be extended up to 150 min using encapsulated PLBs in *Dendrobium* Sonia-28 (Ching et al. 2012). Moreover, 150 min of vitrification yielded with 89.4 % survival in *Dendrobium candidum* Wall ex. Lindl. (Yin and

Hong 2009). In addition, the best regrowth was observed in *Dendrobium chrysanthum* with 100 min of PVS2 incubation (60 %) (Mohanty et al. 2013) and *Dendrobium nobile* with up to 150 min at 25 or 0 °C (75.9 %) (Mohanty et al. 2012).

8.2.3 Physical Dehydration

Dehydration-based methods, in which vitrification of cells and tissues involve physical dehydration of cells and tissues to get rid of extra water content that could form ice crystals during cryopreservation via the usage of air desiccation and silica gel.

Dehydration period is variable depending on the utilization of encapsulated or naked explants. Three-hour desiccation under laminar flow cabinet of encapsulated explants of *Camellia japonica* resulted with 69.2 % survival (Janeiro et al. 1996), while 5–6 h of dehydration yielded 68 % for *Gentiana tibetica* and 83 % for *Gentiana cruciata* encapsulated embryogenic cell suspensions (Mikula et al. 2008). Also, it is possible to dehydrate encapsulated explants to 25 % water content in *Quercus suber* with obtaining more than 90 % viability (Fernandes et al. 2008). For *Clitoria ternatea*, 5 h of air desiccation was found to be optimal for 55 % regeneration (Nair and Reghunath 2008), whereas 1 h of laminar flow desiccation of *Dendrobium nobile* resulted in 50.2 % regeneration (Mohanty et al. 2012). It is also reported that 3 h of desiccation on silica gel was enough for maximal absorbance in TTC test for *Dendrobium* 'Sonia-7' PLBs (Subramaniam et al. 2011). Encapsulation-dehydration methods gave relatively higher recovery in *Melia azedarach* somatic embryos (36 %) in comparison with naked somatic embryos (30 %) (Scocchi et al. 2007).

Unlike encapsulation, naked explants need relatively shorter desiccation time in order to avoid over-dehydration. For instance, 1.5 h of air desiccation yielded more than 66 % viability in *Paeonia lactiflora* Pall. somatic embryos (Kim et al. 2006).

8.2.4 Cooling, Thawing, and Post-treatment

In order to avoid ice crystal formation, rapid cooling and rapid thawing must be applied for the samples since crystal formation is deleterious for the cell membrane integrity. For thawing, approximately 40 °C of water bath about 1–3 min could be used practically. Also high sucrose concentration (e.g., MS medium supplemented with 1.2 M sucrose) post-treatment could be considered to unload cryoprotectants like PVS2 since the mixture contains toxic chemicals like DMSO in vitrification-based cryopreservation techniques.

The percentage of regenerated or viable explants is often used for assessing the viability rate after cryopreservation. Another viability assessment that is being frequently used is 2,3,5-triphenyltetrazolium chloride test (TTC) (Mikula et al. 2008; Antony et al. 2010; Subramaniam et al. 2011; Ching et al. 2012; Poobathy et al. 2013b) which is based on the reduction of TTC salts in mitochondria, thus assessing viability via absorbance of red formazan at 490 nm by spectrophotometer.

8.2.5 Genetic Stability Assessments

Since cryopreservation procedures result in the exposure of tissues to physical, chemical, and physiological stresses that cause cryoinjury, assessment of genetic stability should be performed to validate newly established cryopreservation protocols (Harding 2004). It is generally accepted that cryopreservation procedure itself does not induce any accumulative DNA polymorphisms; however, genetic instabilities may be the result of the whole culture-cryoprotection-regeneration process (Sánchez et al. 2008). Besides, epigenetic variation in chromatin and DNA methylation of gene sequences have been observed in plants after cryopreservation, which suggests altered patterns of gene expression (Harding 2004). Random amplified polymorphic DNA (RAPD, Sánchez et al. 2008), simple sequence repeat (SSR, Fernandes et al. 2008), and amplified fragment length polymorphism (AFLP, Fernandes et al. 2008) are frequently used molecular techniques that are suitable for detecting polymorphisms in cryopreserved somatic embryos of ornamental plants (Kulus and Zalewska 2014). On the other hand, flow cytometry is also being increasingly used to detect genome content or to check ploidy level changes both after cryopreservation and in plant tissue culture.

In *Gentiana* species, cryopreservation did not affect ploidy level; however, *G. tibetica* showed higher DNA content after cryopreservation (Mikula et al. 2008). It was found that the changes in DNA content of non-cryopreserved and cryopreserved samples were minimal (≤0.01 pg/2C) in *Q. suber* (Fernandes et al. 2008). Genetic stability of the cryopreserved material was also confirmed by SSR. However, a few extra AFLP bands were detected in cryopreserved samples that dehydrated to 25 % water content (WC) which revealed that the occurrence of putative small mutations should not be excluded. The authors concluded that the optimized encapsulation-dehydration protocol (dehydration to 35 % WC) was an efficient cryogenic method for recovery, survival, and genetic/morphologic stability of oak somatic embryos. In five out of six lines of *Q. robur*, RAPD profiles of cryopreserved somatic embryos and regenerated plantlets were identical to those of the controls (Sánchez et al. 2008). Moreover, there were no significant differences between the recovery frequencies of samples retrieved from LN after 1 week and 1 year of cryostorage. For *Dendrobium* 'Bobby Messina' cultures, only two polymorphic bands were detected by using RAPD markers and concluded that genetic stability was maintained after cryopreservation (Antony et al. 2012).

8.2.6 Medium Additives

Activated charcoal demonstrates efficient absorption ability; thus, it is frequently used in plant tissue culture to remove phenolic compounds and excess amounts of plant growth regulators. With this ability, activated charcoal is used to promote somatic embryogenesis, rooting, stem induction, etc. (Thomas 2008). Other additives like antioxidants could be used in order to minimize oxidative damage. Addition of 0.6 mM ascorbic acid and 2.0 g/l activated charcoal as additives in recovery medium resulted with 16 % regeneration in *Dendrobium* 'Sonia-28' cultures after cryopreservation (Poobathy et al. 2013b). However, inclusion of 0.6 mM ascorbic acid in every step decreased regeneration of *Dendrobium* Bobby Messina PLBs to 10 % (Antony et al. 2013).

8.2.7 Biochemical and Microscopy Analyses

Cryopreservation procedures involve rapid freezing, thawing, physical and chemical dehydration of cells or tissues, ice crystal formation which could disrupts cell wall and membrane integrity, plasmolysis and nuclear membrane damage. In order to observe these damages, microscopy analyses could be conducted. PLBs of *Dendrobium* 'Sonia-28' freezing injury were observed as rupture in the plasma membrane (Ching et al. 2012). Antony and colleagues (2011) showed higher packed cell volume in cryopreserved cells with more homogenous cell population, mostly cell wall damage and nuclear shrinkage after cryopreservation. Somatic embryo production was also more rapid in cryopreserved cells. However, with scanning electron microscopy analyses of PLBs of *Dendrobium* Bobby Messina, no difference was observed in comparison to stock PLB cultures, and PLBs remained intact (Antony et al. 2013).

This freezing damage effect could also be observed with biochemical tests. Zainuddin and colleagues (2011) noted low level of total chlorophyll, total soluble protein, and antioxidant activities after cryopreservation.

8.3 Concluding Remarks and Future Prospects

Immediate goals of ornamental biotechnology are the production of novel flower color and shape induction, systemic virus and viroid resistance, greenhouse growth and architecture modification, bioreactor-based mass production of plantlets/organs, and secondary metabolite synthesis (Teixeira da Silva 2003). Some of these goals could be achieved via the combined usage of somatic embryogenesis and transgenesis or preferably by cisgenesis (the use of recombinant technology to introduce genes from crossable donor plants, Akdemir et al. 2012) or intragenesis (introduced genes, originate from the same species or a crossable species), but intragenes are hybrid genes which can have genetic elements from different genes and loci (Rommens et al. 2007). Embryogenic cultures are the most commonly utilized target tissue for non-chimeric transgenics recovery (Taylor and Fauquet 2002). However, in order to use cisgenesis, ornamental germplasm should be conserved properly and indefinitely by application of cryogenic technology to introduce genes from crossable wild ornamental plants. Besides, cryopreservation also plays an important role in maintaining transgenic, cisgenic, or intergenic somatic embryos of plant species including ornamentals in a stable way.

References

Akdemir H, Kaya E, Ozden Y (2010) *In vitro* proliferation and minimum growth storage of fraser photinia: influences of different medium, sugar combinations and culture vessels. Sci Hortic 126(2):268–275. doi:10.1016/j.scienta.2010.07.005

Akdemir H, Gago J, Gallego PP, Ozden-Ciftci Y (2012) Recent advances in fruit species transformation. In: Ozden-Ciftci Y (ed) Transgenic plants-advances and limitations. InTech, Crotia, pp 27–62

Antony JJ, Keng CL, Rathinam X, Marimuthu S, Subramaniam S (2010) Preliminary study on cryopreservation of *Dendrobium* Bobby Messina protocorm-like bodies by vitrification. Afr J Biotechnol 9(42):7064–7070

Antony JJ, Keng CL, Rathinam X, Marimuthu S, Subramaniam S (2011) Effect of preculture and PVS2 incubation conditions followed by histological analysis in the cryopreserved PLBs of *Dendrobium* Bobby Messina orchid. Aust J Crop Sci 5(12):1557–1564

Antony JJ, Keng CL, Rathinam X, Marimuthu S, Subramaniam S (2012) Polymorphism analysis of cryopreserved *Dendrobium* Bobby Messina protocorm-like bodies (PLBs) using RAPD markers. POJ 5(5):427–431

Antony JJ, Keng CL, Mahmood M, Subramaniam S (2013) Effects of ascorbic acid on PVS2 cryopreservation of *Dendrobium* Bobby Messina's PLBs supported with SEM analysis. Appl Biochem Biotechnol 171(2):315–329. doi:10.1007/s12010-013-0369-x

Antony JJ, Mubbarakh SA, Mahmood M, Subramaniam S (2014) Effect of plasmolysis on protocorm-like bodies of *Dendrobium* Bobby Messina orchid following cryopreservation with encapsulation-dehydration method. Appl Biochem Biotechnol 172(3):1433–1444. doi:10.1007/s12010-013-0636-x

Bozena P (2008) Employment of encapsulation-dehydration method for liquid nitrogen cryopreservation of ornamental plant explants propagated *in vitro*. Folia Hortic 20(1):61–71

Ching LP, Poobathy AR, Subramaniam S (2012) Encapsulation-vitrification of *Dendrobium* sonia-28 supported by histology. POJ 5(4):345–350

Dela Cruz M, Christensen JH, Thomsen JD, Müller R (2014) Can ornamental potted plants remove volatile organic compounds from indoor air?—a review. Environ Sci Pollut Res 21:13909–13928. doi:10.1007/s11356-014-3240-x

Engelmann F (2010) Use of biotechnologies for the conservation of plant biodiversity. In Vitro Cell Dev Biol Plant 47(1):5–16. doi:10.1007/s11627-010-9327-2

Fernandes P, Rodriguez E, Pinto G, Roldan-Ruiz I, de Loose M, Santos C (2008) Cryopreservation of *Quercus suber* somatic embryos by encapsulation-dehydration and evaluation of genetic stability. Tree Physiol 28:1841–1850

Fjeld T, Veiersted B, Sandvik L, Riise G, Levy F (1998) The effect of indoor foliage plants on health and discomfort symptoms among office workers. Indoor Built Environ 7(4):204–209. doi:10.1177/1420326x9800700404

George EF, Hall MA, De Klerk G-J (2008) Somatic embryogenesis. In: George EF, Hall MA, De Klerk G-J (eds) Plant propagation by tissue culture, 3rd edn. Springer, Dordrecht, Netherlands, pp 335–354

Gogoi K, Kumaria S, Tandon P (2012) A comparative study of vitrification and encapsulation-vitrification for cryopreservation of protocorms of *Cymbidium eburneum* L., a threatened and vulnerable orchid of India. Cryo Letters 33(6):443–452

Halmagyi A, Fischer-Kluver G, Mix-Wagner G, Schumacher HM (2004) Cryopreservation of *Chrysanthemum morifolium* (*Dendranthema grandiflora* Ramat.) using different approaches. Plant Cell Rep 22(6):371–375. doi:10.1007/s00299-003-0703-9

Harding K (2004) Genetic integrity of cryopreserved plant cells: a review. Cryo Letters 25(1):3–22

Hazubska-Przybył T, Bojarczuk K, Chmielarz P, Michalak M (2012) Somatic embryogenesis and cryopreservation of ornamental *Picea* species: modern methods of propagation and long-term storage. Acta Horticult (ISHS) 937:729–735

Hwa TK, Xavier R, Keng CL, Sreeramanan S (2009) An assessment of early factors influencing the PVS2 vitrification method using protocorm-like bodies of *Dendrobium* Sonia 28. Am Eursian J Sustain Agric 3(3):280–289

Janeiro LV, Vieitez AM, Ballester A (1996) Cryopreservation of somatic embryos and embryonic axes of *Camellia japonica* L. Plant Cell Rep 15:699–703

Jekkel Z, Gyulai G, Kiss J, Kiss E, Heszky LE (1998) Cryopreservation of horse-chestnut (*Aesculus hippocastanum* L.) somatic embryos using three different freezing methods. Plant Cell Tissue Organ Cult 52(3):193–197. doi:10.1023/A:1006057819124

Khoddamzadeh AA, Sinniah UR, Lynch P, Kadir MA, Kadzimin SB, Mahmood M (2011) Cryopreservation of protocorm-like bodies (PLBs) of *Phalaenopsis bellina* (Rchb.f.) christenson by encapsulation-dehydration. Plant Cell Tissue Organ Cult 107(3):471–481. doi:10.1007/s11240-011-9997-4

Kim HM, Shin JH, Sohn JK (2006) Cryopreservation of somatic embryos of the herbaceous peony (*Paeonia lactiflora* Pall.) by air drying. Cryobiology 53(1):69–74. doi:10.1016/j.cryobiol.2006.03.012

Kulus D, Zalewska M (2014) Cryopreservation as a tool used in long-term storage of ornamental species – a review. Sci Hortic 168:88–107. doi:10.1016/j.scienta.2014.01.014

Lee YI, Hsu ST, Yeung EC (2013) Orchid protocorm-like bodies are somatic embryos. Am J Bot 100(11):2121–2131. doi:10.3732/ajb.1300193

Li DZ, Pritchard HW (2009) The science and economics of ex situ plant conservation. Trends Plant Sci 14(11):614–621. doi:10.1016/j.tplants.2009.09.005

Martínez MT, Ballester A, Vieitez AM (2003) Cryopreservation of embryogenic cultures of *Quercus robur* using desiccation and vitrification procedures. Cryobiology 46(2):182–189

Mikula A, Olas M, Sliwinska E, Rybczynski JJ (2008) Cryopreservation by encapsulation of *Gentiana* spp cell suspensions maintains regrowth, embryogenic competence and DNA content. Cryo Letters 29(5):409–418

Mikula A, Tomiczak K, Rybczynski JJ (2011) Cryopreservation enhances embryogenic capacity of *Gentiana cruciata* (L.) suspension culture and maintains (epi)genetic uniformity of regenerants. Plant Cell Rep 30(4):565–574. doi:10.1007/s00299-010-0970-1

Mohanty P, Das MC, Kumaria S, Tandon P (2012) High-efficiency cryopreservation of the medicinal orchid *Dendrobium nobile* Lindl. Plant Cell Tissue Organ Cult 109(2):297–305. doi:10.1007/s11240-011-0095-4

Mohanty P, Das MC, Kumaria S, Tandon P (2013) Cryopreservation of pharmaceutically important orchid *Dendrobium chrysanthum* Wall. ex Lindl. using vitrification based method. Acta Physiol Plant 35(4):1373–1379. doi:10.1007/s11738-012-1163-z

Murashige T, Skoog F (1962) A revised medium for rapid growth and bio assays with tobacco tissue cultures. Physiol Plant 15(3):473–497. doi:10.1111/j.1399-3054.1962.tb08052.x

Nair DS, Reghunath BR (2008) Effective use of encapsulation-dehydration technique in cryopreserving somatic embryos of butterfly pea (*Clitoria ternatea* L.). J Herbs Spices Med Plants 13(3):83–95. doi:10.1300/J044v13n03_07

Nishizawa S, Sakai A, Amano Y, Matsuzawa T (1993) Cryopreservation of asparagus (*Asparagus officinalis* L.) embryogenic suspension cells and subsequent plant regeneration by vitrification. Plant Sci 91(1):67–73

Ozudogru EA, Previati A, Lambardi M (2010) In vitro conservation and cryopreservation of ornamental plants. In: Jain SM, Ochatt SJ (eds) Protocols for in vitro propagation of ornamental plants, vol 589, Methods in molecular biology. Humana Press, New York, USA, pp 303–324

Panis M (2009) Cryopreservation of *Musa* germplasm. In: Engelmann F, Benson E (eds) Technical guidelines no. 9, 2nd edn. Bioversity International, Montpellier, France

Panis B, Piette B, Swennen R (2005) Droplet vitrification of apical meristems: a cryopreservation protocol applicable to all *Musaceae*. Plant Sci 168(1):45–55

Poobathy R, Izwa N, Julkifle AL, Subramaniam S (2013a) Cryopreservation of *Dendrobium* sonia-28 using an alternative method of PVS2 droplet freezing. Emir J Food Agric 25(7):531–538. doi:10.9755/ejfa.v25i7.15988

Poobathy R, Sinniah UR, Mahmood M, Subramaniam S (2013b) Refinement of a vitrification protocol for protocorm-like bodies of *Dendrobium* sonia-28. Turk J Bot 37:940–949. doi:10.3906/bot-1208-35

Reed BM, Schumacher L, Wang N, D'Achino J, Barker RE (2006) Cryopreservation of bermudagrass germplasm by encapsulation dehydration. Crop Sci 46(1):6–11. doi:10.2135/cropsci2004.0620

Rommens CM, Haring MA, Swords K, Davies HV, Belknap WR (2007) The intragenic approach as a new extension to traditional plant breeding. Trends Plant Sci 12(9):397–403

Rout GR, Mohapatra A, Jain SM (2006) Tissue culture of ornamental pot plant: a critical review on present scenario and future prospects. Biotechnol Adv 24(6):531–560. doi:10.1016/j.biotechadv.2006.05.001

Sánchez C, Martínez MT, Vidal N, San-José MC, Valladares S, Vieitez AM (2008) Preservation of *Quercus robur* germplasm by cryostorage of embryogenic cultures derived from mature trees and RAPD analysis of genetic stability. Cryo Letters 29(6):493–504

Scocchi A, Vila S, Mroginski L, Engelmann F (2007) Cryopreservation of somatic embryos of paradise tree (*Melia azedarach* L.). Cryo Letters 28(4):281–290

Shibli RA (2000) Cryopreservation of black iris (*Iris nigricans*) somatic embryos by encapsulation-dehydration. Cryo Letters 21(1):39–46

Subramaniam S, Sinniah UR, Khoddamzadeh AA, Periasamy S, James JJ (2011) Fundamental concept of cryopreservation using Dendrobium sonia-17 protocorm-like bodies by encapsulation-dehydration technique. Afr J Biotechnol 10(19):3902–3907

Suzuki M, Tandon P, Ishikawa M, Toyomasu T (2008) Development of a new vitrification solution, VSL, and its application to the cryopreservation of gentian axillary buds. Plant Biotechnol Rep 2(2):123–131. doi:10.1007/s11816-008-0056-5

Swarts ND, Dixon KW (2009) Terrestrial orchid conservation in the age of extinction. Ann Bot 104(3):543–556

Taylor NJ, Fauquet CM (2002) Microparticle bombardment as a tool in plant science and agricultural biotechnology. DNA Cell Biol 21(12):963–977. doi:10.1089/104454902762053891

Teixeira da Silva JA (2003) *Chrysanthemum*: advances in tissue culture, cryopreservation, postharvest technology, genetics and transgenic biotechnology. Biotechnol Adv 21(8):715–766. doi:10.1016/s0734-9750(03)00117-4

Teixeira da Silva JA, Zeng S, Galdiano RF Jr, Dobranszki J, Cardoso JC, Vendrame WA (2014) *In vitro* conservation of *Dendrobium* germplasm. Plant Cell Rep 33(9):1413–1423. doi:10.1007/s00299-014-1631-6

Thomas TD (2008) The role of activated charcoal in plant tissue culture. Biotechnol Adv 26(6):618–631. doi:10.1016/j.biotechadv.2008.08.003

Thomsen JD, Sønderstrup-Andersen HKH, Müller R (2011) People–plant relationships in an office workplace: perceived benefits for the workplace and employees. HortSci 46(5):744–752

Tokuhara K, Mii M (1993) Micropropagation of *Phalaenopsis* and *Doritaenopsis* by culturing shoot tips of flower stalk buds. Plant Cell Rep 13(1):7–11. doi:10.1007/BF00232306

Tsukazaki H, Mii M, Tokuhara K, Ishikawa K (2000) Cryopreservation of *Doritaenopsis* suspension culture by vitrification. Plant Cell Rep 19:1160–1164

Valladares S, Toribio M, Celestino C, Vieitez AM (2004) Cryopreservation of embryogenic cultures from mature *Quercus suber* trees using vitrification. Cryo Letters 25:177–186

Wang GL, Xu CY, Wang GD (2010) Study on vitrification cryopreservation condition of *Anthurium andraeanum* embryonic suspension cells. J Plant Res Environ 19(1):26–31

Wang B, Zhang Z, Yin Z, Feng C, Wang Q (2012) Novel and potential application of cryopreservation to plant genetic transformation. Biotechnol Adv 30(3):604–612. doi:10.1016/j.biotechadv.2011.10.008

Wu YL, Shen XU (2011) Cryopreservation of *Dendrobium wardianum* Warner. protocorms by vitrification. Chin J Cell Bio 33(3):279–287

Yin M, Hong S (2009) Cryopreservation of *Dendrobium candidum* Wall. ex Lindl. protocorm-like bodies by encapsulation-vitrification. Plant Cell Tissue Organ Cult 98(2):179–185. doi:10.1007/s11240-009-9550-x

Zainuddin M, Julkifle AL, Pobathy R, Sinniah UR, Khoddamzadeh AA, Antony JJ, Pavallekoodi, Subramaniam S (2011) Preliminary analysis of cryopreservation of *Dendrobium* Bobby Messina orchid using an encapsulation-dehydration technique with Evans blue assay. Afr J Biotechnol 10(56):11870–11878. doi:10.5897/AJB11.002

9 Role of SERK During Somatic Embryogenesis and Its Interaction with Brassinosteroids

Shonima Talapatra, Poorna Goswami, Subhasree Das, and Sarmistha Sen Raychaudhuri

Abstract

Plants possess a unique property called cellular totipotency which is a series of complex molecular and biochemical steps to produce a complete normal plant. Somatic embryogenesis, the most reliable and useful tool for in vitro plant propagation, uses this property of totipotency to regenerate a whole plant from competent somatic cells in the presence of endogenous or exogenous signals. Development of embryo from somatic cells is regulated by the differential expression pattern of a myriad of genes among which the most important regulator is somatic embryogenesis receptor kinase (*SERK*) gene. Phytohormones like brassinosteroids (BRs) play important roles in directing the plant cells to undergo restructuring program for differentiation, development, and organogenesis. Along with this, BR has been known to elicit a stress response mechanism in plants. SERK, a known LRR-RLK member, can form heterodimer complex with the main BR receptor, brassinosteroid insensitive 1 (BRI1), to induce BR-dependent signaling pathway. The present review encompasses the role of *SERK* in inducing somatic embryogenesis and its interaction with BR during BR signaling.

Keywords

Plant tissue culture • Somatic embryogenesis receptor kinase • Brassinosteroid insensitive 1

9.1 Introduction

Plant embryogenesis can be classified into two groups – zygotic embryogenesis and somatic embryogenesis. Zygotic embryogenesis occurs when two male gametes (generative nuclei) of the pollen grain enters the ovule and double fertilization takes place where one male nucleus

S. Talapatra • P. Goswami • S. Das • S.S. Raychaudhuri (✉)
Department of Biophysics, Molecular Biology and Bioinformatics, University of Calcutta, 92, APC Road, Kolkata 700009, India
e-mail: shonima_talapatra@yahoo.com; poorna.complete@gmail.com; subhasree1947@gmail.com; sarmistharc@gmail.com

A. Mujib (ed.), *Somatic Embryogenesis in Ornamentals and Its Applications*, DOI 10.1007/978-81-322-2683-3_9

fuses with the egg nucleus; this phenomenon is known as syngamy, and the other male nucleus fuses with polar nuclei, leading to the formation of endosperm. Thus, a zygote is formed which through postfertilization modifications results in the formation of embryo and then undergoes cell division and cell differentiation to form a mature sporophyte.

But somatic embryogenesis follows a different mechanism. The plant cells have a unique property called totipotency, where every cell has the capacity to develop into a mature, functional, full grown form. Here all the somatic cells contain the same genetic information for whole plant regeneration. Somatic embryogenesis utilizes this property of totipotency to form embryos from the somatic or sporophytic cells without undergoing gametic fusion. The embryos developed in this way are known as somatic embryos and are generally produced in vitro. It is best used for studying the cellular and molecular mechanisms of embryo formation in plants.

Plant growth regulators (PGRs), supplemented in proper combination and concentration, play a crucial role during the induction of in vitro somatic embryogenesis.

Development of somatic embryos, in vitro, can be achieved by two ways – direct and indirect. Direct somatic embryogenesis does not involve callus formation and can be established from microspores, ovule, zygotic embryos, or even young seedlings to form somatic embryos, whereas indirect somatic embryogenesis requires an intervening callus (undifferentiated mass of cells) formation before giving rise to embryoids.

Sharp et al. (1980) and Evans and Sharp (1983) first proposed that somatic embryogenesis can occur from the explants in two different types of cells – pre-embryogenic determined cells (PEDCs) and induced embryogenic determined cells (IEDCs). PEDCs do not require any prior mitotic divisions and the embryogenic pathway is predetermined. These cells only require the synthesis of an inducer or removal of an inhibitor to undergo mitosis for expressing their embryogenic potentiality. On the contrary, IEDCs follow mitotic division and redifferentiation to develop their embryogenic status. But, both PEDCs and IEDCs have the potential to gain cellular totipotency, thereby generating somatic embryos (Fig. 9.1).

Though the development process of zygotic embryo and somatic embryo undergoes similar morphological changes starting from globular shape to torpedo structure, there do exists some distinct differences among these two. While zygotic embryos during their course of development consist of endosperm, suspensors, and seed coat, the somatic embryos are devoid of it.

Further, in somatic embryogenesis, bipolar embryos (also known as plantlet embryos) are

Fig. 9.1 Schematic diagram showing somatic embryogenesis

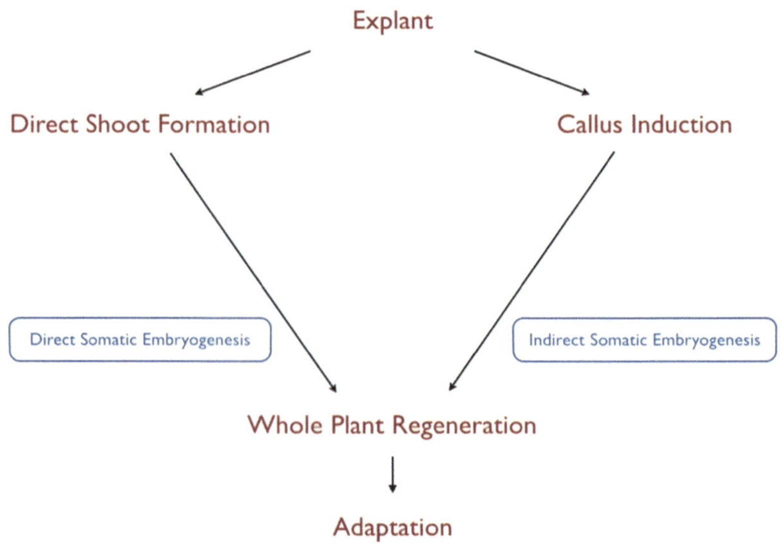

formed with distinct root and shoot apex. Also, the embryos do not need dessication and dormancy, while cotyledons are formed as it is needed during zygotic embryogenesis. The zygotic embryos undergo maturation, dessication, and dormancy as a natural process.

In vitro somatic embryogenesis requires participation of several genes, among which somatic embryogenesis receptor kinase (*SERK*) plays a significant role. The present review is a critique of retrospective as well as recent literature on the multitasking gene *SERK*.

9.2 Gene Expression during Somatic Embryogenesis

Somatic embryogenesis illustrates the perception of plant totipotency by restructuring somatic cells toward embryogenic pathway. The complex developmental pathway of somatic embryogenesis is regulated by the differential expression pattern of different genes (Chugh and Khurana 2002). A number of genes have been known to play significant role in different stages of embryo differentiation. These genes are identified at various developmental stages of somatic embryogenesis and grouped according to their function. Some of the important genes involved at various stages of somatic embryogenesis are listed below (Table 9.1).

Somatic embryogenesis receptor kinase, also known as *SERK*, is one of those genes which are involved in the regulation of somatic embryogenesis at molecular level. It was first isolated from the carrot embryogenic cells and termed as *Daucus carota SERK* (*DcSERK*) (Schmidt et al. 1997). SERK protein is localized on plasma membrane as membrane receptor having an extracellular and an intracellular domain for the transduction of extracellular signals. SERK protein belongs to the receptor-like protein kinase (RLK) family. RLKs are involved in different signal transduction pathways during plant development. SERK falls in the category of leucine-rich repeat RLKs (LRR-RLKs) which comprises the largest group of RLKs in the plant kingdom (Becraft 2002). On analyzing different sequences of SERK protein, it was observed that these proteins evolved via gene duplication which gave rise to two existing precursors – SERK1 and SERK2 and SERK3, SERK4, and SERK5. These precursors, over time, are further duplicated and mutated to generate five SERK members (Albrecht et al. 2008). *SERK* genes are upregulated by the plant growth regulators auxin and cytokinin in various plants (Zhang et al. 2011). Besides playing an important role in somatic embryogenesis, *SERK* gene also possesses diverse range of functions in plant physiology and development. The *SERK* gene is expressed both in embryogenic and non-embryogenic callus cultures under in vitro conditions and also associated with embryogenic competence (Sharma et al. 2008a, b; Savona et al. 2012). *SERK* is also known for its involvement in responses against biotic and abiotic stress conditions and plays an important role in brassinosteroid (BR) signaling (Nolan et al. 2003; Santos and Arago 2009). Discovery of different *SERK* genes isolated from different plant species is tabulated in chronological order in (Table 9.2).

9.3 Functions of *SERK*

SERK is one of the known genes which play an important role in the acquisition of embryogenic competence in plant cells. *SERK* genes are upregulated by auxin in some plant species and by both auxin and cytokinin in others (Zhang et al. 2011). Expression of *SERK* gene starts from the induction of embryogenic stage to the globular stage of somatic embryos, but not in the non-embryogenic stages. *SERK* acts as molecular marker for distinguishing embryogenically competent and non-competent cells. The fusion of *SERK* promoter with the luciferase (*LUC*) reporter gene demonstrated that the elongated cells in carrot, expressing *SERK*, indeed have the ability to undergo somatic embryo formation (Shah et al. 2001a). The five members of *SERK* family in *Arabidopsis thaliana* (*AtSERK1–5*) have specific expression in the nucellus, the megaspore, and embryo sac alongside in the stages of somatic embryogenesis (Shah et al. 2001b). A connection between gene expression of *SERK1* and embryogenic

Table 9.1 Different genes involved in somatic embryogenesis in plants

Gene	Function	Species studied	Reference
ASET1, ASET2, and ASET3	Signal transduction genes. Expressed at early embryogenic stages	*Medicago sativa*	Giroux and Pauls (1997)
AtECP31 and AtECP63	ABA-responsive genes. Expressed in the torpedo stage of somatic embryo	*Arabidopsis thaliana*	Yang et al. (1996) and Yang et al. (1997)
CAM genes	Signal transduction genes. Expressed at different stages of embryogenesis	*Daucus carota* and other plant species	Ling et al. (1991) and Periera and Zielinski (1992)
Cdc2MS	Cell cycle gene. Expressed in shoots and auxin-induced suspension cultures	*Medicago sativa*	Hirt et al. (1991)
CEM1	Housekeeping gene. A globular embryo-specific cDNA encodes for elongation factor-1α found in actively dividing cells	*Daucus carota*	Sato et al. (1995)
CGS103	Cell cycle gene. Glutamine synthetase. Expressed only in later stages of seed development	*Daucus carota*	Higashi et al. (1998)
CGS102 and *CGS201*	Cell cycle genes. Glutamine synthetase. Expressed during the early stages of somatic embryogenesis and developing seeds	*Daucus carota*	Higashi et al. (1998)
Dcarg-1	Auxin-responsive gene. Heat shock protein. Detected specifically during early induction period	*Daucus carota*	Kitamiya et al. (2000)
DcECP31, DcECP40, and DcECP63	ABA-responsive genes. Expressed in the torpedo stage of somatic embryo	*Daucus carota*	Chugh and Khurana (2002), Kiyosue et al. (1992) and Zhu et al. (1997)
H3-1 and *H3-11*	Housekeeping genes. Involved in somatic embryogenesis in response to auxin treatment	*Medicago sativa*	Kapros et al. (1992)
SERK	Signal transduction gene. Plays a role in acquisition of embryogenic competence in plant cells	Different plant species	Schmidt et al. (1997)
Small auxin upregulated genes (pJCW1 and pJCW2)	Auxin-responsive genes. Expressed during embryogenesis in soybean	*Glycine max*	Hagen et al. (1984)
topI	Housekeeping gene. Expressed during torpedo stage of somatic embryogenesis	*Daucus carota*	Balestrazzi et al. (1996, 2001)

competence was observed in *Arabidopsis thaliana* seedling. The callus derived from the seedling has been found to overexpress *AtSERK1* as well as exhibited 3–4 times higher embryogenic competence as compared with the wild type which indicated that the protein encoded by *AtSERK1* can confer embryogenic competence in cultures (Hecht et al. 2001). However, in rice plants, *SERK* expression was observed in non-embryogenic tissues (Ito et al. 2005). *SERK* gene also helps in shoot embryogenesis in sunflower (Thomas et al. 2004), apomixes in *Poa pratensis* (Albertini et al. 2005), and formation of protocorm-like bodies in rose (Kedong et al. 2011). Apart from having a role in somatic embryogenesis, *SERK* genes are considered to play a significant role in stress response. *OsSERK* responds against fungal infections (Hu et al. 2005), whereas *MaSERK* imparts disease resistance response in banana (Huang et al. 2010) (Table 9.3).

Table 9.2 *SERK* genes isolated from different plant species in chronological order

Plant	Types of SERK genes	Reference
Daucus carota	*DcSERK*	Schmidt et al. (1997)
Dactylis glomerata	*DgSERK*	Somleva et al. (2000)
Arabidopsis thaliana	*AtSERK1–AtSERK5*	Hecht et al. (2001)
Zea mays	*ZmSERK1–ZmSERK3*	Baudino et al. (2001)
Medicago truncatula	*MtSERK1–MtSERK6*	Nolan et al. (2003)
Helianthus annuus	*HaSERK*	Thomas et al. (2004)
Ocotea catharinensis	*Ocotea SERK*	Santa-Catarina et al. (2004)
Citrus unshiu	*CitSERK*	Shimada et al. (2005)
Poa pratensis	*PpSERK*	Albertini et al. (2005)
Oryza sativa	*OsSERK1–OsSERK2*	Ito et al. (2005)
Theobroma cacao	*TcSERK*	Santos et al. (2005)
Triticum aestivum	*TaSERK1–TaSERK3*	Singla et al. (2008)
Solanum tuberosum	*StSERK*	Sharma et al. (2008a, b)
Cocos nucifera	*CnSERK*	Pérez-Núñez et al. (2009)
Vitis vinifera	*VvSERK1–VvSERK2*	Maillot et al. (2009)
Musa acuminata	*MaSERK*	Huang et al. (2010)
Rosa hybrida (ornamental)	*RhSERK1–RhSERK4*	Zakizadeh et al. (2010)
Glycine max	*GmSERK*	Yang et al. 2011
Rosa canina (ornamental)	*RcSERK*	Kedong et al. (2011)
Cyclamen persicum Mill.(ornamental)	*CpSERK1–CpSERK2*	Savona et al. (2012)
Cyrtochilum loxense (ornamental)	*ClSERK*	Cueva et al. (2012)
Ananas comosus	*AcSERK1–AcSERK2*	Ma et al. (2012a, b)
Garcinia mangostana	Mangosteen *SERK*	Rohani et al. (2012)
Momordica charantia	*McSERK*	Talapatra et al. (2014)
Cucurma alismatifolia Gagnep (ornamental)	*CaSERK*	Sucharitakul et al. (2014)
Prunus salicina	*PsSERK1*	Jayanthi et al. (2014)
Prunus persica	*PpSERK1*	Jayanthi et al. (2014)
Adiantum capillus-veneris (ornamental)	*AcvSERK1*	Li et al. (2014)

9.4 Enhancement of Somatic Embryogenesis and Expression of *SERK* by Additive Supplementation In Vitro

Several investigations have been carried out to observe the effect of external physical and chemical factors on somatic embryogenesis. Casein hydrolysate (CH) and coconut water (CW) are two most significant external additives used in plant tissue culture. CH in plant tissue culture is known to provide a useful source of amino acids to the plant cells. It is usually added in the media within a concentration range of 0.05–5 %. On the other hand, CW is an undefined media containing mixture of different amino acids, nitrogenous compounds, inorganic elements, organic acids, sugars and their derivatives, vitamins, and many other components. It is mostly added in the range of 5–20 %.

In *Plantago ovata*, it has been observed that CH and CW impart a positive effect in enhancing somatic embryogenesis (Das Pal and Sen Raychaudhuri 2001). It was also seen that somatic embryogenesis and polyphenol content increased with the use of CH and CW (Talapatra and Sen Raychaudhuri 2012). Use of exogenous additives during somatic embryogenesis in *Momordica charantia* showed an increase in the mass of embryogenic callus, numbers of somatic embryos in callus tissue, and SERK expression at both

Table 9.3 Specific functions of *SERK* homologs (*SERK1*, *SERK2*, *SERK3*, *and SERK4*)

Type of *SERK*	Plant species	Functions
SERK 1	*Zea mays*	Expressed in embryogenic as well as non-embryogenic in vitro culture with broader expression in reproductive tissue (Baudino et al. 2001)
	Medicago truncatula	Somatic embryogenesis, in vitro rhizogenesis, formation of primary meristem in plants (Nolan et al. 2003; 2009; Wang et al. 2011)
	Arabidopsis thaliana	Male sporogenesis (Albrecht et al. 2005); tapetum development and microspore maturation (Colcombet et al. 2005)
		Expressed in procambium and immature vascular cells (Kwaaitaal and de Vries 2007); regulates organ separation in flower (Lewis et al. 2010)
	Cyclamen persicum	Helps in maintaining stem cells in pluripotent conditions leading to totipotency and somatic embryo formation (Savona et al. 2012)
SERK 2	*Zea mays*	Expressed in embryogenic as well as non-embryogenic in vitro culture, with expression in all tissues including vegetative ones (Baudino et al. 2001)
	Arabidopsis thaliana	Male sporogenesis (Albrecht et al. 2005); tapetum development and microspore maturation (Colcombet et al. 2005)
	Ananas comosus	Organ development and abiotic stress response (Ma et al. 2012a, b)
SERK 3	*Arabidopsis thaliana*	Involved in brassinosteroid-dependent and brassinosteroid-independent signaling pathway (Albrecht et al. 2008); helps in innate immunity (Albrecht et al. 2008); cell death control with *SERK 4* (Albrecht et al. 2008; Li 2010)
SERK 4	*Arabidopsis thaliana*	Cell death control with *SERK 3* (Albrecht et al. 2008; Li 2010)

mRNA and protein levels (Talapatra et al. 2014). This provides evidence that SERK is involved in somatic embryogenesis. CH and CW, both the additives, were found to enhance somatic embryogenesis as well as *SERK* expression. Talapatra (2014) have used three different concentrations of both CH (1 g/L, 2 g/L, and 3 g/L) and CW (5 %, 10 %, and 15 %) and observed that 2 g/L CH and 5 % CW concentration was most effective in enhancing somatic embryogenesis as well as expression of *SERK*. The enhanced expression of *SERK* gene and protein in CH- and CW-treated callus validated the use of external additives for the production of elite variety of *M. charantia* with effective regeneration method (Talapatra 2014).

Addition of exogenous polyamines enhances somatic embryogenesis. The three most common polyamines found in higher plants are putrescine, spermidine, and spermine (Paul et al. 2009). The effects of exogenous polyamines in the callus cultures of *M. charantia* have been extensively studied by Paul et al. (2009). When polyamines were added to the embryogenic callus, it resulted in the increase of fresh weight and number of somatic embryos with putrescine imparting maximum effect. Hence, addition of exogenous polyamines has a positive correlation with the enhancement of somatic embryogenesis.

9.5 Expression of *SERK* Gene during Somatic Embryogenesis

Calluses, the dedifferentiated somatic cells (derived from the explant), are capable of exhibiting meristematic activities and, hence, are able to give rise to totipotent embryogenic cells which can lead to either organogenesis or formation of somatic embryo by somatic embryogenesis. Some of the cells of the callus have the potency to form plant stem cells in plant tissue culture as they are formed de novo and are capable of self-renewal leading to the formation of several derivative cells with different developmental fate. The embryo transition takes place from these transient stem cell aggregates having meristematic features and thus can be called as pluripotent in nature. The stem cell aggregates depending upon suitable environment, can give rise to any of the two possible pathways: either embryogenesis or organogenesis.

In the embryogenic line, stem cell aggregates form totipotent pre-embryogenic aggregates (PEAs). These PEAs further develop into somatic embryos retaining the meristematic features of the initial stem cells present in the callus. In organogenic line, the stem cell aggregates having the meristematic activities give rise to meristemoids. These are clumps of cells randomly located within the callus containing a distinct nucleus and vacuoles and are smaller in comparison to other cells in the callus. Unlike the embryogenic callus, not all the cells in the meristemoids possess meristematic features. Only those meristemoids which have grown into polarized structures retain their meristematic activity. These polarized structures further develop into shoot and root primordia. The meristemoids initially retain the characteristics of the stem cell from which they are derived. In course of time, the overall meristematic property is lost and becomes restricted only toward root and shoot formation, together termed as organogenesis. Thus, both PEAs and meristemoids show the characteristics of trans-amplifying cells (TACs) as they have derived from common stem cell aggregates and are capable of developing into somatic embryos and plant organs, respectively (Savona et al. 2012) (Fig. 9.2).

SERK gene has been known to be a marker of somatic embryogenesis in several model plant species as it enhances the embryogenic competence of the somatic cells. In *Daucus carota*, *SERK* gene was found to make somatic cells competent for the formation of somatic embryos (Schmidt et al. 1997). *SERK* gene expression in *Dactylis glomerata* was observed during the induction of embryogenic cells and can be used as a molecular marker to track embryogenic cell formation. Transient *SERK* gene expression was seen in leaf explant and suspension culture during early embryo development. Also shoot apical meristem, scutellum, coleoptiles, and coleorhiza showed *SERK* expression in later embryonic

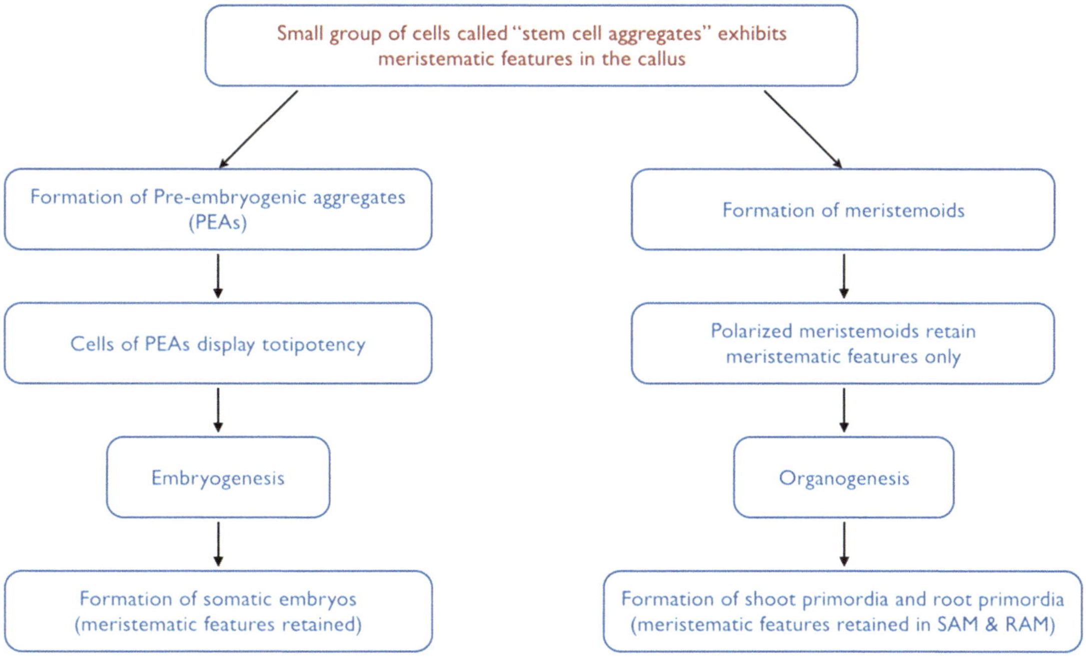

Fig. 9.2 Schematic diagram showing embryogenic and organogenic line during somatic embryogenesis (Savona et al. 2012)

stages. Though there is no conclusive answer, it can be assumed that there may be the presence of a member of *SERK* gene family which gives rise to embryonic competence to certain non-embryogenic tissues during developmental stages (Somleva et al. 2000). *AtSERK1* expression in *Arabidopsis thaliana* occurred in the cells which acquired embryogenic competence. It is expressed in somatic embryos and ovules before fertilization and during embryo development (Hecht et al. 2001). *CnSERK* expression in *Cocos nucifera* suggested that it is a potential marker of cell competence in somatic embryogenesis and helps in the induction of somatic embryo formation (Pérez-Núñez et al. 2009). While monitoring *SERK* gene expression during somatic embryogenesis, it has been noticed that expression of *SERK* increased during the callus induction phase in *Cyclamen persicum* and *Solanum tuberosum.* *SERK* expression did not show any variation during embryo transition phase; however, in somatic embryos, *SERK* expression was upregulated. Apart from that, different organs of the plant like seeds, microtubers, leaves, and flower buds showed varied expressions of *SERK* gene. So it can be inferred that *SERK* gene not only helps in embryogenesis but also serves as a marker of organogenesis. SERK1 and SERK2 are usually the two transmembrane proteins of the SERK family which help the somatic cells to acquire embryogenic competence (Sharma et al. 2008a, b; Savona et al. 2012). In *Adiantum capillus-veneris*, an ornamental fern species, expression of *AcvSERK* was observed in the whole embryo development process but decreased during shoot formation. In situ hybridization of *AcvSERK1* during green globular body (GGB)-derived somatic embryogenesis and in the meristematic callus suggests that it can be a marker for somatic embryogenesis and are capable to form embryos (Li et al. 2014). Talapatra et al. (2014) established somatic embryogenesis in *M .charantia* from leaf explants. Expression of *SERK* gene was determined by densitometry and real-time PCR during different developmental stages of somatic embryogenesis. It was observed that *SERK* gene expressed in all stages of somatic embryogenesis; but its expression was high during the initiation of this process. To investigate the role of SERK protein in *M. charantia*, these authors custom designed primary antibody against SERK protein. Expression analysis by flow cytometry revealed that *McSERK* expression was highest during the initial stages (globular stage) of embryogenesis. *McSERK* was found to be localized in the plasma membrane of the embryogenic cells as was observed in confocal microscopic imaging study (Talapatra 2014).

It is still not clear how SERK helps in acquiring embryogenic competence at molecular level, but according to Aker et al. (2006), it has been found SERK interacts with CDC48A protein, a cell cycle regulation protein. These two proteins co-localize and interact in certain parts of the plasma membrane and endoplasmic reticulum (ER) facing the plasma membrane. Misfolded SERK1 receptors bind to the N-domain of CDC48A protein and are degraded by proteasomal pathway. There are many hypotheses regarding the interaction of SERK with CDC48. One of them suggests that under normal condition, CDC48A is phosphorylated by SERK1 kinase and helps in cytokinesis and re-localization of CDC48A to the cell plate, thereby helping the cells to gain embryogenic competence during cell division (Rienties et al. 2005).

9.6 Structure of Somatic Embryogenesis Receptor Kinase (SERK)

The complete sequence of *McSERK* gene from *M. charantia* was documented by Talapatra et al. (2014). This gene shows high sequence similarity with *AtSERK1* as homolog and with *Cucumis SERK1* and *MtSERK1* as ortholog. Since *McSERK* shows similarity with these gene sequences, it could be conjectured that *McSERK* has a positive influence during somatic embryogenesis.

SERK belongs to the Leucine Rich Repeat Receptor like Kinase (LRR-RLK) superfamily. It is known that SERK has autophosphorylation activity. *McSERK* contains an N-terminal signal peptide (SP), a leucine zipper (LZ), five leucine-rich repeats (LRRs), a distinct serine-proline-proline (SPP) domain, a transmembrane domain in the extracellular region followed by a kinase domain, and C-terminal domain in the cytosolic region. Since SERK is present in the transmembrane region, this LRR kinase transmits signal to the downstream molecules by forming homo- or heterodimer with other RLKs on ligand binding (Talapatra et al. 2014). LRR kinase is considered to play a crucial role in protein-protein interaction (Schmidt et al. 1997). This dimerization causes the phosphorylation of the kinase domain, and the signal proceeds to the next molecules in the signal cascade.

9.7 Molecular Modeling of McSERK

The calculated molecular mass of predicted McSERK protein was found to be 68.870 KD. The three-dimensional model predicted that McSERK is a monomeric protein. Its LRR region was found to be made up of β-sheets and SPP domain as a coiled structure. The TM region showed a high degree of hydrophobicity as compared to the C-terminal end of the protein. This modeling provided an insight into the hypothetical three-dimensional structure of McSERK, and it was found to be highly similar to that of BRI1-associated receptor kinase (BAK1) (Talapatra et al. 2014).

Since it is known that BRI1-associated receptor kinase/somatic embryogenesis receptor kinase 3 (BAK1/SERK3) forms heterodimer with brassinosteroid insensitive 1 (BRI1), it may be suggested that *SERK* genes play some important role in BR signaling.

9.8 Interaction of SERK with Brassinosteroids

Brassinosteroids (BRs) are plant polyhydroxylated steroid hormones involved in several plant physiological processes like seed development and germination, cell division, differentiation of tracheary elements, and enhancement of photosynthesis by increasing carbon dioxide assimilation and Rubisco activity (Li et al. 2005; Vriet et al. 2012; Gruszka 2013). Further it has a broad response toward stress physiology especially abiotic stress like oxidative, salt, drought, and thermotolerance (Oh et al. 2012; Gruszka 2013).

In BR signaling, BR ligand is received by a transmembrane polypeptide receptor, brassinosteroid insensitive 1 (BRI1), which forms a hetero-receptor complex with SERK and leads to abiotic stress response. BRI1, belonging to the large family of LRR-RLKs, is a 1196 amino acid-long protein comprising N-terminal signal peptide, extracellular LRR domain, single-pass transmembrane domain, cytoplasmic juxtamembrane domain (also having kinase activity domain), and a C-terminal domain in the cytoplasmic side (Friedrichsen et al. 2000; Geldner et al. 2007). It has been observed that full activation and functioning of BRI1 on ligand binding requires formation of heterodimers with any of the four members of SERK family, particularly with BRI1-associated receptor kinase/somatic embryogenesis receptor kinase (BAK1/SERK3) (Albrecht et al. 2008; Gou et al. 2012). Usually SERK5 is not involved in the interaction with BRI1 as there is a mutation in one of the amino acids residue which deters its kinase activity for phosphorylation (Gou et al. 2012).

When a BR ligand binds, autophosphorylation event takes place in the activation loop within the kinase domain. Further, BR activity is enhanced by phosphorylation in the juxtamembrane and C-terminal domain (Tang et al. 2010). Also a conformational change is induced in the extracellular LRR domain of BRI1, thereby revealing the dimerization interface so that the co-receptor can bind in the kinase domain for transphosphorylation to take place. All these events lead to the removal of auto-inhibition of BRI1 C-terminal domain and release BKI1 (BRI kinase inhibitor 1) protein allowing the formation of receptor complex with BAK1/SERK3. On binding to the BRI1, BAK1/SERK3 gets activated through phosphorylation in their activation loop by BRI1. The SERK in turn phosphorylates the juxtamembrane and C-terminal domain of BRI1, thereby allowing full functional activation of the kinase domain, leading to the expression of BR-responsive genes. In the absence of BR ligand, BRI1 forms a homodimer where the cytoplasmic domain interacts with the membrane-anchored protein BKI1 and hence restricts the interaction of the kinase domain and reciprocal phosphorylation between BRI1 and BAK1/SERK3. Without BAK1, BR signaling does not take place as BRI1 alone is incapable of phosphorylating the downstream molecules. Hence, two important proteins, BES1 and BZR1, responsible for the expression of BR-responsive genes, remain inactive (through phosphorylation), being unable to enter the nucleus and failing to regulate the expression of BR-responsive genes (Gou et al. 2012) (Fig. 9.3).

Figure 9.3 gives us a bird's-eye view of interaction between SERK and brassinosteroids. Interaction of SERK1 and SERK3, also known as BAK1 (brassinosteroid insensitive 1-associated kinase 1) along with BRI1 (brassinosteroid insensitive 1) in the presence of BR, comes under BR-dependent pathway (Li et al. 2002; Nam and Li 2002; Karlova et al. 2006). But in the case of BR-independent pathways, association between SERK3/BAK1 and FLS2 (flagellin-sensing 2), a binding receptor of flg22 in FLS2 pathway, provides innate immunity to plants (Chinchilla et al. 2007; Heese et al. 2007). In other BR-independent pathways, interaction of SERK3 (BAK1)/SERK4 (BKKI) and BIR1 (BAK1-interacting receptor-like kinase) that activated signal transduction pathways leads to cell death control (He et al. 2007; Gao et al. 2009), and involvement of SERK1/SERK2 with EMS1 (ethyl methanesulfate 1)/EXS mediates another development and controls male sporogenesis (Canales et al. 2002; Albrecht et al. 2005; Zhao et al. 2002; Colcombet et al. 2005).

BR-responsive genes are involved in several plant growth and development processes. Some of the genes are also related to plant stress responses. According to Yuxin et al. (2001), genes

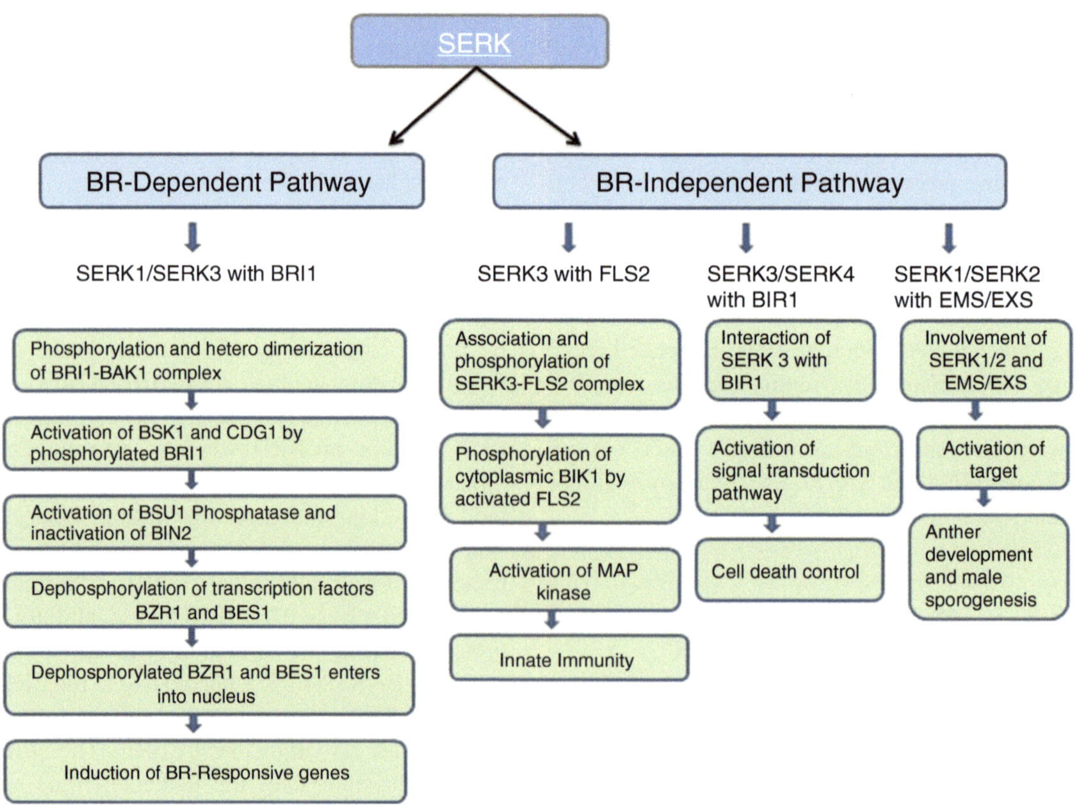

Fig. 9.3 Role of SERK in BR-dependent and BR-independent signaling pathway

which are involved in stress physiology are cyclophilin genes, putative protein kinase, and heat shock proteins (HSP90s). These genes are also associated with transcriptional regulation and BR signal transduction. Recent investigations have shown that HSP90 is directly involved with this pathway. Two isoforms of HSP90 found to be associated with other proteins of the pathway are HSP90.1 and HSP90.3. Both of them interact with brassinosteroid insensitive 2 (BIN2) in the cytoplasm upon BR ligand binding, thereby inactivating BIN2 (Samakovli et al. 2014). Shigeta et al. (2014) observed that HSP90.3 can also form a complex with BRI1 EMS SUPPRESSOR1 (BES1) in vitro. These facts suggest that HSPs play a crucial role in the regulation of BR signal transduction pathway (Samakovli et al. 2014; Shigeta et al. 2014).

9.9 Conclusion

Somatic embryogenesis is massively used for the propagation of plants with high commercial value and their use in research for plant breeding and crop improvement. It is a well-coordinated series of developmental changes, governed by a group of factors which changes the fate of a somatic cell to form embryo followed by the formation of a complete plant. Somatic embryogenesis is associated with a number of factors and the active participation of various regulatory genes. It is very clear from the present review that plant growth regulators, when supplemented in a proper combination and concentration in the culture medium, induces somatic embryogenesis receptor kinase (*SERK*) gene that switches on the signal for embryo formation. *SERK* is a member of the receptor kinase gene family and is reported to be the prime member for initiating the restructuring program of somatic embryogenesis in a variety of plant species. *SERK* remains as a gene family consisting of five different members in plants. It has been observed that *SERK* expression is increased during the initiation of embryo formation and reduces with the maturation of embryo. SERK protein also forms complex with other receptor kinases like brassinosteroid insensitive 1 (BRI1) and gets involved in the developmental and stress-responsive pathways. BRI1 mainly forms heterodimer with SERK3 and after the full functional activation of the kinase domain induces the expression of brassinosteroid-responsive genes.

References

Aker J, Borst JW, Karlova R, de Vries SC (2006) The *Arabidopsis thaliana* AAA protein CDC48A interacts *in vivo* with the somatic embryogenesis receptor-like kinase 1 receptor at the plasma membrane. J Struct Biol 156:62–71. doi:10.1016/j.jsb.2006.03.004

Albertini E, Marconi G, Reale L, Barcaccia G, Porceddu A, Ferranti F, Falcinelli M (2005) *SERK* and *APOSTART*: candidate genes for apomixis in *Poa pratensis*. Plant Physiol 138:2185–2199

Albrecht C, Russinova E, Hecht V, Baaijens E, de Vries SC (2005) The *Arabidopsis thaliana* somatic embryogenesis receptor-like kinases1 and 2 control male sporogenesis. Plant Cell 17:3337–3349

Albrecht C, Russinova E, Kemmerling B, Kwaaitaal M, de Vries SC (2008) *Arabidopsis* somatic embryogenesis receptor kinase proteins serve brassinosteroid-dependent and -independent signaling pathways. Plant Physiol 148:611–619

Balestrazzi A, Toscano I, Bernacchia G, Luo M, Otte S, Carbonera D (1996) Cloning of a cDNA encoding DNA *topoisomerase I* in *Daucus carota* and expression analysis in relation to cell proliferation. Gene 183:183–190

Balestrazzi A, Bernacchia G, Pitto L, Luccarni G, Carbonera D (2001) Spatial expression of DNA *topoisomerase I* genes during cell proliferation in *Daucus carota*. Eur J Histochem 45:31–38

Baudino S, Hansen S, Brettshneider R, Hecht VFG, Dresselhaus T, Lors H, Dumas C, Rogowsky PM (2001) Molecular characterization of two novel maize LRR receptor-like kinase, which belong to the *SERK* gene family. Planta 213:1–10

Becraft PW (1998) Receptor kinase in plant development. Trends Plant Sci 3:384–388

Becraft PW (2002) Receptor kinase signaling in plant development. Annu Rev Cell Dev Biol 18:163–192

Canales C, Bhatt AM, Scott R, Dickinson H (2002) EXS, a putative LRR receptor kinase, regulates male germline cell number and tapetal identity and promotes seed development in *Arabidopsis*. Curr Biol 12:1718–1727

Chinchilla D, Zipfel C, Robatzek S, Kemmerling B, Nurnberger T, Jones JD, Felix G, Boller T (2007) A flagellin-induced complex of the receptor FLS2 and BAK1 initiates plant defence. Nature 448:497–500

Chugh A, Khurana P (2002) Gene expression during somatic embryogenesis: recent advances. Curr Sci 83(6):715–730

Colcombet J, Boisson-Dernier A, Ros-Palau R, Vera CE, Schroeder JI (2005) *Arabidopsis* somatic embryogenesis receptor kinases1 and 2 are essential for tapetum development and microspore maturation. Plant Cell 17:3350–3361

Cueva A, Concia L, Cella R (2012) Molecular characterization of a *Cyrtochilum loxense* Somatic Embryogenesis Receptor-like Kinase (*SERK*) gene expressed during somatic embryogenesis. Plant Cell Rep 31:1129–1139

Das (Pal) M, Raychaudhuri SS (2001) Enhanced development of somatic embryos of *Plantago ovata* Forsk by additives. In Vitro Cell Dev Biol Plant 37:568–571

Evans DA, Sharp WR (1983) Single gene mutations in tomato plants regenerated in tissue culture. Science 221:949–951

Friedrichsen DM, Joazeiro CAP, Li J, Hunter T, Chory J (2000) Brassinosteroid-insensitive 1 is a ubiquitously expressed leucine-rich repeat receptor serine/threonine kinase. Plant Physiol 123:1247–1255

Gao M, Wang X, Wang D, Xu F, Ding X, Zhang Z, Bi D, Cheng YT, Chen S, Li X, Zhang Y (2009) Regulation of cell death and innate immunity by two receptor-like kinases in *Arabidopsis*. Cell Host Microbe 6:34–44

Geldner N, Hyman DL, Wang X, Schumacher K, Chory J (2007) Endosomal signaling of plant steroid receptor kinase BRI1. Genes Dev 21:1598–1602

Giroux RW, Pauls KP (1997) Characterization of somatic embryogenesis-related cDNAs from alfalfa (*Medicago sativa* L.). Plant Mol Biol 33:393–404

Gou X, Yin H, He K, Du J, Yi J, Xu S, Lin H, Clouse SD, Li J (2012) Genetic evidence for an indispensable role of somatic embryogenesis receptor kinases in Brassinosteroid signaling. PLoS Genet 8(1):e1002452

Gruszka D (2013) The Brassinosteroid signaling pathway-new key players and interconnections with other signaling networks crucial for plant development and stress tolerance. Int J Mol Sci 14:8740–8774

Hagen G, Kleinschmidt A, Guilfoyle TJ (1984) Auxin-regulated gene expression in intact soybean hypocotyl and excised hypocotyl sections. Planta 162:147–153

He K, Gou X, Yuan T (2007) BAK1 and BKK1 regulate brassinosteroid-dependent growth and brassinosteroid-independent cell-death pathways. Curr Biol 17:1109–1115

Hecht V, Vielle-Calzada JP, Hartog MV, Schmidt ED, Boutilier K, Grossniklaus U, de Vries SC (2001) The *Arabidopsis* somatic embryogenesis receptor kinase 1 gene is expressed in developing ovules and embryos and enhances embryogenic competence in culture. Plant Physiol 127:803–816

Heese A, Hann DR, Gimenez-Ibanez S, Jones AM, He K, Li J, Schroeder JI, Peck SC, Rathjen JP (2007) The receptor-like kinase SERK3/BAK1 is a central regulator of innate immunity in plants. Proc Natl Acad Sci 104:12217–12222

Higashi K, Shiota H, Kamada H (1998) Patterns of expression of the genes for glutamine synthetase isoforms during somatic and zygotic embryogenesis in carrot. Plant Cell Physiol 39:418–424

Hirt H, Pay A, Gyorgyey J, Bako L, Nemet K, Bogre L, Schweyen RJ, Heberle-Bors E, Dudits D (1991) Complementation of a yeast cell cycle mutant by an alfalfa cDNA encoding a protein kinase homologous to p34cdc2. Proc Natl Acad Sci 88:1636–1640

Hu H, Xiong L, Yang Y (2005) Rice *SERK1* gene positively regulates somatic embryogenesis of cultured cells and host defense response against fungal infection. Planta 222:107–117

Huang X, Lu XY, Zhao JT, Chen JK, Dai XM, Xiao W, Chen YP, Chen YF, Huang XL (2010) *MaSERK1* gene expression associated with somatic embryogenic competence and disease resistance response in banana (*Musa* spp.). Plant Mol Biol Report 28:309–316

Ito Y, Takaya K, Kurata N (2005) Expression of *SERK* family receptor-like protein kinase genes in rice. Biochim Biophys Acta 1730:253–258

Jayanthi M, Jerard A, Sherif S, Jayasankar S (2014) Molecular characterization of somatic embryogenesis receptor-like kinase (SERK) genes from plum (Prunus salicina) and peach (Prunus persica). Indian J Hortic 71(4):560–563

Kapros T, Bögre L, Németh K, Bakó L, Györgyey J, Wu SC, Dudits D (1992) Differential expression of histone *H3* gene variants during cell cycle and somatic embryogenesis in Alfalfa. Plant Physiol 98(2):621–625

Karlova R, Boeren S, Russinova E, Aker J, Vervoort J, de Vries SC (2006) The *Arabidopsis* somatic embryogenesis receptor-like kinase1 protein complex includes brassinosteroid-insensitive1. Plant Cell 18:625–638

Kedong X, Qinglin L, Huifang Y, Li Z, Lili D, Fengluan L, Ling B, Nan M, Liangjun Z (2011) Isolation and molecular characterization of RcSERK1: a Rosa canina gene transcriptionally induced during initiation of protocorm-like bodies. Afr J Biotechnol 10(20):4011–4017

Kitamiya E, Suzuki S, Sano T, Nagata T (2000) Isolation of two genes that were induced upon initiation of somatic embryogenesis on carrot hypocotyls by high concentrations of 2,4-D. Plant Cell Rep 19:551–557

Kiyosue T, Yamaguchi-Shinozaki K, Shinozaki K, Higashi K, Satoh S, Kamada H, Harada H (1992) Isolation and characterization of a cDNA that encodes *ECP31*, an embryogenic-cell protein from carrot. Plant Mol Biol 19:239–249

Kwaaitaal MA, de Vries SC (2007) The *SERK1* gene is expressed in procambium and immature vascular cells. J Exp Bot 58:2887–2896

Lewis MW, Leslie ME, Fulcher EH, Darnielle L, Healy P, Youn JY, Liljegren SJ (2010) The SERK1 receptor-like kinase regulates organ separation in Arabidopsis flowers. Plant J 62(5):817–828

Li J, Wen J, Lease KA, Doke JT, Tax FE, Walker JC (2002) BAK1, an *Arabidopsis* LRR receptor-like protein kinase, interacts with BRI1 and modulates brassinosteroid signaling. Cell 110:213–222

Li L, Xu J, Xu ZH, Xue HW (2005) Brassinosteroids stimulate plant tropisms through modulation of polar auxin transport in *Brassica* and *Arabidopsis*. Plant Cell 17:2738–2753

Li J (2010) Multi-tasking of somatic embryogenesis receptor like-protein kinases. Curr Opin Plant Biol 13(5):509–514

Li X, Fang YH, Han JD, Bai SN, Rao GY (2014) Isolation and Characterization of a novel somatic embryogenesis receptor kinase gene expressed in the fern *Adiantum capillus-veneris* during shoot regeneration *in vitro*. Plant Mol Biol Report. doi:10.1007/s11105-014-0769-2

Ling V, Perera I, Zielinski RE (1991) Primary structures of *Arabidopsis* calmodulin isoforms deduced from the sequences of cDNA clones. Plant Physiol 96:1196–1202

Ma J, He Y, Hu Z, Xu W, Xia J, Guo C, Lin S, Cao L, Chen C, Wu C, Zhang J (2012a) Characterization and expression analysis of *AcSERK2*, a somatic embryogenesis and stress resistance related gene in pineapple. Gene 500:115–123

Ma J, He Y, Wu C, Liu H, Hu Z, Sun G (2012b) Cloning and molecular characterization of a *SERK* gene transcriptionally induced during somatic embryogenesis in *Ananas comosus* cv. Shenwan. Plant Mol Biol Report 30:195–203

Maillot P, Lebel S, Schellenbaum P, Jacques A, Walter B (2009) Differential regulation of *SERK*, *LEC*-like and pathogenesis related genes during indirect secondary somatic embryogenesis in grapevine. Plant Physiol Biochem 47:743–752

Nam KH, Li J (2002) BRI1/BAK1, a receptor kinase pair mediating brassinosteroid signaling. Cell 110:203–212

Nolan KE, Irwanto RR, Rose RJ (2003) Auxin up-regulates *MtSERK1* expression in both *Medicago truncatula* root-forming and embryogenic cultures. Plant Physiol 33:218–230

Nolan KE, Kurdyukov S, Rose RJ (2009) Expression of the somatic embryogenesis receptor-like kinase1 (*SERK1*) gene is associated with developmental change in the life cycle of the model legume *Medicago truncatula*. J Exp Bot 60:1759–1771

Oh MH, Clouse SD, Huber SC (2012) Tyrosine phosphorylation of the BRI1 receptor kinase occurs via a post-translational modification and is activated by the juxtamembrane domain. Front Plant Sci 3:1–14

Paul A, Mitter K, Raychaudhuri SS (2009) Effect of polyamines on *in vitro* somatic embryogenesis in *Momordica charantia* L. Plant Cell Tissue Organ Cult 97:303–311

Pérez-Núñez MT, Souza R, Sáenz L, Chan JL, Zúñiga-Aguilar JJ, Oropeza C (2009) Detection of a *SERK*-like gene in coconut and analysis of its expression during the formation of embryogenic callus and somatic embryos. Plant Cell Rep 28:11–19

Periera IY, Zielinski R (1992) Structure and expression of the *Arabidopsis CaM-3* calmodulin gene. Plant Mol Biol 19:649–664

Rienties IM, Vink J, Borst JW, Russinova E, de Vries SC (2005) The *Arabidopsis* SERK1 protein interacts with the AAA-ATPase AtCDC48, the 14-3-3 protein GF14lambda and the PP2C phosphatase KAPP. Planta 221(3):394–405

Rohani ER, Ismanizan I, Noor NM (2012) Somatic embryogenesis of mangosteen. Plant Cell Tissue Organ Cult 110:251–259

Samakovli D, Margaritopoulou T, Prassinos C, Milioni D, Hatzopoulos P (2014) Brassinosteroid nuclear signaling recruits HSP90 activity. New Phytol 203:743–757

Santa-Catarina C, Hanai LR, Dornelas MC, Viana AM, Floh EIS (2004) *SERK* gene homolog expression, polyamines and amino acids associated with somatic embryogenic competence of *Ocotea catharinensis* Mez. (Lauraceae). Plant Cell Tissue Organ Cult 79:53–61

Santos MO, Arago FJL (2009) Role of SERK genes in plant environmental response. Plant Signal Behav 4(12):1111–1113

Santos MO, Romano E, Yotoko KSC, Tinoco MLP, Dias BBA, Araga˜o FJL (2005) Characterisation of the cacao somatic embryogenesis receptor-like kinase (*SERK*) gene expressed during somatic embryogenesis. Plant Sci 168:723–729

Sato S, Toya T, Kawahara R, Whittier RF, Fukuda H, Komamine A (1995) Isolation of a carrot gene expressed specifically during early-stage somatic embryogenesis. Plant Mol Biol 28:39–46

Savona M, Mattioli R, Nigro S, Falasca G, Della Rovere F, Costantino P, De Vries SC, Ruffoni B, Trovato M, Altamura MM (2012) Two SERK genes are markers of pluripotency in *Cyclamen persicum* Mill. J Exp Bot 63(1):471–488

Schmidt ED, Guzzo F, Toonen MA, de Vries SC (1997) A leucine rich repeat containing receptor-like kinase marks somatic plant cells competent to form embryos. Development 124:2049–2062

Shah K, Schmidt EDL, Vlak JM, de Vries SC (2001a) Expression of the *Daucus carota* somatic embryogenesis receptor kinase (DcSERK) protein in insect cells. Biochimie 83:415–421

Shah K, Gadella TW Jr, van Erp H, Hecht V, de Vries SC (2001b) Subcellular localization and oligomerization of the *Arabidopsis thaliana* somatic embryogenesis receptor kinase 1 protein. J Mol Biol 309:641–655

Sharma SK, Millam S, Hedley PE, McNicol J, Bryan GJ (2008a) Molecular regulation of somatic embryogenesis in potato: an auxin led perspective. Plant Mol Biol 68:185–201

Sharma SK, Millam S, Hein I, Bryan GJ (2008b) Cloning and molecular characterisation of a potato *SERK* gene transcriptionally induced during initiation of somatic embryogenesis. Planta 228:319–330

Sharp WR, Sondahl MR, Caldas LS, Maraffa SB (1980) The physiology of *in-vitro* asexual embryogenesis. Hortic Rev 2:268–310

Shigeta T, Zaizen Y, Asami T, Yoshida S, Nakamura Y, Okamoto S, Matsuo T, Sugimoto Y (2014) Molecular evidence of the involvement of heat shock protein 90 in brassinosteroid signaling in *Arabidopsis* T87 cultured cells. Plant Cell Rep 33:499–510. doi:10.1007/s00299-013-1550-y

Shimada T, Hirabayashi T, Endo T, Fujii H, Kita M, Omura M (2005) Isolation and characterization of the somatic embryogenesis receptor-like kinase gene homologue, (*CitSERK1*) from *Citrus unshiu* Marc. Sci Hortic 103:233–238

Singla B, Khurana JP, Khurana P (2008) Characterization of three somatic embryogenesis receptor kinase genes from wheat, *Triticum aestivum*. Plant Cell Rep 27:833–843

Somleva MN, Schmidt EDL, de Vries SC (2000) Embryogenic cells in *Dactylis glomerata* L. (Poaceae) explants identified by cell tracking and by *SERK* expression. Plant Cell Rep 19:718–726

Sucharitakul K, Rakmit R, Boonsorn Y, Leelapon O, Teerakathiti T, Bunnag S, Chanvivattana Y (2014) Isolation and expression analysis of a SOMATIC EMBRYOGENESIS RECEPTOR-LIKE KINASE (SERK) gene in Curcuma alismatifolia Gagnep. J Agric Sci. doi:10.5539/jas.v6n10p207

Talapatra S (2014) Studies on biochemical changes and expression of SERK gene during somatic embryogenesis in Momordica charantia L. Ph.D. thesis, University of Calcutta, India

Talapatra S, Ghoshal N, Raychaudhuri SS (2014) Molecular characterization, modeling and expression analysis of a somatic embryogenesis receptor kinase (*SERK*) gene in *Momordica charantia* L. during somatic embryogenesis. Plant Cell Tissue Organ Cult 116(3):271–283

Tang W, Deng Z, Wang ZY (2010) Proteomics shed light on the brassinosteroid signaling mechanisms. Curr Opin Plant Biol 13:2727–2733

Thomas C, Meyer D, Himber C, Steinmetz A (2004) Spatial expression of a sunflower *SERK* gene during induction of somatic embryogenesis and shoot organogenesis. Plant Physiol Biochem 42:35–42

Vriet C, Russinova E, Reuzeau C (2012) Boosting crop yields with plant steroids. Plant Cell 24:842–857

Wang XD, Nolan KE, Irwanto RR, Sheahan MB, Rose RJ (2011) Ontogeny of embryogenic callus in *Medicago truncatula*: the fate of the pluripotent and totipotent stem cells. Ann Bot 107:599–609

Yang H, Saitou T, Komeda Y, Harada H, Kamada H (1996) Late embryogenesis abundant protein in *Arabidopsis thaliana* homologous to carrot *ECP31*. Physiol Plant 98:661–666

Yang H, Saitou T, Komeda Y, Harada H, Kamada H (1997) *Arabidopsis thaliana ECP63* encoding a LEA protein is located in chromosome 4. Gene 184:83–88

Yang C, Zhao T, Yu D, Gai J (2011) Isolation and functional characterization of a *SERK* gene from soybean (*Glycine max* (L.) Merr.). Plant Mol Biol Report 29:334–344

Yuxin H, Yuxin W, Yonghong W, Fang B, Ning L, Zhenhua P, Jiayang L (2001) Identification of brassinosteroid responsive genes in *Arabidopsis* by cDNA array. Sci China Ser C 44(6):477–484

Zakizadeh H, Stummann BM, Lutken H, Muller R (2010) Isolation and characterization of four somatic embryogenesis receptor-like kinase (*RhSERK*) genes from miniature potted rose (*Rosa hybrida* cv. Linda). Plant Cell Tissue Organ Cult 101:331–338

Zhang S, Liu X, Lin Y, Xie G, Fu F, Liu H, Wang J, Gao S, Lan H, Rong T (2011) Characterization of a *ZmSERK* gene and its relationship to somatic embryogenesis in a maize culture. Plant Cell Tissue Organ Cult 105:29–37

Zhao DZ, Wang GW, Speal B, Ma H (2002) The excess microsporocytes1 gene encodes a putative leucine-rich repeat receptor protein kinase that controls somatic and reproductive cell fate in the *Arabidopsis* anther. Genes Dev 16:2021–2031

Zhu C, Kamada H, Harada H, He M, Hao S (1997) Isolation and characterization of a cDNA encoded an embryogenic cell protein-63 related to embryogenesis from carrot. Acta Bot Sin 39:1091–1098

10 Somatic Embryogenesis, Cryopreservation, and In Vitro Mutagenesis in Cyclamen

Ehsan Mohammad Tagipur, Gamze Seker,
Jaime A. Teixeira da Silva, and Yesim Yalcin Mendi

Abstract

The cyclamen is a member of the family Primulaceae and consists of 22 species, originated from the Mediterranean Basin. Due to inbreeding depression, abortive zygotic embryo formation, and varying ploidy levels, vegetative production is frequently selected as the preferred form of propagation by breeders. Somatic embryogenesis is a highly effective method to produce cyclamen plants vegetatively. Somatic embryo formation in cyclamen is affected by genotype, source of explant, medium content, plant growth regulators, and their concentrations and culture conditions. Somatic embryos can be multiplied efficiently by repetitive or secondary embryogenesis on MS without plant growth regulators. The obtained somatic embryos can be used to produce synthetic seeds for clonal plant propagation and to protect valuable or rare cyclamen genetic resources.

Mutation breeding is a useful method to improve crops. Heavy-ion beam irradiation in mutation breeding can be used to produce new varieties of cyclamen. However, ion beams have not been sufficiently characterized in terms of mutagens for plant. Cryopreservation is also becoming a very important way for long-term storage of plant genetic resources, and effective cryopreservation protocols have been developed for many of plant species. Cryopreservation requires a minimum of space and low level of maintenance. In this chapter, we discussed and reviewed the progress of in vitro research on cyclamen.

E. Mohammad Tagipur
Department of Biotechnology, Institute of Natural and Applied Sciences, University of Cukurova, Balcalı, Adana 01330, Turkey

G. Seker • Y. Yalcin Mendi (✉)
Department of Horticulture, Faculty of Agriculture, University of Cukurova, Balcalı, Adana 01330, Turkey
e-mail: ymendi@gmail.com

J.A. Teixeira da Silva (✉)
Miki-cho Post Office, PO Box 7, Ikenobe 3011-2, Kagawa-ken 761-0799, Japan
e-mail: jaimetex@yahoo.com

A. Mujib (ed.), *Somatic Embryogenesis in Ornamentals and Its Applications*,
DOI 10.1007/978-81-322-2683-3_10

Keywords
Cryopreservation • *Cyclamen* • In vitro mutagenesis • Somatic embryogenesis

10.1 Introduction

The genus *Cyclamen* was traditionally classified in the Primulaceae family, was briefly reclassified as a Myrsinaceae family member (Grey-Wilson 2003), but has returned to the Primulaceae (The Plant List 2015a). All 116 species (The Plant List 2015b) except for one have been well established in cultivation, and some are easy to cultivate (Grey-Wilson 2003). *Cyclamen persicum* Mill. is an important pot flower (Schwenkel and Winkelmann 1998). It is traditionally propagated by seed despite the lack of homogeneity of some cultivars, the existence of inbreeding depression, and the high cost of hybrid seeds (Ruffoni et al. 2000). Mass propagation by tissue culture is thus an effective propagation method for *Cyclamen* (Takamura and Miyajima 1997). Micropropagation of *Cyclamen* through organogenesis and somatic embryogenesis is established reported protocols (Schwenkel and Grunewaldt 1998; Kreuger et al. 1995; Dillen et al. 1996; Schwenkel and Winkelmann 1998).

As with other ornamental plants, cyclamen breeding is performed by classical hybridization methods and phenotypic selection, based on the identification of novel morphological characters (Çürük et al. 2015), but these methods take more time to develop new cultivars (Sugiyama et al. 2008). Classical cyclamen breeding is very difficult because of inbreeding depression, different ploidy levels, and abortive embryo formation (Jalali et al. 2012). Conventionally, cyclamen is propagated by seeds that are produced by manual pollination and are therefore expensive (up to 0.20 €/seed) (Kocak et al. 2014). Due to inbreeding depression (Affre and Thomson 1999), the production of an F_1 cyclamen hybrid from parental lines is difficult (Jalali et al. 2012). Therefore, breeders pay great attention to vegetative production in breeding programs. Also, if carried out effectively and economically, vegetative propagation can allow breeders to produce clonal varieties that allow for the selection of plants with superior characteristics (Jalali et al. 2012). Cyclamen tends to be mass propagated by cuttings or separation of tubers; therefore, in vitro tissue culture provides an important alternative to traditional vegetative reproduction, and the development of new *Cyclamen* varieties has been possible through the use of biotechnological methods (Jalali et al. 2012).

Somatic embryogenesis (SE) is a valuable tool in plant biotechnology and can be utilized in a number of ways (Zimmerman 1993; Yang and Zhang 2010): (1) for large-scale clonal propagation of elite cultivars, it can serve as a substitute approach to conventional micropropagation (2) as a tool in genetic improvement programs, SE via callus or secondary SE can be an effective receptor tissue for application in genetic transformation, and (3) it provides a model system for expanding knowledge on molecular, regulatory, and morphogenetic events related to plant embryogenesis.

10.2 Somatic Embryogenesis

SE is the process by which somatic cells develop through several stages of embryogeny, resulting in whole plants. An efficient vegetative propagation in cyclamen is possible by using in vitro culture, but reports are restricted to the wild species *C. persicum* (Karam and Al-Majathoub 2000a, b) and its cultivars (Hohe et al. 2001). Although SE and/or organogenesis in *C. persicum* has been reported by several groups (Kiviharju et al. 1992; Takamura et al. 1995; Takamura and Tanaka 1996; Schwenkel and Winkelmann 1998; Seyring et al. 2009; Jalali et al. 2010a, b; Yang and Zhang 2010), the frequency of plant regeneration from somatic embryos is low. Hybrids of a number of species combinations have been obtained by using embryo rescue (Ishizaha 2008 and references therein), which is able to overcome hybrid inviability by cultivating hybrid embryos in vitro to prevent abortion. Another aim of SE is to establish a protocol for the production of synthetic seeds that can replace generative propagation (Lyngved et al. 2008; Sharma et al. 2013). Other more specialist applications include the production of high-value compounds by somatic embryos grown in bioreactors (Dunwell et al. 2008). Seyring et al. (2009) investigated SE in *C. africanum* Boiss. and Reut., *C. cilicium* Boiss. and Heldr., *C. coum* Mill., *C. hederifolium* Ait., *C. persicum* Mill., and *C. purpurascens* Mill. using the protocol developed by Schwenkel and Winkelmann (1998). After an 8–12 week period, callus was observed from all explant types (young leaves, petioles, flower buds, and peduncles), and after 3–7 months structures formed, some of which were somatic embryo-like structures (SELS). Somatic embryos developed in all species, but not all explants of all species could induce somatic embryos.

10.2.1 Direct and Indirect Somatic Embryogenesis

Somatic embryos can differentiate either directly from the explant without an intervening callus phase or indirectly after a callus phase, referred to as direct somatic embryogenesis (DSE) and indirect somatic embryogenesis (ISE), respectively (Sharp et al. 1980; Yang and Zhang 2010). It is believed that both processes are extremes of one continuous developmental pathway (Carman 1990). The main factors involved in each pathway depend on the source and physiological state of the explant employed and the type and concentration of the plant growth regulators (PGRs). Explants from which DSE is most likely to occur include microspores (microsporogenesis), ovules, zygotic and somatic embryos, and seedlings (Malik et al. 2007; Yang and Zhang 2010; Żur et al. 2015). Among the two forms of SE, ISE has been the most extensively studied and used in transformation and somatic hybridization (Li et al. 2006; Yang and Zhang 2010).

During ISE in *Cyclamen*, both embryogenic and non-embryogenic callus are present. It is usually easy to distinguish between both callus forms on the basis of morphology and color. Embryogenic callus has nodular features and a smooth surface, and the embryogenic cells that form somatic embryos are characterized generally as small and isodiametric in shape. These cells have large and densely stained nuclei and nucleoli and are densely cytoplasmic with small vacuoles, thick cell walls, and a higher metabolic activity, while non-embryogenic callus is rough, friable, and translucent (Jimenez and Bangerth 2001; Yang and Zhang 2010). Somatic embryos in *Cyclamen* species can be easily distinguished from adventitious shoots because embryos do not have any connection to callus (Nadja et al. 2010). In our unpublished studies, we observed differences between *Cyclamen* species in terms of the shape and color of callus. Callus obtained from *C. hederifolium* and *C. persicum* is soft, and the structure of callus is more transparent than that of *C. coum* and *C. cilicium*. On the other hand, *C. coum* and *C. cilicium* callus is more compact and opaque than callus from *C. hederifolium* and *C. persicum*.

Terakawa et al. (2008) observed that somatic embryos formed directly from the leaf segments when placed onto somatic embryo formation medium consisting of MS (Murashige and Skoog 1962) supplemented with 4 mg/L 2,4-dichlorophenoxy acetic acid (2,4-D), 0.1 mg/L kinetin, 3 % (w/v) sucrose (pH 5.8), and 0.2 % Gelrite and incubated in the dark at 25 °C. After 3 months, somatic embryos that formed directly on the surface of the leaf segments were transferred to test tubes containing PGR-free MS liquid medium supplemented with 3 % (w/v) sucrose (Fig. 10.1; Tables 10.1, 10.2, and 10.3).

10.2.2 Developmental Stages of Somatic Embryos

Theoretically, the development of a zygotic embryo can be divided into two stages: an initial morphogenetic stage and a second metabolic

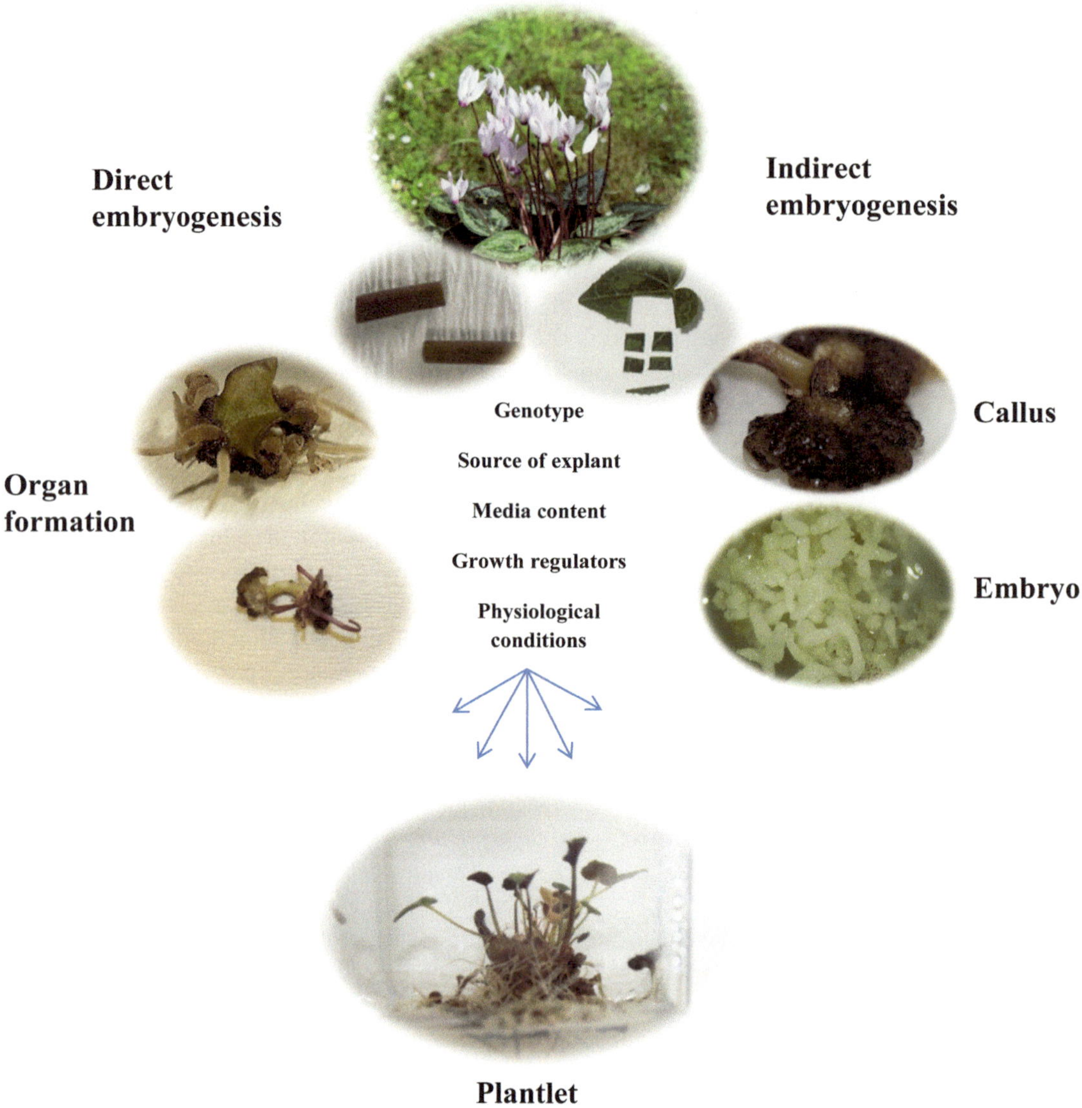

Fig. 10.1 Direct and indirect somatic embryogenesis in *Cyclamen persicum*

stage characterized by biochemical activities (Yang and Zhang 2010). In addition, during the morphogenetic stage, zygotic embryo development in dicots can be divided into four general sequential stages: globular-shaped, heart-shaped, torpedo-shaped, and cotyledonary stages (Yang and Zhang 2010). As the zygotic coequal, somatic embryos pass through various morphological stages: the proembryogenic masses organize into a typical globular embryo, which progresses through heart- and torpedo-shaped stages. At the cotyledonary stage, instead of entering developmental arrest, the embryos develop a shoot meristem and the seedling grows. Unlike their zygotic counterpart, somatic embryos are easily separable, culture conditions can be controlled, and the lack of material is not a restricting factor for experiments (Kawahara and Komamine 1995; Yang and Zhang 2010). SE formation in *C. persicum* takes place in 3 months, globular-shaped somatic embryos form in 4 months, heart-shaped somatic embryos form

Table 10.1 In vitro shoot induction in *Cyclamen* spp. in different media and PGR conditions

Species	Type of explant	Basal medium	PGR	Growth condition(s)	References
C. persicum Mill. (Vuurbaak)	Etiolated petiole	1/3 MS	NAA (0.037 mg l^{-1}) + BA (0.23 mg l^{-1})	25 °C, dark or light (2300 lux), 20 °C, and 16 h	Ando and Murasaki (1983)
C. persicum Mill. (Vuurbaak) and (Victoria)	Etiolated petiole	1/3 MS	NAA (0.037 mg l^{-1}) + BA (0.23 mg l^{-1})	20 °C, dark, shoot and root formation in 16-h photoperiod, 2500 lux	Murasaki and Tsurushima (1988)
C. persicum Mill.	Lamina, petiole, and peduncle	1/3 MS or Nitsch + 15–30 g l^{-1} sucrose + 8 g l^{-1} agar	IAA or NAA (0–1 mg l^{-1}) + BAP or Kin (0.5–1 mg l^{-1})	20 °C, 12-h photoperiod, 2000 lux	Schwenkel and Grunewaldt (1988)
C. persicum Mill.	Young leaf	MS + 60 g l^{-1} sucrose + 8 g l^{-1} agar	BA (4 mg l^{-1}) + Kin (4 mg l^{-1}) + NAA (3 mg l^{-1})	18 °C, dark for shoot induction, 5000–6000 lux for rooting	Dillen et al. (1996)
Wild *C. persicum* Mill.	Leaf disc, petiole, petal, and peduncle	1/3 MS + 30 g l^{-1} sucrose + 8 g l^{-1} agar	NAA (0.1 mg l^{-1}), TDZ (0.022 or 0.22 mg l^{-1} depending on explant type)	22 °C, dark	Karam and Al-Majathoub (2000b)
Wild *C. persicum* Mill.	Aseptic in vitro leaf from microshoot	1/2 MS + 8 g l^{-1} agar	NAA (0.1 mg l^{-1}) + TDZ (0.22 mg l^{-1})	22 °C, dark	Karam and Al-Majathoub (2000a)
C. persicum Mill. (concert)	Tuber, leaf, and petiole	MS + 30 g l^{-1} sucrose + 8 g l^{-1} agar	NAA (1 mg l^{-1}) + BA (1 mg l^{-1}) or TDZ (0.9 mg l^{-1})	22–24 °C, 16-h photoperiod, 50 mol $m^{-2}s^{-1}$ light	Abu-Qaoud (2004)
C. mirabile lamina	Petiole, root, intact tuber, and tuber section	MS, 1/2 MS, 30 g l^{-1} sucrose + 8 g l^{-1} agar	IAA (0.1 mg l^{-1}) + Kin (0.5 mg l^{-1}) or NAA (0.5 mg l^{-1}) + Kin (0.5 mg l^{-1})	24 °C, dark, or 16-h photoperiod	Yamaner and Erdag (2008)

Table 10.2 Tuber formation in various species of *Cyclamen*

Species	Type of explant	Basal medium	PGR	Growth conditions	References
C. persicum Mill. (Vuurbaak)	Cotyledon	1/2 MS	NAA (1.9 mg l^{-1})	Undefined	Fukui et al. (1988)
Wild *C. persicum* Mill.	Root explant	1/2 MS + 30 g l^{-1} sucrose	BA (1 mg l^{-1})	22 °C, dark	Karam and Al-Majathoub (2000a)
C. persicum Mill. (concert)	Leaf explant	MS	TDZ (0.4–0.9 mg l^{-1})	22–24 °C, dark	Abu-Qaoud (2004)
C. mirabile	Tuber	MS, 1/2 MS	BA (2 mg l^{-1}) + NAA (0.1 mg l^{-1})	24 ± 1 °C, 16-h photoperiod	Yamaner and Erdag (2008)

Fig. 10.2 Somatic embryogenesis in *Cyclamen persicum*. (**a**) Globular somatic embryo, (**b**) heart-shaped somatic embryo, (**c**) torpedo-shaped somatic embryo, (**d**) plantlets regenerated *via* somatic embryogenesis

1 month later (fifth month), and for torpedo, it took for an additional 1 month (sixth month), while cotyledonary stage formation is complete by the end of seventh month (Fig. 10.2, Table 10.4).

10.3 Factors Affecting Somatic Embryogenesis

10.3.1 Genotype

The genotype is most likely the most important factor affecting SE in *Cyclamen*. Kocak et al. (2014) observed embryogenic potential of different genotypes of *C. persicum* when cultured on MS supplemented with 2.0 mg/L 2,4-D and 0.8 mg/L 6-(y,y-dimethylallylamino)purine (2iP). In cyclamen, the in vitro response also depends on the genotype (Takamura and Tanaka 1996; Schwenkel and Winkelmann 1998) suggesting the genetic control of regeneration ability. Takamura and Miyajima (1997) observed the differentiation of somatic embryos in four out of 13 cyclamen cultivars, while Schwenkel and Winkelmann (1998) tested 30 genotypes for their in vitro reaction and found that 29 genotypes were able to regenerate somatic embryos (Tables 10.2, 10.3, and 10.4).

10.3.2 Source of Explant

In *Cyclamen*, a variety of explants (leaves, petioles, ovules, ovaries, etiolated petioles, and leaf callus) have been used to induce SE. The choice of explant is an important factor in the induction of SE. Karam and Al-Majathoub (2000a)

Table 10.3 Somatic embryogenesis of *Cyclamen persicum* Mill.

Cultivar	Type of explant	Sterilization protocol	Callus induction media	Somatic embryogenesis media	Callus formation	Somatic embryo formation	References
C. persicum Mill.(mini rose)	Leaf	Ethanol 70 % (30 s) + NaOCl 5 % (5 min)	LS + 1 mg l^{-1} 2,4-D + 0.1 mg l^{-1} Kin + 0.2 % Gelrite, dark, and 26 °C	PGR-free medium	60 % callus	Seven (46 %) embryogenic calli	Otani and Shimada (1991)
C. persicum Mill. (Anneke)	Cotyledon, petiole, tuber, and root	Aseptic seedlings used (no chemical treatment)	MS + 1 mg l^{-1} 2,4-D + 0.1 mg l^{-1} Kin + 0.2 % gellan gum (dark and light)	PGR-free medium	80 % cotyledon, 65 % petiole, 40 % tuber, 35 % root	Number of somatic embryos per organ, 200/cotyledon, 100/petiole, 60/tuber, 200/root	Takamura and Miyajima (1997)
C. persicum Mill.	Ovule	Ethanol 70 % (30s) + NaOCl 2.6 % (20 min)	1/2 MS + 2 mg l^{-1} 2,4-D + 0.8 mg l^{-1} 2iP + 3.7 g l^{-1} Gelrite in dark and 22–25 °C	PGR-free medium (1/2 MS) formed callus	47–58 % of ovules	0–30 % callus formed somatic embryos	Schwenkel and Winkelmann (1998)
C. persicum Mill. (Halios)	Ovule	Ethanol 70 % + NaOCl 1.2 % (20 min)	1/2 MS + 2 mg l^{-1} 2,4-D + 0.8 mg l^{-1} 2iP + 4 g l^{-1} Phytagel, dark, 24 °C	PGR-free medium (1/2 MS)	Not specified	Seven somatic embryos/callus induced from two ovules	Savona et al. (2007)
C. persicum Mill. (Halios)	Leaf	Ethanol 70 % (60 s) + $HgCl_2$ 1 % (9 min) + NaOCl 1 % (8 min)	1/2 MS + 4 mg l^{-1} 2,4-D + 0.1 mg l^{-1} Kin + 8 g l^{-1} agar, dark, 25 °C	PGR-free medium (1/2 MS)	66 % callus	Ten somatic embryos/cm^3 callus	Jalali et al. (2010b)
C. persicum Mill. 15 genotypes	Leaf, ovule, petiole	Ethanol 70 % (1 min) + $HgCl_2$ 1 % (10 min) + NaOCl 30 % (20 min)	1/2 MS + 2 mg l^{-1} 2,4-D + 0.8 mg l^{-1} 2iP + 4 g l^{-1}	PGR-free medium (MS)	74 % callus	31 % embryos	Kocak et al. (2014)

observed that the greatest regeneration percentage (88 %) of cyclamen somatic embryos was obtained on leaves with midrib tissue. Ando and Murasaki (1983) observed that the regenerative potential of petioles in cyclamen was superior to those of normal, non-etiolated petioles. Petiole explants followed by ovary explants formed the callus on callus induction medium (Kocak et al. 2014). However, ovary and ovule explants tended to take longer to induce callus than the leaves and petioles (Kocak et al. 2014). Kiviharju et al. (1992) showed that petioles and peduncles produced very homogenous callus, in terms of color and morphology. Karam and Al-Majathoub (2000a) noted that peduncle and petal explants formed most callus on half-strength MS medium containing 0.22 mg/L thidiazuron after 8 weeks of culture. Jalali et al. (2010b) showed that most callus formed from leaf explants on half-strength MS containing 2 mg/L 2,4-D and 0.1 mg/L kinetin. Kocak et al. (2014) compared the embryogenic potential of different *C. persicum* explants when cultured on MS with 2.0 mg/L 2,4-D and 0.8 mg/L 2iP (Fig. 10.3). Petioles formed more callus than the ovule and leaf explants, with average callus formation as follows: 34.3 %, 30.16 %, 26.6 %, and 15.6 % from petioles, ovaries, ovules, and leaves, respectively (Kocak et al. 2014) (Tables 10.2, 10.3, and 10.4).

Fig. 10.3 Regeneration stages in *Cyclamen persicum*. Leaves (**a**), petioles (**b**), ovules (**c**), and ovaries (**d**); somatic embryo formation on each explant (**a1**, **b1**, **c1**, and **d1**, somatic embryos); **a2**, **b2**, **c2**, and **d2**, plantlets

Table 10.4 Somatic embryogenesis of wild *Cyclamen* species

Species	Type of explants	Callus induction media	Somatic embryogenesis media	Growth conditions for embryo induction	Reference
C. cilicium, *C. coum*, *C. graecum*, *C. hederifolium*, *C. persicum*, *C. purpurascens*, and *C. rohlfsianum*	Petiole, leaf	(1) LS + 50.0 g l^{-1} sucrose + 2.0 g l^{-1} gellan gum + 2,4-D (1.0 mg l-1) + Kin (0.1 mg l-1) (2) MS + 60.0 g l^{-1} sucrose + 3.0 g l^{-1} gellan gum + 2,4-D (5.0 mg l^{-1}) + Kin (0.1 mg l^{-1})	Same medium without PGRs	–	Furukawa et al. (2001)
18 wild *Cyclamen* species	Petiole and leaf	MS + 30 g l^{-1} sucrose + 3 g l^{-1} gellan gum	Same medium without PGRs	–	Furukawa et al. (2002)
C. africanum Boiss. and Reut., *C. cilicium* Boiss. and Heldr., *C. coum* Mill., *C. hederifolium* Ait., *C. persicum* Mill., *C. purpurascens* Mill.	Young leaf, petiole, flower bud, peduncle	1/2 MS + 2 mg l^{-1} 2,4-D + 0.8 mg l^{-1} 2iP + 3.7 g l^{-1} Gelrite, dark, and 25 °C	PGR-free MS medium	16 weeks, dark, 23 °C	Seyring et al. (2009)
C. coum	Leaf	1/2 MS + 0.8 mg l^{-1} 2,4-D + 0.5 mg l^{-1} BA for 4–6 months and later 1/2 MS + 2 mg l^{-1} 2,4-D + 0.8 mg l^{-1} 2iP	Six media with different PGRs, calcium, activated charcoal, and NH_4NO_3 content	4 weeks, dark, 24 °C	Prange et al. (2010b)
C. graecum, *C. mirabile*, *C. alpinum*	Cotyledon, tuber, and root from aseptic seedling	1/2 MS + 2.0 mg l^{-1} 2,4-D + 0.8 mg l^{-1} 2iP and 3.7 g l^{-1} Gelrite in dark at 24 ± 1 °C	Six media differing in PGRs, calcium, activated charcoal, and NH_4NO_3 content	8 weeks, dark, 22–24 °C	Prange et al. (2010a)

10.3.3 Preparation of Explants

Explants must be disinfected from microorganisms to provide effective and uncontaminated growth during tissue culture. For this purpose, different sterilization protocols must be developed for *Cyclamen* species. In the protocol developed by Kocak et al. (2014) for *C. persicum*, explants were washed under tap water for 20 min, then placed in 0.1 % $HgCl_2$ for 5 min, sterilized in 70 % ethanol for 1 min, rinsed in sterile distilled water (SDW), dipped in 30 % Domestos® (NaOCl, 4.5 % v/v) for 20 min, and finally rinsed five times and maintained in SDW until explant preparation.

In order to sterilize all explants from *C. hederifolium*, *C. coum*, *C. pseudibericum*, *C. mirabile*, *C. cilicium*, and *C. graecum*, explants were washed under tap water for 20 min; *C. hederifolium* was treated with 0.1 % $HgCl_2$ for 10 min; *C. pseudibericum*, *C. mirabile*, *C. cilicium*, and *C. graecum* were treated with 0.1 % $HgCl_2$ for 20 min; and *C. coum* was treated with 0.1 % $HgCl_2$ for 30 min. Then explants were washed with SDW and then treated with 0.4 % $AgNO_3$ for 10 min. Explants were rinsed three times with SDW, dipped in 70 % ethanol for 1 min, rinsed in SDW, dipped in 30 % Domestos® (NaOCl, 4.5 % v/v) for 20 min, and then finally rinsed five times and maintained in SDW until explant preparation (unpublished data).

10.3.4 Plant Growth Regulators

SE formation on cultured *Cyclamen* explants is dependent upon the exogenous application of an auxin and its concentrations. Winkelmann observed most effective SE formation from the leaves and petioles of *C. persicum* on ½ MS containing 0.8 mg/L 2iP and 2 mg/L 2,4-D. In our unpublished research, the capacity of SE of *C. cilicium*, *C. pseudibericum*, *C. hederifolium*, *C. parviflorum*, and *C. mirabile* was investigated using leaf and petiole explants. The most effective PGR concentration for both explants in *C. cilicium* was 2 mg/L 2,4-D and 1.5 mg/L 2iP. The most effective SE of *C. pseudibericum* was 2 mg/L 2iP and 2 mg/L 2,4-D for leaves and 2.5 mg/L 2iP and 1 mg/L 2,4-D for petioles. The most effective SE in *C. mirabile* and *C. hederifolium* was possible on ½ MS added with 2 mg/L 2iP and 2 mg/L 2,4-D.

Abscisic acid (ABA) has been reported to play an active role in SE through the acquisition of embryogenic competence and in developing embryo (Kikuchi et al. 2006), i.e., in facilitating somatic embryo maturation (Othmani et al. 2009; Sholi et al. 2009), improving somatic embryo quality by increasing desiccation tolerance, and preventing precocious germination in higher plants (Rai et al. 2011). ABA may serve as one of the interesting alternatives in increasing the rate of induction in cyclamen SE (Table 10.2, 3, 4).

10.4 Proteomic Analysis of Somatic Embryogenesis

Proteomic studies are a powerful tool for monitoring the physiological status of cells and tissues under specific developmental conditions (Rose et al. 2004). A proteomic analysis of zygotic and somatic embryos of *C. persicum* was performed by Winkelmann et al. (2006) in which the influence of storage protein composition of zygotic and somatic embryos on seed and germination physiology was studied. A total of 200 protein spots were detected in zygotic and somatic embryos (IPG, pH 3–10), and 26 proteins in 83 spots were identified by nano-LC-MS/MS (Bian et al. 2010). To understand the function of abundant proteins in the proteomes of somatic embryos in *Cyclamen*, identified proteins were categorized into groups either by gene ontology or by sequence and conserved domain homology with other known proteins.

10.5 Mutation Breeding

Mutation breeding has a very important and established place in agriculture. Moreover, 3,000 mutant crop varieties have been produced over the past 60 years (Mutant Variety Database, Joint FAO/IAEA Programme). Most of those varieties were created using ionizing radiation, especially gamma rays. High-energy ion beams that are generated by accelerators have been shown to have greater biological effects than gamma rays. Increasing evidence suggests that ion beams can efficiently induce mutations (Tanaka et al. 2010).

Mutation breeding can induce a forceful change in the target organism and shorten the period required for breeding new varieties (Kozgar and Khan 2012). Mutations created by irradiation, gamma rays, X-rays, and neutrons have been used mainly for mutation breeding (Sugiyama et al. 2008). Heavy-ion beam irradiation has also attracted increasing attention as a new mutation method as it induces plant mutants at a high frequency, shows broad mutation spectrum, and produces novel mutants (Tanaka 1999). Fast neutron radiation is also a technique that has not yet been applied to *Cyclamen* but may be a useful form of inducing variants (Falusi et al. 2013).

The effect of $^{12}C_6$ ion beam irradiation on SE from *Cyclamen persicum* callus was examined by Sugiyama et al. (2008) using scapes obtained from potted plants. Callus was exposed to a $^{12}C_6$ ion beam at doses ranging between 10 and 80 Gy. The fresh weight of somatic embryos produced was determined. The initial trial showed that mutant callus, somatic embryos, and plantlets could not be obtained. In a second trial, tubers were irradiated at 4–16 Gy to examine the effect of radiation dose on tuber viability. Two petal color mutants were obtained at 12 Gy indicating that mutations induced by heavy-ion beam irradiation can be a useful means of obtaining variation in flower characteristics in a short period of time.

10.6 Cryopreservation

Cryopreservation allows for the storage of living cells, tissues, or organs at ultralow temperatures, usually in liquid nitrogen at −196 °C, ceasing cell division and metabolic activity, and the preservation of important germplasm without the loss of genetic stability (Benson 2008). The storage of ornamental plant material via cryopreservation has multiple advantages including low maintenance costs, reduced space, minimum loss of materials by contamination and human error, and a longer storage period (Teixeira da Silva et al. 2015). Preserving genetic resources by cryopreservation is a beneficial way to protect endangered or wild species.

To date, two cyclamen cryopreservation protocols exist. Winkelmann et al. (2004) cryopreserved embryogenic suspension cultures of *C. persicum*. Different cryoprotectants (0.09, 0.2, 0.4, or 0.6 M sucrose or 0.4 M sorbitol) were tested during pre-culture, and 0.6 M sucrose resulted in the highest rate of regrowth (75 %). Additional pretreatment with 0.6 M sucrose and 10 % DMSO for 1 day positively affected regrowth. The optimal time of exposure to 0.6 M sucrose was 2–4 days, and regrowth rates of cryopreserved and unfrozen callus were similar.

In another study by Winkelmann et al. (2004), with the aim of producing synthetic embryos from globular somatic embryos of *C. persicum*, two encapsulation systems (conventional alginate beads and alginate hollow beads) were tested. The percentage germination of alginate beads was as high as that for control embryos (97 %), but germination was delayed. Hollow beads showed lower germination percentage (71 %). When stored at 4 °C, the viability of encapsulated and nonencapsulated somatic embryos was reduced although the speed of germination was increased by incorporating medium into the capsules in both capsule types.

10.7 Conclusion

Early research on cyclamen SE generally focused on hormonal regulation, and new methods were developed to produce a wider range of *Cyclamen* species using SE. SE studies on cyclamen were initiated in the 1990s. In addition, mutation breeding, classic breeding, identification of secondary metabolites, and molecular characterization (e.g., Çürük et al. 2015) have also been studied. SE in cyclamen usually consists of two culture steps: (1) induction of embryogenic callus and somatic embryos in medium with an auxin or with an auxin and a low concentration of cytokinin and (2) germination of somatic embryos from callus in medium without PGRs. Detailed protocols for SE of *C. persicum* are now routine (Prange et al. 2010a, b; Winkelmann 2010; Jalali et al. 2012). Although relatively widespread research has been conducted on germination, encapsulation and desiccation tolerance of somatic embryos (Winkelmann et al. 2003), genetic instability in somatic embryos (Borchert et al. 2007), and secondary SE (You et al. 2011; Jalali et al. 2012), asynchronous development and malformation (Rode et al. 2011; Jalali et al. 2012) are still problems that need to be overcome.

References

Abu-Qaoud H (2004) Direct regeneration in *Cyclamen persicum* Mill. using seedling tissues. An-Najah Univ J Res (Nat Sci) 18:147–156

Affre L, Thomson J (1999) Variation in self-fertility, inbreeding depression and levels of inbreeding in four *Cyclamen* species. J Evol Biol 12:113–122

Ando T, Murasaki K (1983) *In vitro* propagation of *Cyclamen* by the use of etiolated petioles. Tech Bull Fac Chiba Univ 32:1–5

Benson EE (2008) Cryopreservation of phytodiversity: a critical appraisal of theory & practice. Crit Rev Plant Sci 27:141–219

Bian F, Zheng C, Qu F, Gong X, You C (2010) Proteomic analysis of somatic embryogenesis in *Cyclamen persicum* Mill. Plant Mol Biol Rep 28:22–31

Borchert T, Fuchs J, Winkelmann T, Hohe A (2007) Variable DNA content of *Cyclamen persicum* regenerated via somatic embryogenesis: rethinking the concept of long-term callus and suspension cultures. Plant Cell Tissue Organ Cult 90:255–263

Carman JG (1990) Embryogenic cells in plant tissue cultures: occurrence and behavior. In Vitro Cell Dev Biol Plant 26:746–753

Çürük P, Sogut Z, Bozdoğan E, İzgü T, Sevindik B, Tagipor E, Teixeira da Silva JA, Serce S, Aka Kaçar Y, Yalcin Mendi Y (2015) Morphological characterization of *Cyclamen* sp. grown naturally in Turkey: part I. South Afr J Bot 100:7–15

Dillen W, Dijkstra I, Oud J (1996) Shoot regeneration in long-term callus cultures derived from mature flowering plants of *Cyclamen persicum* Mill. Plant Cell Rep 15:545–548

Dunwell JM, Gibbings JG, Mahmood T, Naqvi SMS (2008) Germin and germin-like proteins: evolution, structure and function. Crit Rev Plant Sci 27:342–375

Falusi OA, Abejide DR, Liman MM, Teixeira da Silva JA (2013) Effect of fast neutron irradiation (FNI) on pollen germinability of three Nigerian sesame varieties. Int J Biotechnol Res 1(6):87–90

Fukui H, Yamamoto T, Asano T, Nakamura M (1988) Effect of plant growth regulators on in vitro organogenesis of cyclamen (*Cyclamen persicum* Mill.). Res Bull Fac Agr Gifu Univ 53:139–145

Furukawa K, Kakihara F, Kato M (2001) Somatic embryogenesis in seven wild cyclamen species. J SHITA 13:270–278

Furukawa K, Kakihara F, Kato M (2002) Somatic embryos produced from aseptic seedlings of wild Cyclamen species. J SHITA 14:71–80

Grey-Wilson C (2003) *Cyclamen*: a guide for gardeners, horticulturists and botanists, Newth edn. Batsford, London

Hohe A, Winkelmann T, Schwenkel HG (2001) Development of somatic embryos of *Cyclamen persicum* Mill. in liquid culture. Gartenbauwissenschaft 66(5):219–224

Ishizaha H (2008) Interspecific hybridization by embryo rescue in genus Cyclamen. Plant Biotechnol 25:511–519

Jalali N, Naderi R, Babalar M, Mirmasoumi M (2010a) Somatic embryogenesisin Cyclamen with two explants and combinations of plant growth regulators. Hortic Environ Biotechnol 51:445–448

Jalali N, Naderi R, Teixeira da Silva J A, Babalar M, Mirmasoumi M (2010b) Influence of salt concentration of media and plant growth regulator combination on callus formation and somatic embryogenesis of *Cyclamen persicum* Mill. Floricult. Ornam Biotechnol 4(1):84–87

Jalali N, Naderi R, Shahi-Gharahlar A, Teixeira da Silva JA (2012) Tissue culture of *Cyclamen* spp. Sci Hortic 137:11–19

Jimenez VM, Bangerth F (2001) Endogenous hormone concentrations and embryogenic callus development in wheat. Plant Cell Tissue Organ Cult 67:37–46

Joint FAO/IAEA Programme, Mutant Variety Database (2015) http://mvgs.iaea.org/default.aspx

Karam NS, Al-Majathoub M (2000a) Direct shoot regeneration and microtuberization in wild *Cyclamen persicum* Mill. using seedling tissue. Sci Hortic 86:235–246

Karam NS, Al-Majathoub M (2000b) In vitro shoot regeneration from mature tissue of wild *Cyclamen persicum* Mill. Sci Hortic 86:323–333

Kawahara R, Komamine A (1995) Molecular basis of somatic embryogenesis. In: Bajaj YPS (ed) Biotechnology in agriculture and forestry, vol 30, Somatic embryogenesis and synthetic seed I. Springer, Berlin, pp 30–40

Kikuchi A, Sanuki N, Higashi K, Koshiba T, Kamada H (2006) Abscisic acid and stress treatment are essential for the acquisition of embryogenic competence by carrot somatic cells. Planta 223:637–645

Kiviharju E, Tuominen U, Tormala T (1992) The effect of explant material on somatic embryogenesis of *Cyclamen persicum* Mill. Plant Cell Tissue Organ Cult 28:187–194

Kocak M, Izgu T, Sevindik B, Tutuncu M, Curuk P, Simsek O, Kacar YA, Teixeira da Silva JA, Mendi YY (2014) Somatic embryogenesis of Turkish *Cyclamen persicum* Mill. Sci Hortic 172:26–33

Kozgar MI, Khan S (2012) Induced mutagenesis in crop plants. Biorem Biodiv Bioavail 6 (Special Issue 1):118 pp

Kreuger M, Postma E, Brouwer Y, Van Holst GJ (1995) Somatic embryogenesis of *Cyclamen persicum* in liquid medium. Physiol Plant 94:605–612

Li ZT, Dhekeny S, Dutt M, van Aman M, Tattersall J, Kelley KT, Gray DJ (2006) Optimizing *Agrobacterium*-mediated transformation of grapevine. In Vitro Cell Dev Biol Plant 42:220–227

Lyngved R, Snipen LG, Iversen TH, Hvoslef-Eide AK (2008) Influence of potential growth factors on the production of proembryogenic masses of *Cyclamen persicum* Mill. in bioreactors. Sci Hortic 118:53–59

Malik MR, Wang F, Dirpaul JM, Zhou N, Polowick PL, Ferrie AMR, Krochko JE (2007) Transcript profiling and identification of molecular markers for early microspore embryogenesis in *Brassica napus*. Plant Physiol 144:134–154

Murasaki K, Tsurushima H (1988) Improvement on clonal propagation of cyclamen in vitro by the use of etiolated petioles. Acta Hortic 226:721–724

Murashige T, Skoog F (1962) A revised medium for rapid growth and bio assays with tobacco tissue cultures. Physiol Plant 15:473–497

Otani M, Shimada T (1991) Somatic embryogenesis and plant regeneration from *Cyclamen persicum* Mill. leaf culture. Plant Tiss Cult Lett 8:121–123

Othmani A, Bayoudh C, Drira N, Marrakchi M, Trifi M (2009) Somatic embryogenesis and plant regeneration in date palm *Phoenix dactylifera* L cv. Boufeggous is significantly improved by fine chopping and partial desiccation of embryogenic callus. Plant Cell Tissue Organ Cult 97:71–79

Prange ANS, Bartsch M, Serek M, Winkelmann T (2010a) Regeneration of different Cyclamen species via somatic embryogenesis from callus, suspension cultures and protoplasts. Sci Hortic 125:442–450

Prange ANS, Serek M, Bartsch M, Winkelmann T (2010b) Efficient and stable regeneration from protoplasts of

Cyclamen coum Miller via somatic embryogenesis. Plant Cell Tissue Organ Cult 101:171–182

Rai MK, Shekhawat NS, Harish GAK, Phulwaria M, Kheta R, Jaiswal U (2011) The role of abscisic acid in plant tissue culture: a review of recent progress. Plant Cell Tissue Organ Cult 106:179–190

Rode C, Gallien S, Heintz D, Van Dorsselaer A, Braun HP, Winkelmann T (2011) Enolases: storage compounds in seeds, evidence from a proteomic comparison of zygotic and somatic embryos of *Cyclamen persicum* Mill. Plant Mol Biol 75:305–319

Rose JKC, Bashir S, Giovannoni JJ, Jahn MM, Saravanan RS (2004) Tackling the plant proteome: practical approaches, hurdles and experimental tools. Plant J 39:715–733

Ruffoni B, Semeria L, Profumo P, Bisio A (2000) *Cyclamen persicum* Mill. somatic embryos developed in suspension cultures: Histological analysis and conversion to plant. Acta Hortic 520:8390

Savona M, Ruffoni B, Giovannini A, Altamura MM (2007) *Cyclamen persicum* Mill. cv. 'Halios': somatic embryogenesis and phenotypic analysis of somatic embryo-derived plants. Acta Hortic 743:91–98

Seyring M, Ewald A, Mueller A, Haensch KT (2009) Screening for propagation suitability *in vitro* of different Cyclamen species. Electron J Biotechnol 12:4–7

Schwenkel HG, Grunewaldt J (1988) In vitro propagation of *Cyclamen persicum* Mill. Acta Hortic 226:659–662

Schwenkel HG, Winkelmann T (1998) Plant regeneration via somatic embryogenesis from ovules of *Cyclamen persicum* Mill. Plant Tissue Cult Biotechnol 4:28–34

Sharma S, Shahzad A, Teixeira da Silva JA (2013) Synseed technology – a complete synthesis. Biotechnol Adv 31(2):186–207

Sharp WR, Sondahl MR, Caldas LS, Maraffa SB (1980) The physiology of *in vitro* asexual embryogenesis. Hortic Rev 2:268–310

Sholi NJY, Chaurasia A, Agrawal A, Sarin NB (2009) ABA enhances plant regeneration of somatic embryos derived from cell suspension cultures of plantain cv. Spambia (*Musa* sp.). Plant Cell Tissue Organ Cult 99:133–140

Sugiyama M, Saito M, Ichida H, Hayashi Y, Ryuto H, Fukunishi N, Terakawa T, Abe T (2008) Biological effects of heavy-ion beam irradiation on *Cyclamen*. Plant Biotechnol 25:101–104

Takamura T, Miyajima I, Matsuo E (1995) Somatic embryogenesis of *Cyclamen persicum* Mill. 'Anneke' from aseptic seedlings. Plant Cell Rep 15:22–25

Takamura T, Tanaka M (1996) Somatic embryogenesis from the etiolated petiole of cyclamen (Cyclamen *persicum* Mill.). Plant Tissue Cult Lett 13:43–48

Tanaka A (1999) Mutation induction by ion beams in *Arabidopsis*. Gamma Field Symp 38:19–27

Tanaka A, Shikazono N, Hase Y (2010) Studies on biological effects of ion beams on lethality, molecular nature of mutation, mutation rate, and spectrum of mutation phenotype for mutation breeding in higher plants. J Rad Res 51:223–233

Teixeira da Silva JA, Kim H-Y, Engelmann F (2015) Chrysanthemum low-temperature storage, cryopreservation and synseed technology. Plant Cell Tissue Organ Cult 120:423–440

Terakawa T, Yamamura T, Murayama T (2008) Improvement of regeneration and transformation systems for *Cyclamen persicum* using somatic embryo culture. Plant Biotechnol 25:77–80

The Plant List (2015a) Primulaceae. http://www.theplantlist.org/1.1/browse/A/Primulaceae/. Last accessed 20 July 2015

The Plant List (2015b) Cyclamen. http://www.theplantlist.org/tpl1.1/search?q=Cyclamen. Last accessed 20 July 2015

Winkelmann T (2010) Clonal propagation of *Cyclamen persicum* via somatic embryogenesis. In: Jain SM, Ochatt SJ (eds) Protocols for *in Vitro* propagation of ornamental plants, methods in molecular biology. Humana Press, Dordrecht, pp 81–290

Winkelmann T, Meyer L, Serek M (2003) Maturation and desiccation of somatic embryos of *Cyclamen persicum*. Acta Hortic 612:27–34

Winkelmann T, Mobmann V, Serek M (2004) Cryopreservation of embryogenic suspension cultures of *Cyclamen persicum* Mill. Plant Cell Rep 23:1–8

Winkelmann T, Heintz D, Van Dorsselaer A, Serek M, Braun HP (2006) Proteomic analyses of somatic and zygotic embryos of *Cyclamen persicum* Mill. Reveal new insights into seed and germination physiology. Planta 224:508–519

Yamaner O, Erdag B (2008) Direct shoot formation and microtuberization from aseptic seedlings of *Cyclamen mirabile* Hildebr. Biotechnology 7:328–332

Yang X, Zhang X (2010) Regulation of somatic embryogenesis in higher plants. Crit Rev Plant Sci 29:36–57

You CR, Fan TJ, Gong XQ, Bian FH, Liang LK, Qu FN (2011) A high-frequency cyclic secondary somatic embryogenesis system for *Cyclamen persicum* Mill. Plant Cell Tissue Organ Cult 107:233–242

Zimmerman JL (1993) Somatic embryogenesis. Plant Cell 5:1411–1423

Żur I, Dubas E, Krzewska M, Janowiak F (2015) Current insights into hormonal regulation of microspore embryogenesis. Front Plant Sci 6:424

Somatic Embryogenesis and *Agrobacterium*-Mediated Genetic Transformation in *Rosa* Species

11

Yuxiao Shen, Wen Xing, Meng Ding, Manzhu Bao, and Guogui Ning

Abstract

Rose (*Rosa* spp.) is not only an important ornamental plant, but also an economical crop. It contains thousands of cultivars and has been grown throughout the world. As most rose cultivars suffer from biotic and abiotic stress, it is necessary and significant to improve the ornamental or agronomic traits. Genetic engineering provides an efficient and convenient way for overcoming these problems. Somatic embryogenesis is widely utilized for plant regeneration and genetic transformation; it is also usually utilized to perform the genetic manipulation in plant molecular breeding. Up to now, somatic embryogenesis and genetic transformation have been applied successfully in many plant species including woody and herbaceous plants. There is some information reporting transgenic roses with altered flower colors, increased disease resistance, or with modified agronomic traits. This chapter discusses somatic embryogenesis, the factors influencing somatic embryogenesis, and the protocol to realize plant regeneration via somatic embryogenesis in rose. The chapter also reviews *Agrobacterium*-mediated genetic transformation, the factors controlling transformation, and the overall protocol to obtain transgenic mediated by cyclic secondary somatic embryogenesis. This somatic embryogenesis and *Agrobacterium*-mediated genetic transformation may provide a feasible effective method to gene functional studies and gene engineering breeding in *Rosa sp.* in future.

Y. Shen • W. Xing • M. Ding • M. Bao • G. Ning (✉)
Key laboratory of Horticultural Plant Biology, Ministry of Education, College of Horticulture & Forestry Sciences, Huazhong Agricultural University, Wuhan 430070, P.R. China

College of Plant Science & Technology, Huazhong Agricultural University, Wuhan 430070, P.R. China
e-mail: 971197465@qq.com; 215519980@qq.com; 453611162@qq.com; mzbao@mail.hzau.edu.cn; whnobleman2004@aliyun.com

A. Mujib (ed.), *Somatic Embryogenesis in Ornamentals and Its Applications*,
DOI 10.1007/978-81-322-2683-3_11

Keywords
Rose • Somatic embryogenesis • Plant regeneration • Genetic transformation

11.1 Introduction

Rose, distributing widely around the world, is not only an important ornamental plant but also economical crop. It includes many species, such as *Rosa hybrida*, *Rosa rugosa*, *Rosa multiflora*, and several others. Roses, with different species, have their own particular advantages and have general applications to our life, such as medicinal functions, esthetically attractive flowers, and often with powerful fragrance. It also has some deficiencies like short flowering, thorniness, susceptible to disease, etc. The traditional breeding ways have great limitations in eliminating these weaknesses, which need a long time and high cost. Traditionally, rose improvement has depended on sexual hybridization and selection, together with identification of natural mutations. Most desirable "modern rose cultivars" do not belong to a single rose species, but are complex hybrids derived from multiple species through conventional breeding (Gudin 2000). However, breeding of roses using these traditional methods is restricted due to the limited gene pool, distant interspecific hybridization with incompatibility issue and varying ploidy among species. Furthermore, the balance of factors determining plant growth and development are commonly altered by sexual crossing and selection procedures, thereby giving rise to unpredictable progeny types (Bao et al. 2012).

The genomic era is now under way for Rosaceae. In the past decade, advances in genomics and bioinformatics of Rosaceae have provided new options to identify useful genes responsible for disease resistance, recurrent flowering, fragrance production, and flower color development (Hattendorf and Debner 2006; Ahmadi et al. 2008; Zhang et al. 2012; Verde et al. 2013). These resources also provide a range of opportunities for rose improvement by transgenic approaches. Genetic engineering technology, a highly desirable breeding strategy for the genetic improvement of plant genotype, provides an opportunity to overcome the restrictions of conventional breeding as the technique can facilitate the introduction and modification of valuable traits without altering the overall quality of the target variety within a short time frame. Therefore, there was a great excitement and expectation when the first establishment of genetic transformation system of rose was reported (Firoozabady et al. 1994). Since the first report, several genetic transformation protocols using reporter genes or genes of interest have been reported in various rose varieties (Dohm et al. 2001, 2002; Li et al. 2002b; Kim et al. 2004; Katsumoto et al. 2007; Chen et al. 2010; Vergne et al. 2010; Zakizadeh et al. 2013).

Among various in vitro tissue culture systems, somatic embryogenesis is the efficient tool for plant regeneration. It has widely been used in germplasm preservation (de Sliva et al. 2014), rapid propagation (de Wit et al. 1990; Murali et al. 1996; Bao et al. 2012), genetic transformation (Vergne et al. 2010; Xing et al. 2014a, b; Wu et al. 2015), and in vitro mutant induction (Isabel et al. 1996). Somatic embryogenesis regeneration pathway provides a more integrated and resilient material than organogenesis-obtained plants (Rout et al. 1991). Furthermore, somatic embryo cells have the totipotency, which can be easier to differentiate into complete plants, so somatic embryo usually is utilized as the better genetic transformation materials than the other plant tissues, especially for plants with high difficulty in transformation.

In this chapter, the somatic embryogenesis, the factors controlling somatic embryogenesis, and the protocol to realize plant regeneration via somatic embryogenesis in rose are reviewed. *Agrobacterium*-mediated genetic transformation, various factors influencing the success of transformation, and the protocol in obtaining transgenic are also discussed. This somatic embryogenesis and *Agrobacterium*-mediated genetic transformation may provide a feasible effective method to gene functional studies and gene engineering breeding in *Rosa sp.*

11.2 Somatic Embryogenesis and Plant Regeneration

Somatic embryogenesis is usually developed via a direct or indirect pathway. In direct somatic embryogenesis, primary embryos develop directly on the surface of organized organ or tissue. Alternatively, indirect somatic embryogenesis occurs through an intermediate step involving callus formation or a cell suspension culture (Bao et al. 2012). Somatic embryogenesis has been regarded to be an efficient method in plant regeneration and genetic engineering in several woody species (Gaj 2004; Tian et al. 2008; Prakash and Gurumurthi 2010; Vergne et al. 2010; Paul et al. 2011). Based on existing documentations, somatic embryogenesis is the potential effective method of micropropagation in *Rosa*. Successful establishment of regeneration systems by somatic embryogenesis has been reported in many rose species in the past years (Li et al. 2002a; Estabrooks et al. 2007; Vergne et al. 2010; Bao et al. 2012; Xing et al. 2014a). Although the somatic embryos have been induced, the efficiency of somatic embryogenesis is low; it is essential and necessary to establish a highly efficient system of somatic embryogenesis. A cycle of secondary somatic embryogenesis has been proved to be a convenient method to make secondary somatic embryo proliferation and germination; it also provides a good source of raw materials for further experiments. Secondary somatic embryo induction is also easier and simpler than primary embryos (Li et al. 2002a; Bao et al. 2012). In addition, several experimental analyses showed that secondary somatic embryos are the best acceptor for genetic transformation in gene engineering breeding.

11.2.1 Factors Affecting Primary Somatic Embryogenesis

Somatic embryogenesis in vitro depends tightly on different factors, such as genotypes, type of plants, environment (including light, temperature, sugars, PGRs, agar, and other additives), and other factors. All these factors affect primary somatic embryo development (shown in Fig. 11.1, step 1).

11.2.1.1 Genotype

Somatic embryogenesis has been reported to be induced in varied plant species, while most of them still limited to genotype. Even in the same genera, there is great difference in the induction frequency of somatic embryos and optimal culture conditions are needed (Norma and Chen 1989; Marchant et al. 1996; Maximova et al. 2002; Bao et al. 2012; Xing et al. 2014a). In *Rosa* species, significant difference also exists among the different cultivars. de Wit et al. (1990) reported that somatic embryogenesis was successfully induced in only 2 cultivars out of 7 different cultivars tested. Murali et al. (1996) also reported 2 cultivars of successful somatic embryogenesis although 22 different cultivars of rose were utilized. The difference in somatic embryogenesis may be due to complexity of genetic background, although reports demonstrate varied somatic embryogenesis ability with appropriate condition.

11.2.1.2 The Types of Explants

Various types of explants have been confirmed to have the capability of somatic embryogenesis in specific rose cultivar. The age and developmental stage and the physiological state of the explants significantly affect somatic embryogenesis. Generally, the explants with low level of differentiation are easier to induce somatic embryos than those with high level. The induction of somatic embryos occurs directly or indirectly from roots, leaves, stems, and immature zygotic embryos; petioles and petals are also proved to be efficient in many *Rosa* species (Rout et al. 1991; Sarasan

Fig. 11.1 Schedule map of the strategy to generate transgenic plant in *Rosa rugosa* or *Rosa bybrida* developed in our lab. Step 1: The development of primary somatic embryos from in vitro culture rose young leaf. Step 2: Repeated cycle proliferation of secondary somatic embryos. Step 3: Transformed secondary embryos harboring *GUS* reporter gene. Step 4: Plantlets regenerated from non-transformed or transformed secondary embryos. Note: Part of photos derived from our previous reports (Xing et al. 2014a, b)

et al. 2001; Li et al. 2002a; Kim et al. 2003, 2009b; Kaur et al. 2006). The comparison studies using different types of explants illustrate that young leaf is the best material for direct induction of somatic embryos (Bao et al. 2012; Xing et al. 2014a).

11.2.1.3 The Plant Growth Regulator (PGR)

It is universally acknowledged that PGRs are the essential factor of somatic embryogenesis in many plant species. Only a limited number of cases reveal somatic embryo induction directly on specific explants without PGRs (Evans et al. 1981; Kim et al. 2003). The auxin and cytokinin were commonly applied in promoting somatic embryos; the types and concentration of PGRs are, however, different in different cultivars and explants. In *Rosa* species, 2,4-D, NAA, pCPA, and 2,4,5-T are usually used as auxin, while commonly used cytokinin are BA, kinetin, and zeatin. Among those, high concentrations of 2,4-D are usually applied to induce embryonic tissues in rose (Marchant et al. 1996; Hsia and Korban 1996; Kaur et al. 2006; Kim et al. 2009a). Other studies indicated that the combination of high concentration of auxin (2,4-D) with cytokinin (BA, kinetin, etc.) can promote the formation of embryogenic calluses, while 2,4-D used alone produced somatic embryos directly on leaflet. In addition, the concentration of PGRs needed to be decreased in the maturation of somatic embryos (Bao et al. 2012; Xing et al. 2014a).

11.2.1.4 Light and Temperature

In rose somatic embryogenesis, darkness is noted to be superior for primary somatic embryo induction than light condition (Murali et al. 1996; Kim et al. 2003; Kaur et al. 2006; Estabrooks et al. 2007). Chen et al. (2014) found that red light treatment (7.2 μmol $m^{-2}s^{-1}$) generated more numbers of embryos from embryogenic calli, promoted storage protein synthesis, and increased somatic embryogenesis in China rose. Under light, explants turned brown and inclined to form calli; light also stimulated embryo proliferation and maturation of secondary somatic embryos. The optimal light condition has been confirmed is 14–16 h photoperiod, 20–40 μmol m^{-2} s^{-1} intensity. There are few reports that discussed the influence of temperature on embryogenesis. It was found that low-temperature treatment enhanced the germination of somatic embryos in some cultivars of rose (Rout et al. 1991; Roberts et al. 1995; Marchant et al. 1996).

11.2.1.5 Carbohydrate

Carbohydrate not only provides the requirements of carbon sources but also influences the osmotic pressure of the culture medium. High concentration of carbohydrate imposes a high osmotic stress, which has been reported to promote somatic embryos formation (Agarwal et al. 2004; Bao et al. 2012). Glucose and sucrose are usually used to induce somatic embryos in *Rosa sp.* Glucose promoted the formation of somatic embryos, and it is superior to sucrose, while medium supplemented with sucrose is advantageous to secondary embryo germination and the plantlet growth (Bao et al. 2012).

11.2.1.6 Basal Medium

In previous studies, MS (Murashige and Skoog basal medium), SH (Schenk and Hildebrandt basal medium), and WPM (Woody plant medium) medium were all employed as the basal medium in the induction and development of somatic embryos in *Rosa sp.* (Marchant et al. 1996; Estabrooks et al. 2007; Kim et al. 2009a, b; Bao et al. 2012). Of which, MS medium has been widely used in somatic embryogenesis, when leaflets, petioles, zygotic embryos, and immature zygotic embryos were used as explants (de Wit et al. 1990; Kim et al. 2003, 2009a; Kaur et al. 2006). The comparative analysis of MS and WPM medium indicated that somatic embryo was only induced in MS on leaflet and petiole explants (Estabrooks et al. 2007). Marchant et al. (1996) successfully induced somatic embryos when culturing petiole and root explants on SH medium, and the induction effect of SH was noted to be superior to MS. Kim et al. (2009b) also obtained embryogenic calluses and somatic embryos from root explants on SH medium. These demonstrated that SH medium is suitable for somatic embryogenesis with root explants, and MS medium is more responsive with leaflet, petiole, and zygotic embryo explants.

11.2.1.7 The Additives

Antioxidants (dithioerythritol, charcoal, silver nitrate) and amino acids (L-proline) are commonly considered to promote the induction of somatic embryos. The browning of explants decreased with the presence of antioxidants; the antioxidants also reduced somatic embryogenesis efficiency in other investigated cases (Pinto et al. 2008). L-proline promotes somatic embryogenesis, especially the primary somatic embryogenesis through promoting organic nitrogen supply and in resisting against possible stress. L-proline also promotes the formation of abnormal embryos in later stage of embryo maturation (Marchant et al. 1996; Kaur et al. 2006).

11.2.2 Maintenance and Germination of Secondary Somatic Embryos

Cycle secondary somatic embryogenesis provides an effective method to maintain embryos, growing for long term, and the secondary somatic embryos still have the ability of regeneration in rose. Li et al. (2002a) first reported the successful proliferation and germination of somatic embryos in *Rosa hybrida*. Subsequently, the proliferation and maturation of secondary embryos succeeded in many rose species (Vergne et al. 2010; Bao et al. 2012; Xing et al. 2014a). In addition, secondary somatic embryos also proved to be the optimal material for genetic transformation in rose. In our laboratory, repeated secondary somatic embryos of rose have been successfully used in genetic transformation after cycle proliferation for about 3–4 years (shown in Fig. 11.1, step 2).

PGRs play an important role in secondary somatic embryogenesis. Low concentration of auxins alone or in combination with cytokinins is suitable for somatic embryo proliferation and germination. Several studies demonstrated that auxin-free or low concentration of 2,4-D contributes to the development and germination of somatic embryos. BA is the most potent PGR for somatic embryo germination in many rose cultivars. GA_3, IAA, and TDZ are also important in stimulating the emergence of buds (Rout et al. 1991; Kintzios et al. 1999; Bao et al. 2012). Sarasan et al. (2001) indicated that BA and methyl laurate both promoted the development and increased frequency of bipolar embryo germination in hybrid rose. After shoot bud appearance, shoot elongation was observed on those medium amended with BA and NAA or PGR-free medium. Rooting was achieved by culturing on low PGR or PGR-free medium (Table 11.1).

11.2.3 Molecular Mechanism of Somatic Embryogenesis

It is generally known that somatic embryo has the cellular totipotency. Somatic embryogenesis involves dedifferentiation and redifferentiation of cells through the complex molecular interactions. The procedure of somatic embryogenesis is always accompanied by the changes of the endogenous substances and the expression of embryonic related gene. Advanced molecular techniques lead to the identification of some master genes in somatic embryogenesis with the aid of bioinformatic tools. Li et al. (2014) found that somatic embryogenesis receptor kinase (*SERK)*, arabinogalactan proteins (AGPs), and members of the WUS-related homeobox protein family may play important roles in the early stage of somatic embryogenesis of *Picea balfouriana* by transcriptome analysis. In cotton, the hormone responsive genes, signal transduction pathway (SERKs), and transcription factors (AP2/ERF, WUS, AGL15) are found associated with somatic embryogenesis (Yang and Zhang 2010). Despite the fact that many genes and pathways are involved in SE, very few genes were functionally characterized. Beside the level of PGRs, soluble protein and other endogenous substances are also altered along with somatic embryogenesis. The addition of arabinogalactan proteins with PGRs to the somatic embryogenesis medium may be very efficient in inducing or promoting somatic embryos.

Some candidate genes of somatic embryogenesis had been identified and assigned to specific chromosome regions in some plants; it lead us to summarize that genotype is the decisive factor for the induction of primary somatic embryos. The optimal environment promoting those embryo-related gene expressions is also the key factor in inducing and improving somatic embryogenesis efficiency. Although most species of plants have the potential to produce somatic embryos, the sensitivity and difficulty level of induction of somatic embryos are different. In rose plants, the embryos are induced successively demonstrating that the optimal conditions activate the expression of embryo-related genes, which later stimulate the induction of somatic embryos.

11.2.4 Efficient Protocol of Somatic Embryogenesis in *Rosa* sp.

11.2.4.1 Establishing of Sterile Shoots

In our previous experiments, the somatic embryos had been induced directly from the in vitro leaves of *R. rugosa*, *R. hybrida*, and *R. multiflora*, respectively. The sterile shoots, which provided the reliable source of leaves for induction of somatic embryos, should be proliferated first. The optimal protocols of proliferating shoots are the same for the three kinds of *Rosa sp.* developed in authors' laboratory. The robust and vigorous stems without leaves were selected and

Table 11.1 The type and concentration of PGRs used in plant regeneration of representative *Rosa* spp. via somatic embryogenesis

Species	Explants type	Embryogenic callus formation	Primary somatic embryogenesis	Secondary somatic embryogenesis	Somatic embryos germination	Shoots elongation	Rooting	References
Rosa hybrida 'Landora'	Leaves, stem	2.2 μM BA, 5.4 μM NAA, 2.2–9.0 μM 2,4-D	2.2 μM BA, 0.05 μM NAA, 0.3 μM GA_3, 600 mg /l L-proline	2.2 μM BA, 0.05 μM NAA, 0.3 μM GA_3	2.2 μM BA, 0.3 μM GA_3, 24.7 μM adenine sulfate (8 ± 1 °C)	–		Rout et al. (1991)
Rosa hybrida 'Trumpeter'	Leaves, petioles, roots	4.0 mg/l 2,4-D	5.0 mg/l 2,4-D, 300 mg/l L-proline	3.0 mg/l 2,4-D, 300 mg/l L-proline	0.1 mg/l IBA, 1.0 mg/l BA(4 ± 1 °C)	0.1 mg/l BA		Marchant et al. (1996)
Rosa hybrida 'Carefree Beauty'	Leaves, stems	1 μM or 10 μM 2,4-D	23 μM TDZ, 3 μM GA_3		PGR-free			Hsia and Korban (1996)
Rosa hybrida 'Arizona'	Petals	0.5–2.0 mg/ l BA, 1.0–2.0 mg/l kinetin, 1.0 mg/l dicamba	0.02 mg/l 2.4-D, 2.45 mg / l 2iP	0.25 mg/l BA	0.25 mg/l ABA	0.5 mg/l NAA, 100 mg/l phloroglucinol		Murali et al. (1996)
Rosa hybrida 'Soraya'	Leaves, stems	53.5 μM pCPA, 4.6 μM Kinetin	53.5 μM pCPA, 4.6 μM kinetin, L-proline		5.2 μM BA, 5.7 μM IAA	–		Kintzios et al. (1999)
Rosa hybrida 'Sumpath'	Immature zygotic embryos	1.36 μM 2,4-D, 4.44 μM BA	4.44 μM BA (BA alone, direct somatic embryos formation)		PGR-free			Kim et al. (2003)
Rosa hybrida 'Livin' Easy'	Leaflet, petiole	10 μM 2,4,5-T	4 μM 2,4,5-T		PGR-free			Estabrooks et al. (2007)
Rosa hybrida 'Charming'	root	3–11 mg/l 2,4-D	PGR-free	1.0 mg/l 2,4-D	0.5 mg/l BA		0.1 mg/l NAA	Kim et al. (2009b)
Rosa hybrida 'Samantha'	Leaves	Not this process	3.0 mg/l 2,4-D	1.0 mg/l 2,4-D, 0.05 mg/l BA	1.0 mg/l TDZ, 0.5 mg/l BA	0.05 mg/l NAA		Bao et al. (2012)
Rosa rugosa	Mature zygotic embryos	2.26 μM2,4-D	0.5 mg/l NAA+0.1 mg /l BA		PGR-free			Kim et al. (2009a)

(continued)

Table 11.1 (continued)

Species	Explants type	Embryogenic callus formation	Primary somatic embryogenesis	Secondary somatic embryogenesis	Somatic embryos germination	Shoots elongation	Rooting	References
Rosa rugosa 'Bao white'	Leaflet	Not this process	4.0 mg/l 2,4-D, 0.05 mg/l kinetin	1.0 mg/l 2,4-D, 0.01 mg/l BA	1.0 mg/l BA	0.5 mg/l BA, 0.01 mg/l NAA	0.5 mg/l IBA	Xing et al. (2014a)
Rosa bourboniana	Immature zygotic embryos	15 μM 2,4-D	15 μM 2,4-D	15 μM 2,4-D, 800 mg/l L-proline	5.0 μM BA	10 μM IBA (liquid medium)		Kaur et al. (2006)
Rosa chinensis 'Old Blush'	Leaves	8.06 or 10.74 μM NAA, 18.24 μM zeatin	8.06 μM NAA, 6.84 μM zeatin and 3 μM GA_3		0.1 g/ l Fe-EDDHA, 0.02 μM NAA, 0.29 μM GA_3, 4.44 μM BA		11.4 μM IAA	Vergne et al. (2010)

Note: *NAA* a-Naphthalene acetic acid, *2,4-D* 2,4-dichlorophenoxyacetic acid, *2,4,5-T* 2,4,5-trichlorophenoxyacetic acid, *TDZ* Thidiazuron, *BA* 6-benzyladenine, *IAA* 3-indoleacetic acid, *IBA* indole-3-butyric acid, GA_3 gibberellic acid, *ABA* abscisic acid, *pCPA* p-chlorophenoxyacetic, *2iP* N^6-(2-isopentenyl) adenine

washed under running water for 30 min. The stems were cut into single nodal sections (approx. 4 cm in length), then subsequently immersed in 70 % (v/v) ethanol for 30 s, 0.1 % (w/v) mercuric chloride for 12 min, and rinsed three times with sterile distilled water. One (1) cm length of stem fragment was cut from each end of the nodal sections and discarded, and the remaining explants were placed onto MS (Murashige and Skoog basal medium) medium supplemented with 0.5 mg/l BA, 0.01 mg/l NAA, 30 g/l sucrose, and 7.5 g/l agar. After sprouting, the shoots were excised and subcultured every 4 weeks on the same fresh medium to maintain the growth and proliferation of shoots. All the cultures were incubated at 24±2 °C, under a 14 h photoperiod with a light intensity of 20–40 μmol m^{-2} s^{-1} provided by cool white fluorescent lights (detail reference to Bao et al. 2012).

11.2.4.2 Primary Somatic Embryogenesis

Unexpanded leaves, applied for primary somatic embryo induction directly, were excised from the in vitro culture sterile shoots. Then, the compound leaves were divided into three or five separate leaflets with the respective petioles present. All the abaxial surface of leaflets was placed down onto the MS medium containing 30 g/l glucose and 3.0 mg/l Phytagel. The optimal medium supplemented with 2,4-D alone stimulated somatic embryos formation for *R. hybria* and *R. multiflora*. The medium with combinations of kinetin and 2,4-D is suitable for the induction of somatic embryos of *R. rugosa* directly. Supplementation of kinetin (0.05 mg/l) and 2,4-D increased the frequency of somatic embryo induction than using 2,4-D alone for *R. rugosa*. However, the high concentration of 2,4-D (3.0–4.0 mg/l) is indispensability in the formation of somatic embryos of the three *Rosa spp.* from leaflets. All the cultures were incubated at 24±2 °C, under darkness for 2 months, and subcultured onto the same fresh medium every 4–5 weeks (shown in Fig. 11.1, step1).

11.2.4.3 Proliferation and Maintenance of Secondary Somatic Embryos

Although the rate of primary somatic embryo initiation was low (4–5 %), secondary somatic embryos could be efficiently induced from primary embryos. Cultured on MS medium supplemented with 30–45 g/l glucose, 1.0 mg/L 2,4-D, and 0.01–0.05 mg/L BA, secondary somatic embryos at all stages like globular, heart, and cotyledonary somatic embryos were observed without browning and abnormal morphology. Generally, the added BA promoted secondary somatic embryo proliferation, but usually the proliferated somatic embryos turned green and germinated as the concentration of BA increasing. In addition, it was observed that cyclic secondary somatic embryos developed vigorously and proliferated well in the weak light (light intensity of 10–20 μmol m^{-2} s^{-1}). The strong light causes a negative effect on embryo proliferation like easier germination and abnormal embryos in *R. multiflora* and *R. Rugosa.* All the cultures were maintained for about 3–4 years in this medium by continuously subculturing on same fresh medium every 4–5 weeks at 24±2 °C (shown in Fig. 11.1, step 2).

11.2.4.4 Secondary Somatic Embryos Germination and Plantlets Regeneration

Maturated and germinated somatic embryos were observed by transferring the somatic embryos to germination medium. In the three *Rosa* species, BA is the essential PGR for the germination of somatic embryos. The combination of 0.5 mg/l BA and 1.0 mg/l TDZ promoted the germination of somatic embryos than used BA alone in *R. hybrida* and *R. multiflora* species. In medium containing 1/2 MS and BA alone, the highest germination frequencies of somatic embryos were obtained in *R. rugosa.* When the cotyledonary secondary somatic embryos were transferred to the germination medium, the cotyledons or the whole embryo turned green and showed sprouting. The embryo regenerated plantlets of *Rosa sp.* were cultured under light for 2–3 months on the shoots elongation and rooting medium (Table 11.2), later transferred to the mixture soil consisting of turf soil + garden soil + sand (2:2:1; v/v/v) and then grown in the greenhouses.

Table 11.2 MS medium added with various PGR levels for somatic embryo germination and plant development in rose

Species	Medium type	Composition
R. rugosa 'Bao white'	Germination medium	1/2 MS + 1.0 mg/L BA + 30 g/L glucose + 3.0 g/L Phytagel
	Shoots elongation medium	MS + 0.5 mg/L BA + 0.01 mg/L MAA + 30 g/L glucose + 7.5 g/L agar
	Rooting medium	1/2MS + 0.5 mg/L IBA + 30 g/L glucose + 7.5 g/L agar
R. hybrida 'Samantha'	Germination medium	MS + 0.5 mg/L BA + 1.0 mg/L TDZ + 30 g/L glucose + 3.0 g/L Phytagel
	Shoots elongation medium	1/2 MS + 0.05 mg/L NAA + 30 g/L glucose + 7.5 g/L agar
	Rooting medium	
R. multiflora var. carnea	Germination medium	MS + 0.5 mg/L BA + 1.0 mg/L TDZ + 30 g/L glucose + 3.0 g/L Phytagel
	Shoots elongation medium	MS + 0.5 mg/L BA + 0.01 mg/L NAA + 30 g/L glucose + 3.0 g/L Phytagel
	Rooting medium	1/2MS + 0.5 mg/L IBA + 30 g/L glucose + 7.5 g/L agar

11.3 Genetic Transformation via Somatic Embryos

Genetic transformation is a convenient and efficient method for the studies of gene function, and it improves the specific characters without changing other traits. In the past studies, the effective systems of genetic transformation had been established in many herbal and woody plants.

11.3.1 Genetic Transformation in *Rosa* sp.

The *Agrobacterium*-mediated transformation is proved to be the most efficient and successful method used widely. In *Rosa* species, the transgenic plants have been obtained from *R. rugosa, R. hybrida, R. chinensis* and other rose through *Agrobacterium*-mediated transformation (Table 11.3). Though microprojectile bombardment (Marchant et al. 1998a) and PEG-mediated protoplast transformation (Schum and Hofmann 2001) were documented, the efficiency of transformation is low. *Agrobacterium*-mediated transformation is proved to be the most effective approach for rose.

Marchant et al. (1998a) successfully performed transformation using embryogenic callus of rose by microprojectile bombardment. It resulted in production of transgenic plants and the efficiency of transformation is 15.6 % higher than the efficiency of *Agrobacterium*-mediated transformation. While in this experiment, 2–6 copies of target gene had been inserted to the rose genome. The first rose *Agrobacterium*-mediated transformation was reported by Firoozabady et al. (1994) via somatic embryogenesis pathway in *Rosa hybrida* cv. Royalty. Since then, the successful establishment of *Agrobacterium*-mediated transformation system had been reported continuously in rose. In addition, the

Table 11.3 Transformation systems used in representative rose plant

Species	Explants	Agrobacterium strains	Transformed gene	Efficiency of transformation	References
R. hybrida 'Careful Beauty'	Embryogenic callus	GV3101 (*A. tumefaciens*)	*GUS*	9 %	Li et al. (2002b)
R. hybrida 'Royalty'	Embryogenic callus	LBA4404 (*A. rhizogenes*)	*GUS*	Over 100 transgenic plants obtained	Firoozabady et al. (1994)
R. hybrida 'Anny'	Somatic embryos	LBA 4404 (*A. tumefaciens*)	*Phosphinothricin acetyltransferase (PAT)*	10 transgenic plantlets obtained	Borissova et al. (2005)
R. hybrida 'Saltze Gold'				3 transgenic plantlets obtained	
R. hybrida 'Tineke'	Embryogenic callus	LBA 4404 (*A. tumefaciens*)	*GFP*	6.6–12.6 % calluses with GFP expression	Kim et al. (2004)
R. hybrida 'Linda'	Embryogenic callus	AGL1 (*A. tumefaciens*)	P_{SAG12}-*ipt*	10 %	Zakizadeh et al. (2013)
Rosa chinensis 'Old Blush'	Somatic embryos	EHA105 (*A. tumefaciens*)	*GUS*	3–9 %	Vergne et al. (2010)
R. rugosa 'Bao white'	Somatic embryos	EHA105 (*A. tumefaciens*)	*GUS*	11.40 %	Xing et al. (2014a)
			FT	Transgenic plantlets obtained	Xing et al. (2014b)
R. canina	PLBs (protocorm-like bodies)	GV3101 (*A. tumefaciens*)	*GUS*	Transgenic plantlets obtained	Bi et al. (2012)

method of PEG-mediated protoplast transformation had been reported by Schum and Hofmann (2001) using *GFP*-gene as reporter and obtained stable transformation. Currently, many attempts have been made to improve the efficiency of transformation, and some target genes have been inserted to the rose genome in improving the agronomic traits.

11.3.2 Factors Affecting *Agrobacterium*-Mediated Transformation

The two aspects are mainly involved in *Agrobacterium*-mediated transformation: (1) transforming ability of a gene into target genome by *Agrobacterium* and (2) regenerating ability of a transformed cell (Juturu et al. 2015). Factors involved in these two steps include genotype, the type, and stage of acceptors, the *Agrobacterium* strain, the method of regeneration (somatic embryogenesis or organogenesis), pre-cultivation and cocultivation method, infection time, the concentration of bacteria, and additives (such as osmotic treatment reagent, acetosyringone). Establishing a high-efficient plant regeneration system is the important prerequisite of successful genetic transformation, and then finding out the optimal combination condition of the above factors is also helpful to improve the efficiency of transformation.

11.3.2.1 Type of Acceptors

Somatic embryos and embryogenic callus are the best receptors applied for genetic transformation because of easy regeneration and less mosaic nature. In fact, the plant tissues (leaf, stem, petiole, root, flower, etc.) have been used for transformation, but attempts were found unsuccessful in obtaining normal transgenic plants with stable expression of target gene (Firoozabady et al. 1994; Li et al. 2002b). Li et al. (2002b) first reported on developing a transformation protocol for rose using a secondary somatic embryogenesis system. In this experiment, embryogenic callus, leaf, and undifferentiated callus were used as explants for transformation. Ultimately, the groups obtained 10 transgenic rose plants via embryogenic callus transformation. The result clearly indicated that the use of embryogenic callus as explant is superior to leaf, callus, and other tissues in rose.

11.3.2.2 Antibiotics in Transformation

Antibiotics like cephalosporin, kanamycin, and hygromycin are commonly used for select resistant plants, and the concentration of antibiotics affected transformation efficiency (Xing et al. 2014a, b). In rose transformation, antibiotics may have an inhibitory effect on callus induction from plant organs under long exposure of antibiotics selection pressure, and the developed putative transgenic callus may lose the ability to differentiate into embryogenic callus and unable to convert into a plantlet. Li et al. (2002b) noted that carbenicillin completely inhibited shoot regeneration from leaf, undifferentiated callus, and embryogenic callus in *Rosa hybrida* 'Carefree Beauty'. Using suitable carbenicillin and cefotaxime antibiotics foe specific rose species, Vergne et al. (2010) established a transformation system of *R. chinensis* using primary somatic embryo explants, and transformation efficiency ranged between three and nine percent. Xing et al. (2014a) first developed a stable and efficient *Agrobacterium*-mediated transformation protocol using secondary somatic embryo explants in *R. rugosa*; the transformation efficiency was 11.4 %. The results indicated that secondary somatic embryos as explants could improve the efficiency of transformation with its potent ability of regeneration.

11.3.2.3 *Agrobacterium* Strains, Concentration, and Infection Time

The use of *Agrobacterium* strains like GV3101 (Li et al. 2002b), LBA4404 (Borissova et al. 2005), AGL1 (Zakizadeh et al. 2013), and EHA105 (Xing et al. 2014a, b) has been reported in the development of transgenic rose (Table 11.3). These results have shown that the diversity of *Agrobacterium* strains and plasmids have no significant difference in transformation. The *Agrobacterium* density and cocultivation time played an important role in the frequency of transformation. The high frequency of resistant embryos was obtained with *Agrobacterium* density (OD_{600}) 0.2~0.6, of which the somatic embryos were inoculated for 30–60 min in bacterial suspension, then cocultivated for 48 h (Li et al. 2003; Borissova et al. 2005; Xing et al. 2014a). The long time of inoculation and coculture with bacteria may lead to decrease transformation efficiency. In rose transformation, the cocultivation medium was usually supplemented with acetosyringone (AS) to enhance the transformation efficiency. There are many studies that show that phenolic substance like AS can activate the vir gene and guide the T-DNA transferred to plant genome.

Many rose cultivars are known to be recalcitrant to transformation, so finding a solution to improve the transformed efficiency is necessary. Micro trauma treatment (Georgiev et al. 2011), drying treatment (Pollin et al. 2006), antioxidants (Dan 2008), surfactant utilization (Liu et al. 2008), and other methods had been reported to improve transformed efficiency in many species. Till today, all transformation protocols published in roses proved that embryogenic callus and somatic embryos both are appropriate explants for transgenic plant production. No matter which explants are used, induction of secondary somatic embryogenesis is especially important for transformation and can be applied for a wide range of rose species.

11.3.3 *Agrobacterium*-Mediated Transformation via Somatic Embryos in *Rosa* spp.

In our present study, an efficient *Agrobacterium*-mediated transformation system via repetitive secondary somatic embryogenesis had been developed in *R. hybrida* 'Samantha' and *R. rugosa* 'Bao white'(Fig. 11.1). The effective methods of genetic transformation are recommended as follows (Ning et al. 2012; Xing et al. 2014a).

11.3.3.1 Bacterial Preparation

Bacterial cells are cultured in liquid LB (Luria-Bertani) medium supplemented with antibiotic (kanamycin is usually used) and then incubated in shaker incubator at 28 °C and 200 rpm for about 8–12 h until the optimal OD_{600} of bacteria (0.4–0.6). Bacterial cells are harvested in liquid MS supplemented with 100–200 μM/L acetosyringone and then incubated about 2 h.

11.3.3.2 *Agrobacterium* Infection

The globular stage somatic embryo clumps were immersed in bacterial suspension liquids for 40–60 min. Subsequently after dried on a sterilized filter paper, the tissues were cocultivated with *Agrobacterium* for 2–3 d in the cocultivation medium under darkness. The cocultivation medium is the secondary somatic embryo proliferation medium supplemented with 100–200 μM/L acetosyringone.

11.3.3.3 Selection and Regeneration

After cocultivation, somatic embryos were transferred into selection and proliferation medium. After 2 months slection on proliferation mediumon, the antibiotic-resistant somatic embryos were seperatedand transferred to germination medium for 2 months. The antibiotic-resistant shoots were subcultured onto selection and shoot elongation medium for 1 month. Vigorously growing antibiotic-resistant shoots were transferred to selection rooting medium for 1 month for rooting. When the positive plantlets rooted, transferred them to the soil and grown in the greenhouse. During selection process, explants were subcultured to a fresh selection medium once every 2–4 weeks. The medium, used for selection and regeneration, were the same as secondary somatic embryo proliferation, germination, shoot elongation, and root induction medium except for supplementing with different concentrations of antibiotic.

11.3.3.4 Identification of Transgenic Plants

GUS assay, GFP expression assay, and Southern blotting are efficient and frequently used methods in establishing a system of genetic transformation. GUS assay can identify the positive plants directly through histochemical stain when *GUS* is used as the reporter gene. If *GFP* is used as reporter gene, the positive plants could be examined by observation under a fluorescence microscope. Southern blotting has been used to detect the copy number of target gene in plants. The detailed process of identification refers to our previous articles (Xing et al. 2014a, b).

11.3.4 Transgenic Research in *Rosa* spp.

Breeding of rose cultivars with fungal disease resistance is the dream of all the rose breeders. Up to now, several successful rose improvements against fungal disease resistance was published, which focused primarily on black spot and powdery mildew diseases. The black spot disease, powdery mildew, downy mildew, and rust are most widespread and pernicious fungal diseases in *Rosa* species. Marchant et al. (1998b) successfully applied transformation procedure to introduce a rice gene, encoding a basic (class I) chitinase into the black spot-susceptible rose (*Rosa hybrida* L.) cv. Glad Tidings. Bioassays demonstrated that expression of the chitinase transgene reduced the severity of black spot development by 13–43 % (Marchant et al. 1998b). Subsequently, according to *Agrobacterium*-mediated transformation protocol, different combinations of three antifungal resistance genes encoding a class 1 chitinase and a type 1 ricosome inhibiting protein from barley as well as the T4 lysozyme gene were introduced into two garden rose cultivars by Dohm et al. (2001). The results showed that the severity of black spot disease development was reduced by 60 %. On the other hand, in the study of powdery mildew resistance, an antimicrobial protein gene, *Ace-AMP1*, was introduced into *Rosa hybrida* cv. Carefree Beauty via *Agrobacterium*-mediated transformation by Li et al. (2003). Transgenic rose lines inoculated with conidial spores showed enhanced resistance to powdery mildew (Li et al. 2003).

The suitable growth temperature for roses was 20–28 °C. Defoliation will happen and many branches will perish at low temperature. Chen et al (2010) successfully introduced the *MtDREB1C* gene, isolated from *Medicago truncatula*, driven by the Arabidopsis rd29A promoter into China rose. It was observed that the transgenic line exhibited markedly increased freezing tolerance in comparison to untransformed controls, without any obvious morphological or developmental abnormality.

Although modern roses (*Rosa hybrida*) resulted from extensive hybridization of wild rose species have various flower colors, violet to blue flower varieties are not created by rose breeders through traditional methods due to the absence of delphinidin-based anthocyanins (Holton and Tanaka 1994). Now, genetic transformation technology provided the opportunity for breeders to create blue roses. By downregulating the endogenous dihydroflavonol 4-reductase (*DFR*) gene and overexpressing the *Iris × hollandica DFR* gene and the viola *F3'5'H* gene in rose, Katsumoto et al. (2007) created roses with blue hues flowers which could not be achieved by hybridization breeding.

Rose was the most popular cut flower. Leaf senescence and the release of ethylene are some of the significant problems for cut flower preservation during postharvest handling of roses. Transformation of rose plants with fusion gene, ligating SAG12 promoter from *Arabidopsis* to the *ipt* gene from *A. tumefaciens* (*PSAG12-ipt*), resulted in significantly delayed leaf senescence and increased ethylene resistance due to specific upregulation of the *ipt* gene under senescence promoting conditions (Zakizadeh et al 2013).

11.4 Future Perspectives

In recent years, significant progress has been made in the area of rose molecular biology. A large number of genomic, transcriptional, and functional annotated sequences can be obtained from the genome database for Rosaceae (GDR) (http://www.rosaceae.org/node/1). In addition to these, several resources of roses have been developed, including the expressed sequence tag (EST) database, cDNA libraries, and microarrays (Ogata et al. 2005; Foucher 2009; Hirata et al. 2012). These resources provide rose breeders and researchers with a range of opportunities for the rose improvement as well as the analysis of gene function by transgenic approaches. Using efficient transformation systems described above to develop roses with varies agronomic traits appears to be a feasible approach. Transgenic approaches are likely to become routine for the elucidation of gene function by overexpression, suppression, or complementation testing in the appropriate genetic background (Yamada et al. 2012).

Up to now, almost all the successful development of transformation systems published focused on rose cultivars. In *Rosa*, there are many other economically important species, such as *Rosa rugosa* and *Rosa damasica* which are the major material of rose essential oil as well as *Rosa laevigata* and *Rosa roxburghii*, which are important medical materials. To our best acknowledge, the investigation of transformation system of these species have not yet been reported. Therefore, there is great expectation when the successful transformation system of many other rose species will be established, and much work is needed to establish the transformation system for specific rose so that it can be useful to a range of genotypes. Somatic embryogenesis, widely used in transformation system of rose cultivars (Kim et al. 2004; Chen et al. 2010; Katsumoto et al. 2007; Zakizadeh et al. 2013), appears to be a feasible approach to achieve the regeneration of transgenic plants in other rose species.

To fulfill the economical and scientific research requirements, genetic engineering is taking an important role in enhancing rose tolerance to biotic and abiotic stresses, adding desirable traits and determining gene function. In the near future, rose genetic transformation will not be a difficult task. Along with this, evaluation of the potential risk of transgenic plants for environment needs to be made. When the consumers accept genetically engineered roses, transgenic rose commercialization will become a reality, which will introduce more desirable traits into commercially important *Rosa* species.

References

Agarwal S, Kanwar K, Sharma DR (2004) Factors affecting secondary somatic embryogenesis and embryo maturation in *Morus alba* L. Sci Hortic 102:359–368

Ahmadi N, Mibus H, Serek M (2008) Isolation of an ethylene-induced putative nucleotide laccase in miniature roses (*Rosa hybrida* L.). J Plant Growth Regul 27:320–330

Bao Y, Liu GF, Shi XP, Xing W, Ning GG, Liu J, Bao MZ (2012) Primary and repetitive secondary somatic embryogenesis in *Rosa hybrid* 'Samantha'. Plant Cell Tiss Org Cult 109:411–418

Bi L, Liu F, Dong A (2012) Establishment of genetic transformation system in protocorm like-body (PLB) of *Rosa canina*. Acta Agric Nucl Sin 26:270–274

Borissova A, Hvarleva T, Bedzhov I, Kondakova V, Atanassov A, Atanassov I (2005) *Agrobacterium* mediated transformation of secondary somatic embryos from *Rosa hybrida* L. and recovery of transgenic plants. Biotechnol Biotechnol Equip 19:70–74

Chen R, Lv J, Liu R, Xiong X, Wang T, Chen S, Guo L, Wang H (2010) DREB1C from *Medicago truncatula* enhances freezing tolerance in transgenic *M. truncatula* and China Rose (*Rosa chinensis* Jacq.). J Plant Growth Regul 60:199–211

Chen JR, Wu L, Hu BW, Yi X, Liu R, Deng ZN, Xiong XY (2014) The influence of plant growth regulators and light quality on somatic embryogenesis in China Rose (*Rosa chinensis* Jacq.). J Plant Growth Regul 33:295–304

Dan Y (2008) Biological functions of antioxidants in plant transformation. In Vitro Cell Dev Biol Plant 44:149–161

Dohm A, Ludwig C, Schilling D, Debener T (2001) Transformation of roses with genes for antifungal proteins. Acta Hortic 547:27–33

Dohm A, Ludwig C, Schilling D, Debener T (2002) Transformation of roses with genes for antifungal protein to reduce their susceptibility to fungal diseases. Acta Hortic 572:105–111

Estabrooks T, Browne R, Dong ZM (2007) 2, 4, 5-Trichlorophenoxyacetic acid promotes somatic embryogenesis in the rose cultivar 'Livin' Easy' (*Rosa sp.*). Plant Cell Rep 26:153–156

Evans DA, Sharp WR, Flick CE (1981) Growth and behavior of cell cultures. In: Thorp TA (ed) Plant tissue culture – methods and application in agriculture. Academic Press, New York, pp 45–113

Firoozabady E, Moy Y, Courtney-Gutterson N, Robinson K (1994) Regeneration of transgenic rose (*Rosa hybrida*) plants from embryogenic tissue. Nat. Biotech 12:609–613

Foucher F (2009) Functional genomics in Rose, genetics and genomics of *Rosaceae*. Plant Genet Genomics Crop Model 6:381–392

Gaj MD (2004) Factors influencing somatic embryogenesis induction and plant regeneration with particular reference to *Arabidopsis thaliana* (L.) Heynh. Plant Growth Regul 43:27–47

Georgiev MI, Ludwig-Müller J, Alipieva K, Lippert A (2011) Sonication-assisted *Agrobacterium* rhizogenes mediated transformation of *Verbascun xanthophoeniceum* Griseb. for bioactive metabolite accumulation. Plant Cell Rep 30:859–866

Gudin S (2000) Rose: genetics and breeding. Plant Breed Rev 17:59–189

Hattendorf A, Debner T (2006) Cloning and characterization of resistance gene analogs from roses. Abstr Int Rosaceae Genomics Conf 3:72

Hirata H, Ohnishi T, Ishida H, Tomida K, Sakai M, Hara M, Watanabe N (2012) Functional characterization of aromatic amino acid aminotransferase involved in 2-phenylethanol biosynthesis in isolated rose petal protoplasts. J Plant Physiol 169:444–451

Holton TA, Tanaka Y (1994) Blue roses-a pigment of our imagination? Trends Biotechnol 12:40–42

Hsia C, Korban SS (1996) Organogenesis and somatic embryogenesis in callus cultures of *Rosa hybrida* and *Rosa chinensis minima*. Plant Cell Tiss Org Cult 44:1–6

Isabel N, Boivin R, Levasseur C, Charest P, Bousquet J, Tremblay FM (1996) Occurrence of somaclonal variation among somatic embryo-derived white spruces (*Picea glauca*, Pinaceae). Am J Bot 83:1121–1130

Juturu VN, Mekala GK, Kirti PB (2015) Current status of tissue culture and genetic transformation research in cotton (*Gossypium spp.*). Plant Cell Tiss Org Cult 120:813–839

Katsumoto Y, Fukuchi-Mizutani M, Fukui Y, Brugliera F, Holton TA, Karan M, Nakamura N, Yonekura-Sakakibara K, Togami J, Pigeaire A, Tao G, Nehra NS, Lu C, Dyson BK, Tsuda S, Ashikari T, Kusumi T, Mason JG, Tanaka Y (2007) Engineering of the rose flavonoid biosynthetic pathway successfully generated blue-hued flowers accumulating delphinidin. Plant Cell Physiol 48:1589–1600

Kaur N, Pati PK, Sharma M, Ahuja PS (2006) Somatic embryogenesis from immature zygotic embryos of *Rosa bourboniana* Dsep. In Vitro Cell Dev Biol Plant 42:124–127

Kim SW, Oh SC, Liu JR (2003) Control of direct and indirect somatic embryogenesis by exogenous growth regulators in immature zygotic embryo cultures of rose. Plant Cell Tiss Org Cult 74:61–66

Kim CK, Chung JD, Park SH, Burrell AM, Kamo KK, Byrne DH (2004) *Agrobacterium tumefaciens* mediated transformation of *Rosa hybrida* using the green fluorescent protein (*GFP*) gene. Plant Cell Tiss Org Cult 78:107–111

Kim SW, Oh MJ, Liu JR (2009a) Somatic embryogenesis and plant regeneration in zygotic embryo explant cultures of rugosa rose. Plant Biotechnol Rep 3:199–203

Kim SW, Oh MJ, Liu JR (2009b) Plant regeneration from the root-derived embryonic tissues of *Rosa hybrida* L. cv. Charming via a combined pathway of somatic embryogenesis and organogenesis. Plant Biotechnol Rep 3:341–345

Kintzios S, Manos C, Makri O (1999) Somatic embryogenesis from mature leaves of rose (*Rosa sp.*). Plant Cell Rep 18:467–472

Li X, Krasnyanski S, Korban SS (2002a) Somatic embryogenesis, secondary somatic embryogenesis, and shoot organogenesis in *Rosa*. Plant Physiol 159:313–319

Li X, Krasnyanski S, Korban SS (2002b) Optimization of the *uidA* gene transfer into somatic embryos of rose via *Agrobacterium tumefaciens*. Plant Physiol Biochem 40:453–459

Li XQ, Gasic K, Cammue B, Broekaert W, Korban SS (2003) Transgenic rose lines harboring an antimicrobial protein gene, *Ace-AMP1*, demonstrate enhanced resistance to powdery mildew (*Sphaerotheca pannosa*). Planta 218:226–232

Li Q, Zhang S, Wang J (2014) Transcriptome analysis of callus from *Picea balfouriana*. BMC Genomics 15:553

Liu SJ, Wei ZM, Huang JQ (2008) The effect of co-cultivation and selection parameters on Agrobacterium mediated transformation of Chinese soybean varieties. Plant Cell Rep 27:489–498

Marchant R, Davey MR, Lucas JA, Power JB (1996) Somatic embryogenesis and plant regeneration in Floribunda rose (*Rosa hybrida* L.) cvs. Trumpeter and Glad Tidings. Plant Sci 120:95–105

Marchant R, Davey MR, Lucas JA, Lamb CJ, Dixon RA, Power JB (1998a) Expression of a chitinase transgene in rose (*Rosa hybrida* L.) reduces development of blackspot disease (*Diplocarpon rosae* Wolf). Mol Breed 4:187–194

Marchant R, Power JB, Lucas JA, Davey MR (1998b) Biolistic transformation of rose (*Rosa hybrida* L.). Ann Bot 81:109–114

Maximova SN, Alemanno L, Young A, Ferriere N, Traore A, Guiltinan MJ (2002) Efficiency, genotypic variability, and cellular origin of primary and secondary somatic embryogenesis of *Theobroma cacao* L. In Vitro Cell Dev Biol Plant 38:252–259

Murali S, Sreedhar D, Lokeswari TS (1996) Regeneration through somatic embryogenesis from petal-derived calli of *Rosa hybrida* L. Arizona (hybrid tea). Euphytica 91:271–275

Ning GG, Xiao X, Lv HY, Li X, Zuo Y, Bao MZ (2012) Shortening tobacco life cycle accelerates functional gene identification in genomic research. Plant Biol 14:934–943

Norma LT, Chen X (1989) Genotype specificity of the somatic embryogenesis response in cotton. Plant Cell Rep 8:133–136

Ogata J, Kanno Y, Itoh Y, Tsugawa H, Suzuki M (2005) Anthocyanin biosynthesis in roses. Nature 435:757–758

Paul S, Dam A, Bhattacharyya A, Bandyopadhyay TK (2011) An efficient regeneration system via direct and indirect somatic embryogenesis for the medicinal tree *Murraya Koenigii*. Plant Cell Tiss Org Cult 105:271–283

Pinto G, Silva S, Park YS, Neves L, Arauujo C, Santos C (2008) Factors influencing somatic embryogenesis induction in Eucalyptus globulus Labill.: basal medium and anti-browning agents. Plant Cell Tiss Org Cult 95:79–88

Pollin LD, Liang H, Rothrock RE, Nishii M, Diehl DL, Newhouse AE, Nairn CJ, Powell WA, Maynard CA (2006) *Agrobacterium*-mediated transformation of American chestnut (*Castanea dentata* (Marsh.) Borkh.) somatic embryos. Plant Cell Tiss Org Cult 84:69–78

Prakash MG, Gurumurthi K (2010) Effects of type of explant and age, plant growth regulators and medium strength on somatic embryogenesis and plant regeneration in *Eucalyptus camaldulensis*. Plant Cell Tiss Org Cult 100:13–20

Roberts AV, Yoloya K, Walker S, Mottley J (1995) Somatic embryogenesis in *Rosa spp*. In: Jain S, Gupta P, Newton R (eds) Somatic embryogenesis in woody plants, vol 2. Kluwer, Dordrecht, pp 277–289

Rout GR, Debata BK, Das P (1991) Somatic embryogenesis in callus culture of *Rosa hybrida* L. cv. Landora. Plant Cell Tiss Org Cult 27:65–69

Sarasan V, Roberts AV, Rout GR (2001) Methyl laurate and 6-benzyladenine promote the germination of somatic embryos of a hybrid rose. Plant Cell Rep 20:183–186

Schum A, Hofmann K (2001) Use of isolated protoplasts in rose breeding. Acta Hortic 547:35–44

Sliva JAT, Zeng S, Galdiano RF, Dobranszki J, Cardoso JC, Vendrame WA (2014) In vitro conservation of *Dendrobium* germplasm. Plant Cell Rep 33:1413–1423

Tian CW, Chen Y, Zhao XL, Zhao LJ (2008) Plant regeneration through protocorm-like bodies induced from rhizoids using leaf explants of *Rosa spp*. Plant Cell Rep 27:823–831

Verde I, Abbott AG, Scalabrin S, Jung S, Shu S, Marroni F (2013) The high-quality draft genome of peach (*Prunus persica*) identifies unique patterns of genetic diversity, domestication and genome evolution. Nat Genet Nat Genet 45:1–10

Vergne P, Maene M, Gabant G, Chauvet A, Debener T, Bendahmane M (2010) Somatic embryogenesis and transformation of the diploid *Rosa chinensis* cv. Old Blush. Plant Cell Tiss Org Cult 100:73–81

Wit JC, Esendam HF, Honkanen JJ, Tuominen U (1990) Somatic embryogenesis and regeneration of flowering plants in rose. Plant Cell Rep 9:456–458

Wu J, Liu C, Seng S, Khan MA, Sui J, Gong B, Liu C, Wu C, Zhong X, He J, Yi M (2015) Somatic embryogenesis and *Agrobacterium* mediated transformation of *Gladiolus hybridus* cv. 'Advance Red'. Plant Cell Tiss Org Cult 120:717–728

Xing W, Bao Y, Luo P, Bao MZ, Ning GG (2014a) An efficient system to produce transgenic plants via cyclic leave-originated secondary somatic embryogenesis in Rosa rugosa. Acta Physiol Plant 36:2013–2023

Xing W, Wang Z, Wang XQ, Bao MZ, Ning GG (2014b) Over-expression of an FT homolog from Prunus mume reduces juvenile phase and induces early flowering in rugosa rose. Sci Hortic 172:68–72

Yamada T, Takagi K, Ishimoto M (2012) Recent advances in soybean transformation and their application to molecular breeding and genomic analysis. Breed Sci 61:480–494

Yang X, Zhang X (2010) Regulation of somatic embryogenesis in higher plants. Crit Rev Plant Sci 29:36–57

Zakizadeh H, Lutken H, Sriskandarajah S, Serek M, Muller R (2013) Transformation of miniature potted rose (*Rosa hybrida* cv. Linda) with P_{SAG12}-*ipt* gene delays leaf senescence and enhances resistance to exogenous ethylene. Plant Cell Rep 32:195–205

Zhang Q, Chen W, Sun L, Zhao F, Huang B, Yang W, Tao Y, Wang J, Yuan Z, Fan G, Xing Z, Han C, Pan H, Zhong X, Shi W, Liang X, Du D, Sun F, Xu Z, Hao R, Lv Y, Zheng Z, Sun M, Luo L, Cai M, Gao Y, Wang J, Yin Y, Xu X, Cheng T, Wang J (2012) The genome of *Prunus mune*. Nat Commun 3:1318

12 Embryogenesis in Ornamental Monocots: Plant Growth Regulators as Signalling Element

A. Mujib, Muzamil Ali, Dipti Tonk, Tasiu Isah, and Nadia Zafar

Abstract

The process of in vitro embryogenesis has several basic and applied applications especially in fast and mass propagation of elite ornamental germplasm. In this chapter, somatic embryogenesis is described in two important ornamental plants. The select plants *Gladiolus* and *Caladium* are the monocot genera, belonging to the family Iridaceae and Araceae in which in vitro embryogenesis were investigated. Indirect somatic embryo formation was noticed in both the plants on tuber/rhizome callus. Although the synthetic auxin 2,4-D is a good signalling element in inducing embryos, the amendment of NAA and BAP improved embryogenesis and embryo numbers. The single use of BAP (0.5–1.0 mg/) was also effective in triggering embryogenesis; the embryo number was, however, low. The signalling and molecular role of auxin, cytokinin, GA_3 and ABA in regulating somatic embryogenesis is briefly summarised. The origin, embryo structure and development are very similar, so was the requirement of plant growth regulator (PGR). The embryo structure and development was investigated morphologically, by using scanning electron microscopy (SEM) and is presented. The addition of 0.5 mg/l GA_3 was noted to be very efficient in promoting embryo maturity; the influence of ABA in embryo maturation is also discussed. The technique of making synthetic seed in *C. bicolor* was made and is described briefly. The regenerative medium for obtaining plants from synthetic seeds was identified after storage at low temperatures (0 and 4 °C). The somatic embryos/synthetic seeds germinated into plantlets on 0.5 mg/l BAP-added medium and the regenerated plants are very similar to the mother plants. This and similar in vitro embryogenesis studies may help in building phylogenetic tree of near or distantly related taxonomic plants/plant groups.

A. Mujib (✉) • M. Ali • D. Tonk • T. Isah • N. Zafar
Cellular Differentiation and Molecular Genetics Section, Department of Botany, Hamdard University, New Delhi 110062, India
e-mail: amujib3@yahoo.co.in

A. Mujib (ed.), *Somatic Embryogenesis in Ornamentals and Its Applications*,
DOI 10.1007/978-81-322-2683-3_12

Keywords
Direct and indirect embryo formation • Gladiolus • Caladium • Monocot • Ornamentals • Plant growth regulators • Scanning electron microscopy • Secondary embryo • Somatic embryogenesis • Signalling

12.1 Introduction

Ornamentals are large groups of plants comprises of bedding-, flowering-, foliage/house-, and plants producing cut-flowers. These groups of floriculture plants are often annuals and perennials in nature. Beside ornamental applications, plants are cultivated for cosmetic/perfume uses and in pharmaceutical industry. Flowers are used in various social ways: garland-making, decoration and for religious ceremonies. The floral industry is booming at a fast pace with an estimated turnover of over Rs. 1000 crore in India in 2006 (Singh 2011). A total area of over 1.26 lakh ha is under flower production covering Maharashtra, Karnataka, Tamil Nadu, Andhra Pradesh and West Bengal provinces (Singh 2009). The floriculture export mostly comprises of dry flowers, fresh cut flowers, bulbs and foliage plants (Singh 2009). Dry flower is the primary floriculture export materials contain Dahlias, marigold, roses, lilies, helichrysum, etc. (Singh 2009). Some of the flowers are air-dried and used; examples are poppy seeds, roses, *Delphinium*, larkspur, lavender, African marigold, strawflower, globe amaranth, lotus pod, etc. (Sarkar 2010). Flower crops are also grown for essential oil production, which includes rose, jasmine, tuberose, etc. *Rosa damascene* is exclusively cultivated for extraction of essential oils, rose water, attar and gulkand. Marigold pigments are widely used for the extraction of natural dye like xanthophylls exploited in poultry industry to enhance the meat colour and yolk of eggs, used in food and textile industry as well. Besides, seedlings, flower-budded plants, rooted cuttings, bulbs, tubers and rhizomes, annual seeds are used in commercial plant/flower production for home, garden and corporate landscaping purposes.

Gladiolus is a perennial bulbous ornamental plant belonging to the family Iridaceae. The genus is distributed to Asia, Mediterranean Europe, South Africa and tropical Africa. The centre of diversity is in the Cape Floristic Region (Goldblatt and Manning 1998). The genus *Gladiolus* comprises of about 260 species and the World Checklist of Selected Plant Families recognises 276 species (Ohri and Khoshoo 1986). The *Gladiolus* flowers are multi-coloured, pink to reddish, light purple with white, white to cream and orange to red in colour. The giant flower spikes cultivated for commercial applications are the results of hybridisation and selection and perhaps the result of genetic manipulation effort. Gladioli have been extensively hybridised and a wide range of ornamental flower colours are available. The important hybrid groups have been obtained by hybridisation involving four or five different species followed by selection, e.g., Grandiflorus, Primulines and Nanus. These groups of *Gladiolus* produce cut flowers. Cytologically, most of the species of *Gladiolus* are diploid ($2n = 30$ chromosomes), but the hybrids of Grandiflora are tetraploid ($2n = 4x = 60$ chromosomes) as the parental species *Gladiolus dalenii* is tetraploid and produced a wide array of hybrids including Grandiflora (Krahulcová 2003; Elena et al. 2012).

Caladium is an immensely valuable horticultural plant, used as potted and landscape plant. These araceous species are seen as indoor, decorative plant in offices and in houses. The *Caladiums* are widely distributed and the genera are primarily localised to New World Tropics (Bown 1988). It contains 12 different species and with over 2000 cultivars (Wilfret 1986, 1993). The identification of cultivars is often difficult since the leaves show similar morphology, striation and pigment. These variable leaf size, shape and colouration add ornamental value to *Caladiums*. The *Caladium* is recognised to be a poisonous plant, causes nausea and vomiting and diarrhoea, damages cornea and causes severe burning of mouth and throat when parts of the plant are eaten (Hostetler and Schneider 2004; Graeme 2011).

C. bicolor (Ait.) Vent. is quite widespread from Costa Rica to northern Argentina, and it is found in India, Bangladesh, Africa, Indian island and Pacific Oceans and in Caribbean island. In *Caladium*, a high degree of chromosomal diversity was noted and a recent study indicates $2n = 18$, 20, 24, 26, 30, 34 and 38 chromosomes in squashed root tip cells (Cao and Deng 2014). Plant with $2n = 28$ chromosome number was also noted in tissue culture-regenerated *Caladium* (Mujib et al. 2008). These plants are propagated asexually by dividing storage organs (rhizome, corm and bulb) and are multiplied conventionally by corm cutting and leaf clipping (Memon 2012). While propagation, the mother corm produces a limited number of daughter corms, which take a couple of season more to mature and produce natural long flowering spike. The storage underground tissues of ornamentals are sensitive to diseases of diverse nature (Massey 1928; Sinha and Roy 2002). The underground organs also show dormancy that lowers conventional multiplication rate. Fast multiplication of propagules is the most important problem in ornamental bulbous plants. Seed propagation is not practised as it shows sterility, often with poor seed germination (owing to self-incompatibilities), causing heterogeneity and progeny variations. Lately, modern biotechnological approaches/plant tissue culture has been successfully used in multiplying plants for novel elite cultivars, besides generating the possibility of raising new and improved cultivars. *Gladiolus* and *Caladium* are important ornamentals showing a very low rate of multiplication conventionally. In vitro technique is a convenient alternative in augmenting multiplication for which two common methods are used: (a) direct regeneration of plantlets from explants and (b) plant regeneration through callus via organogenesis and/or somatic embryogenesis.

12.2 Choice of Explants

The in vitro plant regeneration is significantly influenced by the following few factors: (a) the composition of culture medium, (b) the choice and appropriate concentrations of plant growth regulators (PGRs), (c) the genetic makeup of mother plant and (d) the choice of explants. In *Gladiolus*, a variety of explants such as shoot tip, meristem, leaf base, inflorescence stalk, cormlet, corm piece, etc., have been used for establishing culture and later for plantlet regeneration (Fig. 12.1a–e, Table 12.1). The corm piece and inflorescence stalk are the choice of explants in several studied experiments (Nagaraju and Parthasarthy 1995; Babu and Chawla 2000; Nhut et al. 2004; Xu et al. 2009).

In *Caladium* the rhizome/tuber and leaf petioles have been used as experimental materials. These explants were dissected into 3–4 mm long pieces before inoculation. Shoot meristem (Ali et al. 2007), mature leaf (Ahmed et al. 2002), etc. were also used as explant for initiating in vitro culture and for large-scale plant regeneration (Table 12.2).

12.3 Callus Induction and Involvement of PGRs

The role of PGRs in inducing callus is well established in different groups of plants. In gladiolus, a monocot genus, MS amended with 2,4-D was observed to be very efficient for inducing callus; the use of NAA, dicamba and picloram was earlier noted to be very efficient in promoting callus (Emek and Erdag 2007; Kamo and Joung 2007; Aftab et al. 2008). In our *Gladiolus* study, corms were used as explant on 2,4-D-amended MS medium. The explants showed early swelling and callus started to appear from the cut end of explants within 10 days of incubation. Of the various concentrations tested 0.5–2.0 mg/l 2,4-D induced profuse callus. This synthetic auxin was noted to be very important in inducing callus in several investigated plants (Su et al. 2009; Feher 2015). The increased activity of 2,4-D has been noted to be due to its better stability during culture process (Covington and Harmer 2007).

Fig. 12.1 (**a**) Embryogenic callus, (**b**) somatic embryos at different stages, (**c**) germinating somatic embryos in MS medium, (**d**) regenerated plantlets and (**e**) transplanted plant in pot (bar: A, B: 0.5 cm; C, D and E:1.0 cm)

The callus was reddish yellow, compact, nodular structure with sticky gumish appearance. The addition of BAP (0.5–1.0 mg/l) in NAA (0.5–1.0 mg/l) supplemented medium was noted to be very promotive for callus induction.

In *C. bicolour*, callus was induced from rhizome and leaf petiole on 0.5–2.0 mg/l 2,4-D added MS with variable intensities. The rhizome-induced callus was yellow, hard, compact and nodular structure showing moderate in vitro growth on NAA/BAP-amended medium. The use of 2,4-D was noted to be ineffective for callus propagation as it showed necrosis on callus surfaces. Leaf-induced callus was friable and fast growing. Callus induction frequency and growth was high on medium amended with 0.5 mg/l NAA and 0.5 mg /l BAP, followed by a treatment with 1.0 mg /l NAA and 0.5 mg/l BAP (data not shown).

12.4 Somatic Embryo Formation

In *Gladiolus*, within 3–4 weeks of incubation, embryo started to appear on callus surface, the embryo formation was quite high on NAA+BAP-added medium, maximum being in 0.5 mg/l NAA+2.0 mg/l BAP (Table 12.3), followed by 0.25 mg/l NAA+1.0 mg/l BAP; the high level of PGRs was observed to be inhibitory. Several individual PGRs like 2,4-D, IAA and NAA were

Table 12.1 Various explants, medium and PGRs used for developing callus and plant regeneration in *Gladiolus*

Explant	Medium and PGRs used	Mode of regeneration (organogenesis/ embryogenesis	References
Axillary buds	MS, 0.5 mg/l BAP	Organogenesis, corm formation	Dantu and Bhojwani (1995)
Terminal and lateral buds	MS, 1.0 mg/l BAP	Shoot multiplication by direct organogenesis	Sen and Sen (1995)
Leaf and lateral bud	MS, 1.0 mg/l BAP	Somatic embryogenesis via callus	Emek and Erdag (2007)
Corm, leaf, whole plants	MS, 2.0 mg/l 2,4-D	Somatic embryogenesis via callus	Stefaniak (1994)
Seed or germinated seedlings	MS, 2.0 mg/l 2,4-D	Somatic embryo from callus	Racosy-Tican et al. (2012)
Cormlets	MS, 3.0 mg/l TDZ+ 0.2 mg/l BAP	Somatic embryo from callus and genetic transformation with *gus* and *hpt* genes	Wu et al. (2015)
Corm, style, ovary and stamen	MS, 6.0 mg/l TDZ	Embryogenic callus and regeneration	Xu et al. (2009)
Corm	MS, 18 μM 2,4-D+ 0.5 μM BA	Callus, suspension and plant regeneration (organogenesis)	Nasir et al. (2013)
Cormlet	MS, 0.5 μM+0.5 zeatin	Callus, cell suspension, somatic embryogenesis and plant regeneration	Remotti (1995)
Inflorescence	MS, 4.0 mg/l BAP	Cormlet and plantlet regeneration	Memon (2012)

Table 12.2 Various explants, medium and PGRs used in producing callus and regeneration in *Caladium*

Explant	Medium and PGRs used	Mode of regeneration (organogenesis/ embryogenesis	References
Rhizomes	MS, 2.0 mg/l 2,4-D;	Callus induction	Mujib et al. (1991, 1996, 2008)
	MS, 0.5 mg/l BAP+1.0 mg/l NAA	Somatic embryogenesis	
Young leaf, petiole	MS, 0.1 mg/l 2,4-D+0.5 mg/l BAP;	Callus	Li et al. (2005)
	MS, 2.0 mg/l BAP+0.2 mg/l NAA	Organogenesis and genetic transformation with *Lc* gene	
Corm, leaf petiole	MS, 0.8 mg/l 2,4-D+1.0 mg/l kinetin	Direct organogenesis	Sakpere and Adebona (2007)
Shoot meristem	MS, 1.0 mg/l BAP+0.25 mg/l BAP	Direct shoot formation organogenesis	Ali et al. (2007)
Explants	MS, 1.0 mg/l BAP+ 1.0 mg/l NAA	Direct organogenesis	Ahmed et al. (2002)
Leaf	MS, 2.0 mg/l BAP +0.6 mg/l NAA	Direct tubercles formation	Li et al. (1981)

Table 12.3 Embryogenesis and embryo numbers in *Gladiolus*, MS medium was added with different PGRs

2,4-D	NAA	IAA	BAP	% embryogenesis	No of embryos/callus mass
0.5	0.0	0.0	0.0	85.62±6.43b	6.28±1.08d
1.0	0.0	0.0	0.0	88.15±4.42a	6.45±1.04d
2.0	0.0	0.0	0.0	41.66±4.76d	3.36±0.08e
0.0	0.5	0.0	0.0	0.0 h	0.0 g
0.0	1.0	0.0	0.0	0.0 h	0.0 g
0.0	2.0	0.0	0.0	0.0 h	0.0 g
0.0	0.0	0.5	0.0	0.0 h	0.0 g
0.0	0.0	1.0	0.0	0.0 h	0.0 g
0.0	0.0	2.0	0.0	0.0 h	0.0 g
0.0	0.0	0.0	0.5	17.66±1.22e	3.33±0.99e
0.0	0.0	0.0	1.0	16.35±1.23f	2.48±0.06f
0.0	0.0	0.0	2.0	7.55±1.89 g	2.46±0.05f
0.0	0.25	0.0	1.0	87.50±5.33a	9.33±1.11b
0.0	0.5	0.0	1.0	75.00±4.99c	7.66±1.20c
0.0	0.5	0.0	2.0	74.89±4.57c	14.33±2.08a

added to medium and tested their ability in forming embryos. Except 2,4-D, none of the NAA and IAA level was noted to be stimulatory in forming embryos. Structurally, the embryos were whitish, globular or roundish structures, often clustered in a common callus mass. The embryo progressed to mature embryo with the formation of a notch. Beside typical embryo structures, variation in embryo forms, i.e. elliptical, ovoid and other shapes, was observed. Some of them had reduced hypocotyls and well-developed shoot axis. In *Gladiolus,* no secondary embryo was formed on primary somatic embryos. The use of 2,4-D, which induced secondary embryos in several investigated plants, was observed to be ineffective in developing secondary embryos. Similarly, in the combination of 0.5 mg/l NAA and 2.0 mg/l BAP, which produced primary somatic embryos in numbers, no secondary embryo formation was noted; the single use of BAP and kinetin was also noted to be ineffective.

In *Caladium*, 3–4 weeks old callus started to induce embryos on medium amended with 0.5–1.0 mg/l 2,4-D; the embryo forming ability was even more on 0.5 mg/l BAP and 0.5 mg/l NAA-added medium (Table 12.4). At initiation stage, the embryos were white globular/ovoid structures; later the embryos took bipolar appearance with green apical, cotyledonary end – and a root axis. The cotyledonary end shows pointed apex with swollen base. In *Caladium,* the secondary embryogenesis programme was switched off as no secondary embryo was formed.

12.4.1 The Somatic Embryo Development Pathway

The origin of somatic embryos takes place in two different ways: in direct method, the embryos are formed directly on cultured explant itself without any development of callus. This type of embryo formation was reported in numerous ornamental plants such as *Chrysanthemum*, lily, *Hippeastrum*, *Eucharis*, carnation, orchid, etc. (Liu et al. 1997; Ma and Liu 2003; Mujib et al. 2007; Jiang et al. 2008; Gow et al. 2009). On the other hand, somatic embryos are developed on embryogenic callus, derived from a variety of explants. This indirect way of formation of embryo on callus is much more frequent and is observed in several investigated plants including ornamentals (Chen et al. 1999; Fereol et al. 2002; Jiang and Zhang 2007). In this present study, somatic embryos were formed indirectly on callus mass both in *Gladiolus* and in *Caladium.*

Table 12.4 Embryogenesis and embryo numbers in C. bicolor, MS medium was added with different PGRs

2,4-D	NAA	BAP	% embryogenesis	No of embryos/callus mass
0.5	0.0	0.0	62.25±3.43b	2.18±0.56e
1.0	0.0	0.0	38.65±2.23d	3.25±1.11d
2.0	0.0	0.0	21.66±4.44e	1.68±0.23f
0.0	0.5	0.0	0.0i	0.0 h
0.0	1.0	0.0	0.0i	0.0 h
0.0	2.0	0.0	0.0i	0.0 h
0.0	0.0	0.5	12.64±1.26f	1.26±0.12 g
0.0	0.0	1.0	13.33±2.22 g	1.88±0.32f
0.0	0.0	2.0	4.25±0.32 h	1.76±0.67f
0.0	0.5	0.5	76.22±3.65a	7.65±1.12a
0.0	1.0	0.5	62.18±4.43b	6.85±2.08b
0.0	0.5	1.0	48.66±3.34c	5.88±1.06c

Values are means±standard errors of three replicates; within each column, means followed by the same letter are not significantly different at $p \leq 0.05$ according to DMRT

12.4.2 Scanning Electron Microscopy of Somatic Embryos

SEM examination of embryo bearing callus of *Gladiolus* revealed somatic embryos of various stages developed on callus surfaces, presented in Fig. 12.2c, d. The induced embryos were formed as a heterogeneous mass; the number of globular embryos was low, however. The somatic embryos were smooth, often clustered. SEM of embryogenic callus also revealed long cylindrical embryo structure without distinct notches. In *Caladium,* SEM investigation also shows the same embryo morphology with pointed cotyledonary end and swollen base and a long root axis. The somatic embryos turned green later and germinated into plantlets.

12.4.3 Signalling Role of Auxin/2,4-D in Somatic Embryogenesis

The influence of PGR is very crucial in impacting somatic embryogenesis. In most investigated cases, auxin and cytokinin are added together. In horticulturally important plants, 2,4-D, one of the important PGRs, has been widely employed and was reported in a number of important ornamental plants like *Begonia* (Castillo and Smith 1997), Lily (Liu et al. 1997), *Euphorbia* (Zhu et al. 2004), *Chrysanthemum* (Mandal and Datta 2005) and *Crinum* (Mujib et al. 2006). In this study, the influence of various PGRs on somatic embryo formation and later on plant regeneration was studied. In *Gladiolus,* various auxins (2,4-D, IAA and NAA) were amended in medium and only 2,4-D at low to moderate levels induced somatic embryos while the other two auxins did not have much influence on in vitro embryogenesis. This synthetic auxin was noted to be very crucial for the induction of callus in both the two plants on which somatic embryos were formed. Somatic embryogenesis has been triggered by several internal and external factors of which 2,4-D, a unique signalling element, amended to medium in initiating embryos is well established (Gaj 2004; Feher 2015). 2,4-D is a synthetic auxin showing strong PGR/auxin-like activities; it also induces stress in culture, which triggers the in vitro embryogenesis process (Song 2013). For embryo development, 2,4-D needs to be withdrawn from the medium, and this removal of auxin (2,4-D) induces YUC gene expression and produces a key enzyme (flavin monooxygenase), which helps in biosynthesis of endogenous IAA in cultured tissues (Bai et al. 2013), and this also favours polar transport of IAA

Fig. 12.2 (**a**) Synthetic seeds of *Caladium* showing germination, (**b**) regenerated plantlets, (**c**) scuetellar stage, (**d**) coleoptilar stage of embryo in *Gladiolus* (bar: A, B = 1 cm; C, D = 200 μm)

(Su et al. 2009). Both the two events are thought to play a central role in developing embryos in culture (Nawy et al. 2008). At the same time, the addition of 2,4-D imposes abiotic stress on tissue, which influence various facets of auxin homeostasis including storage and transport of endogenous auxin and induces synthesising transcription factors (Tognetti et al. 2012; Gliwicka et al. 2013).

12.4.4 Molecular Role of Cytokinins in Somatic Embryogenesis

The relationship between cytokinins with embryogenesis is not always direct as cytokinin has often been amended along with auxin in inducing embryos (Gaj 2004). In ornamental plant *Euphorbia*, a subtle ratio of auxin and cytokinin is necessary in triggering embryogenesis (Zhu et al. 2004). The high auxin-cytokinin ratio promoted somatic embryo formation in *Begonia* (Castillo and Smith 1997) and Cyclamen (Takejiro et al. 1995) whereas a high cytokinin and auxin ratio induced somatic embryogenesis in apple (Da et al. 1996). Two or more number of auxins with cytokinin or multiple cytokinins with single auxin have also been added to obtain somatic embryos in a number of studied materials (Chen and Su 2006; Xin et al. 2006; Jiang and Zhang 2007). Cytokinin alone was also amended in medium and was noted to be active in inducing embryos (Iantcheva et al. 1999; Mujib and Samaj 2006). In these studied plants like *Caladium*, 0.5 mg/l BAP was noted to be very efficient in developing embryos. The embryogenesis percentage and embryo numbers increased on NAA containing BAP-added medium. However, NAA alone in medium (without BAP) did not have any influence in developing embryos. Francis and Sorrell (2001) reported that the used PGRs control cell cycle and expression of genes, the two most important steps of embryogenesis. Cytokinin is also reported to regulate cyclin gene, which controls cell cycle passage during division (Jiménez and Thomas 2006). The role of auxin in expressing gene during embryogenesis is known for quite some times (Zhu and Perry 2005; Su et al. 2009), but cytokinin-regulated gene expression investigation is still not too many. In *Medicago truncatula* leaf somatic embryogenesis, ethylene-mediated transcription factor AP2/ERF synthesis was described in auxin- and cytokinin-added condition (Mantiri et al. 2008). Single use of cytokinins is also reported to be an important signalling element, promoting *WUS* expression during somatic embryogenesis and shoot meristem establishment as WUS-encoded transcription factor was detected in early embryogenic cell fate in investigated cellular systems including *Arabidopsis* (Su et al. 2009; Gordon et al. 2009). Here, in these two plants, a link is established between cytokinin and embryogenesis, but at what level the cytokinin acts is still not known; similar gene expression mechanism may be operational in acquiring embryogenic competenance.

12.4.5 Secondary Somatic Embryogenesis: Is It Genetically Controlled or Acquired?

Adventives or secondary somatic embryos are often formed on primary embryo structures, and this phenomenon has been noticed in a number of plants especially in dicots (Junaid et al. 2007; Karami et al. 2007). In these studied genera, secondary somatic embryos were not formed on primary embryos, amended with PGRs in which primary embryos were routinely induced. The embryo-bearing callus were kept on medium for extended period of time (without regular subculturing), provided a condition of stress, did not induce secondary or adventive embryos. It is therefore believed that the genetic make-up determines secondary embryogenesis process and external PGRs application in medium or stress did not have much influence on secondary embryogenesis unlike primary embryogenesis.

12.4.6 Maturation of Somatic Embryos and PGRs' Involvement

The maturation and fast embryo germination is a key in getting plantlets in numbers, and it partly depends upon the number of embryos induced and the quality of embryos. The formation of aberrant embryos and frequent induction of embryonal budding or secondary embryos restrict embryo maturation and delay embryo germination process. Besides, induced somatic embryos show dormancy, similar to zygotic embryos, which also limits embryo germination rate. In recent time, a large number of compounds/techniques such as abscisic acid (ABA), GA_3, sugars, sugar alcohol, polyethylene glycol (PEG), activated charcoal (AC), low temperature treatment, etc., are amended to medium as to improve embryo maturation and in promoting embryo germination (Lipaska and Konradova 2004; Robichaud et al. 2004). Here, the embryos of *Gladiolus* were cultured on MS medium containing salts, vitamins and inositol but without PGRs for embryo maturation. In other experiments, the medium was amended with 0.25–1.0 mg/l abscisic acid (ABA) and 0.25–1.0 mg/l gibberellic acid (GA_3). The added GA_3 into the medium improved embryo quality (promoted embryo sizes and turned embryos green at the tips), which later germinated into plantlets. The embryo maturation frequency was maximum on 0.5 mg/l GA_3-supplemented medium, followed by 1.0 mg/l GA_3 treatment. Embryo maturity improved with time and about 62. 15 % embryos matured on 0.5 mg/l GA_3-amended medium at the end of 8 weeks. The addition of ABA to MS was noted to be less efficient and had a very little influence on embryos (morphophysiologically). The embryo maturation percentage was relatively low, which varied from 12.15 to 18.75 % range. In hybrid chestnut, similar response, i.e. poor embryo germination, was noted in ABA-added medium (Vieitez 1994). Conversely, the requirement of ABA was known to be essential during somatic embryogenesis especially in *Arabidopsis* (Su et al. 2012; Bai et al. 2013). In some coniferous plants too, ABA improved embryo quality and facilitated embryo maturation process (Lelu-Walter et al. 2008; Maruyama et al. 2007). The medium fortified with GA_3 especially at 0.5–1.0 mg/l significantly improved somatic embryo sizes and facilitated photosynthetic pigment synthesis, a good sign of embryo maturation prior to plantlet development. The synthesised pigments improve photosynthetic ability and produce cellular energy reserves in the form of various sugars, lipid, triglycerides and protein, required during embryo germination. ABA/GA_3 ratio is currently known to be directly interlinked with *LEC* gene expression during embryo/seed maturation (Braybrook and Harada 2008). This *LEC* gene (*LEC1, LEC2*) and other associated genes like *FUS3* form a regulatory network and help to make building storage reserve during the course of embryo maturation and cotyledon development (Harada 2001; Santos-Mendoza et al. 2008).

12.4.7 Somatic Embryo Germination and Plantlets Regeneration

In *Gladiolus*, 2,4-D induced callus and embryos, but plantlet regeneration was not achieved on this medium. The embryo bearing calluses were transferred to a variety of BAP- and NAA-amended media for their ability to germinate into plantlets. Almost in all tested levels somatic embryo germinated into plantlets, maximum plantlet regeneration was noted on 0.5 mg/l BAP containing medium, followed by 1.0 mg/l BAP. Somatic embryo germination was also pretty high on 0.5 BAP- and 0.25–0.5 mg/l NAA-added medium. The individual shoots were later transferred to medium added with IBA and NAA (0.5, 1.0 and 2.0 mg/l) for the induction of roots. Both the two auxins promoted roots, but the influence of IBA was more compared to NAA, and 1.0 and 2.0 mg/l of IBA were noted to be more efficient in promoting roots.

Beside somatic embryogenesis, plant regeneration was achieved in *Caladium* on MS medium added with 0.5–4.0 mg/l BAP and 0.5–2.0 mg/l NAA. Before shoot bud emergence, upper surfaces of rhizome calli showed green patches from which tiny shoot buds formed. The shoot forming ability of callus was quite high and almost all PGR combinations especially amended with NAA and BAP induced shoots at variable numbers and frequency, maximum being at 0.5 mg/l BAP and 1.0 mg/l NAA. Single use of BAP at 0.5 mg/l BAP was equally efficient in producing shoots. Typical somatic embryos are bipolar structure with distinct shoot and root end that later converted into plantlets with no separate root induction step is necessary. Many of the embryos, however, did not have adequate root axes nor had less developed root ends. In such conditions induction of roots is necessary. Of the various PGRs tested, 0.25–2.0 mg/l IBA and NAA (0.25–2.0 mg/l) were observed to be very promotive for developing roots. The embryo-derived plantlets with shoot and root grew normally in medium. The shoots originated through organogenetic way upon transfer on NAA- and IBA-added medium produced roots at variable numbers; the influence of IBA at 2.0 mg/l was more on rooting.

12.5 Encapsulation and Storage of Synthetic Seeds at Low Temperature

The rapid loss of genetic diversity can be protected by fast multiplication of elite, endangered plants with ability of germplasm by employing in situ and in vitro conservation methods. Since 1980s, the encapsulation methodology has been receiving attention as a good alternative for storage of germplasm (Danso and Ford-Lloyd 2003; Rai et al. 2009). Synthetic seeds are prepared by using somatic embryos/other explants with artificial protective layer, which can serve as true seeds (Cangahuala-Inocente et al. 2007). The protocol of synthetic seed production by encapsulating plant parts in *C. bicolor* was made (Fig. 12.2a, b) and is described. The suitable medium for obtaining plantlets from synthetic seeds after short-term storage at low temperatures (0 and 4 °C) was also optimised. A variety of alginate and calcium chloride levels were tested in making synthetic beads and 3.0 % sodium alginate and 100 mM calcium chloride combination was noted to be very efficient in making synthetic bead, followed by 3.0 % sodium alginate + 125 mM calcium chloride treatment. The combinations of NAA and BAP were efficient in recovering plants from synthetic seeds of which 1.0 mg/l BAP+0.5 mg/l NAA produced maximum plantlets, followed by 0.50 mg/l BAP+1.0 mg/l NAA treatment. The encapsulated somatic embryos were cultured and kept at low temperatures (0 and 4 °C) for storage of *Caladium* germplasm. We noted that the preservation at 4 °C was much more efficient than the conservation at 0 °C (data not shown). The encapsulated embryos stored at 4 °C for 6 weeks showed fairly good germination (58.66 %) in developing plants; the germination ability was, however, declined with time. Recovery of plantlets of synthetic seeds was poor, stored at 0 °C, and the germination percentage was very low when the synthetic seeds were kept for longer period of time. The germinated plantlets, however, grew well and are morphologically very similar to parent plants.

12.6 Concluding Remarks

Ornamentals especially the bulbous/tuberous plants are propagated vegetatively, which is a very slow process. By using somatic embryogenesis, the rate of multiplication can be augmented several-fold as was observed in many plant genera like orchids, *Coffea*, etc. (Belarmino and Mii 2000). Comparing to direct regeneration or organogenesis, plants regenerated through somatic embryogenesis show both roots and shoots as the embryos are bipolar structure, avoiding root induction stage.

Somatic embryogenesis is regarded a unique mechanism for studying cellular differentiation. It has several basic and applied applications. The external and internal triggers including the genes/homologues of genes, which regulate in vitro embryogenesis are known at least partially in other groups of plants, but there is very little information in ornamentals, unfortunately. How a somatic cell acquires embryogenic competence in response to PGRs, stress and with related molecular signal is not elucidated clearly in ornamentals.

The present study indicated some similarities in embryogenic processes including origin, development and germination stages. The PGRs requirement is also nearly the same at different stages of embryogenesis. The secondary embryogenesis is not operational in these studied monocot plants unlike dicotyledonous plants. Similar in vitro embryogenesis studies may help in building evolutionary link among plants/plant groups.

The embryos are originated from single cell of cultivated tissues; the genetic manipulation is often simple, easy and ideal for developing transgenic in numbers. High transformation frequency with low level of somaclonal variation in embryo mother cells (with bipolar ability) makes them a good recipient of foreign gene in transgenic experiments (Vergne et al. 2010). This concept has been utilised in genetic information studies in a number of important plants including flowering ornamentals like *Chrysanthemum* (Jiang and Zhang 2007), rose (Li et al. 2002), *Lilium* (Ogaki et al. 2008) and even in gladiolus (Kamo et al. 2009; Wu et al. 2015). With the progress and optimisation of pre-culture, co-culture time, media and explant selection, the transformation experiments can be extended to other groups of plants of monocots, which have been thought to be inadequate/recalcitrant for genetic engineering studies.

In vitro propagated plants especially developed from compact organised structure are true to type, but the plantlets raised from callus/embryogenic suspension or less organised structure like protoplast show somaclonal variation. It is therefore essential to detect culture-induced variation at early stages by using molecular techniques. DNA markers such RAPD (random amplified polymorphic DNA) and ISSR (inter-simple sequence repeat) have recently been used in detecting somaclonal variation in several tissue culture raised ornamental plants (Podwyszynska et al. 2006, 2010; Bondrea et al. 2007). In *Gladiolus* and *Caladium,* however, no such DNA marker techniques are used to test the genetic fidelity of regenerated plants, developed from callus, suspension or somatic embryo.

Important, endangered or rare ornamental germplasm may be conserved by utilising callus, embryonic callus, embryos/embryo mother cells and synthetic seeds in in vitro conditions (Kaviani 2010). Low- and ultralow temperature and identification and optimisation of cryoprotectants may be very important factors in preserving biological materials (Sarmah et al. 2010). The use of minimal, sucrose-free medium or mineral oil overlay may reduce the cost of in vitro conservation. High-alkaloid, perfume, pigment and oil-producing cell lines of ornamentals can also be cryopreserved for short-, medium- and long-term basis.

Acknowledgement The first author is thankful to the Department of Botany, Hamdard University (Jamia Hamdard), for getting various facilities. The first author acknowledges present and past research students' help.

References

Aftab F, Alam M, Afrasiab H (2008) In vitro shoot multiplication and callus induction in *Gladiolus hybridus* Hort. Pak J Bot 40(2):517–522

Ahmed EU, Hayashi T, Zhu Y, Hosokawa M, Yazawa S (2002) Lower incidence of variants in *Caladium bicolor* Ait. Plants propagated by culture of explants from younger tissue. Sci Hortic 96:187–194

Ali A, Munawar A, Naz S (2007) An in vitro study on micropropagation of *Caladium bicolor*. Int J Agric Biol 9(5):731–735

Babu P, Chawla HS (2000) In vitro regeneration of *Agrobacterium* mediated transformation in gladiolus. J Hortic Sci Biotechnol 75(4):400–404

Bai B, Su YH, Yuan J, Zhang XS (2013) Induction of somatic embryos in Arabidopsis requires local *YUCCA* expression mediated by the down-regulation of ethylene biosynthesis. Mol Plant 6:1247–1260

Belarmino MM, Mii M (2000) Agrobacterium-mediated genetic transformation of a *Phalaenopsis* orchid. Plant Cell Rep 19(5):435–442

Bondrea IO, Pamfil D, Van Heusden S, Van Tuyl J, Meijer-Dekens F, Bondrea M, Sestras R, Rusu AR, Lucaci M, Patrascu BI, Balteanu VA (2007) AFLP as a modern technique for DNA fingerprinting and identification *Tulipa* cultivars. Bull USAMV-CN 63–64

Bown D (1988) Aroids, plants of the arum family. Timber Press, Portland, 256 pp

Braybrook SA, Harada JJ (2008) LECs go crazy in embryo development. Trends in Plant Sci 13: 624–630

Cangahuala-Inocente GC, Dal Vesco LL, Steinmacher D, Torrees AC, Guerra MP (2007) Improvements in somatic embryogenesis protocol in Feijoa *(Acca sellowiana* (Berg) Biuret): induction, conversion and synthetic seeds. Sci Hortic 111:228–234

Cao Z, Deng Z (2014) Interspecific genome size and chromosome number variation shed new light on species classification and evolution in *Caladium*. J Am Soc Hortic Sci 139(4):449–459

Castillo B, Smith MAL (1997) Direct somatic embryogenesis from *Begonia gracilis* explants. Plant Cell Rep 16(6):385–388

Chen WS, Su SC (2006) The difference and occurrence of somatic embryogenesis in early varieties of walnut. Sci Tech Inf 16:176–177

Chen LP, Wang BL, Chen MF (1999) Studies on somatic embryogenesis of *Euphorbia pulcherrima in vitro* culture. Plant Physiol Commun 35(6):463–465

Covington MF, Harmer SL (2007) The circadian clock regulates auxin signalling and responses in Arabidopsis. PLoS Biol 5(8), e227

Da KD, Zhang S, Li YZ, Qi ZJ (1996) Direct somatic embryogenesis from *in vitro* leaves of apple. Acta Hortic Sin 23(3):241–245

Danso KE, Ford-Lloyd BV (2003) Encapsulation of nodal cuttings and shoot tips for storage and exchange of cassava germplasm. Plant Cell Rep 21:718–725

Dantu PK, Bhojwani SS (1995) In vitro corm formation and field evaluation of corm-derived plants of *Gladiolus*. Sci Hortic 61:115–129

Elena R-T, Barna B, Ana-Maria S (2012) In vitro culture and medium-term conservation of the rare wild species *Gladiolus imbricatus*. Afr J Biotechnol 11(81):14703–14712

Emek Y, Erdag B (2007) *In vitro* propagation of *Gladiolus anatolicus* (Boiss.) Stapf. Pak J Bot 39(1):23–30

Feher A (2015) Somatic embryogenesis – stress-induced remodeling of plant cell fate. Biochim Biophys Acta 1849(4):385–402

Fereol L, Chovelon V, Causse S (2002) Evidence of a somatic embryogenesis process for plant regeneration in Garlic (*Allium sativum* L.). Plant Cell Rep 21(3):197–203

Francis D, Sorrell DA (2001) The interface between the cell cycle and plant growth regulators: a mini review. J Plant Growth Regul 33:1–12

Gaj MD (2004) Factors influencing somatic embryogenesis induction and plant regeneration with particular reference to *Arabidopsis thaliana* (L.) Heynh. J Plant Growth Regul 43:27–47

Gliwicka M, Nowak K, Balazadeh S, Mueller-Roeber B, Gaj MD (2013) Extensive modulation of the transcription factor transcriptome during somatic embryogenesis in *Arabidopsis thaliana*. PLoS One 8, e69261

Goldblatt P, Manning JC (1998) *Gladiolus* in Southern Africa: systematics, biology, and evolution. Fernwood Press, Cape Town

Gordon SP, Chickarmane VS, Ohno C, Meyerowitz EM (2009) Multiple feedback loops through cytokinin signaling control stem cell number within the *Arabidopsis* shoot meristem. Proc Natl Acad Sci USA 106:16529–16534

Gow WP, Chen JT, Chang WC (2009) Effects of genotype, light regime, explant position and orientation on direct somatic embryogenesis from leaf explants of *Alaenopsis* Orchid. Acta Physiol Plant 31(2):263–269

Graeme KA (2011) Toxic plant ingestions. In: Auerbach PS (ed) Wilderness medicine, 6th edn. Elsevier, Philadelphia, pp 862–870

Harada JJ (2001) Role of Arabidopsis *LEAFY COTYLEDON* genes in seed development. J Plant Physiol 158:405–409

Hostetler MA, Schneider SM (2004) Poisonous plants. In: Tintinalli JE, Kelen GD, Stapczynski JS, Ma OJ, Cline DM (eds) Emergency medicine: a comprehensive study guide, 6th edn. McGraw-Hill, New York, chap 205

Iantcheva A, Barbulova A, Vlahova M, Kondorosi E, Elliott M, Atanassov A (1999) Regeneration of diploid annual medics via direct somatic embryogenesis promoted by thidiazuron and benzylaminopurine. Plant Cell Rep 18:904–910

Jiang XW, Zhang QX (2007) Studies on transgenic acceptor system of groundcover *Chrysanthemum* via indirect somatic embryogenesis. For Res 20(3):328–333

Jiang XW, Chen FJ, Lu M, Cai M, Zhang QX (2008) Direct somatic embryogenesis in Ground-cover *Chrysanthemum*. J Beijing For Univ 30(2):65–70

Jiménez VM, Thomas C (2006) Participation of plant hormones in determination and progression of somatic embryogenesis. In: Mujib A, Samaj J (eds) Somatic embryogenesis. Springer, Heidelberg, pp 103–118

Junaid A, Mujib A, Bhat MA, Sharma MP (2007) Somatic embryogenesis and plant regeneration in *Catharanthus roseus*. Biol Plant 51:641–646

Kamo K, Joung YH (2007) Gladiolus. In: Pua EC, Davey MR (eds) Biotechnology in mass agriculture and forestry: transgenic crops VI, vol 61. Springer, Berlin/Heidelberg, pp 289–298

Kamo K, Joung YH, Green K (2009) GUS expression in *Gladiolus* plants controlled by two Gladiolus ubiquitin promoters. Floricult Ornam Biotechnol 3(1):10–14

Karami O, Deljao A, Pour AM (2007) Repetitive somatic embryogenesis in carnation in picloram supplemented media. J Plant Growth Regul 51:33–39

Kaviani B (2010) Cryopreservation by encapsulation dehydration for long-term storage of some important germplasm: seed of lily [*Lilium ledebourii* (Baker) Bioss.], embryonic axe of Persian lilac (*Melia azedarach* L.), and tea (*Camellia sinensis* L.). Plant Omics J 3:177–182

Krahulcová A (2003) Chromosome numbers in selected monocotyledons (Czech Republic, Hungary and Slovakia). Preslia Praha 75:97–113

Lelu-Walter MA, Bernier-Cardou M, Klimaszeska K (2008) Clonal plant production from self- and cross-pollinated seed families of *Pinus sylvestris* (L.) through somatic embryogenesis. Plant Cell Tiss Org Cult 92:31–45

Li WA, Zhang YH, Wang YQ (1981) Studies on tissue culture of various cultivars of caladium. Acta Phytophysiol Sin 17(3):245–250

Li XQ, Krasnyanski SF, Korban SS (2002) Optimization of the *UidA* gene transfer into somatic embryos of Rose via *Agrobacterium tumefaciens*. Plant Physiol Biochem 40(5):453–459

Li SJ, Deng XM, Mao HZ, Hong Y (2005) Enhanced anthocyanins synthesis in foliage plant *Caladium bicolor*. Plant Cell Rep 23:716–720

Lipaska H, Konradova H (2004) Somatic embryogenesis in conifers: the role of carbohydrate metabolism. In Vitro Cell Dev Biol Plant 40:23–30

Liu XM, Zhou PH, Qu SC, Lu XY, Luo ZM (1997) *In vitro* induction of indefinite bubs and somatic embryos from scale leave of tetraploid 'Longya Lily. Acta Hortic Sin 24(4):353

Ma GH, Liu N (2003) Direct somatic embryogenesis and shoot formation from cultured young leaf of *Kalanchoe blossfeldiana*. Plant Physiol Commun 39(6):625

Mandal AKA, Datta SK (2005) Direct somatic embryogenesis and plant regeneration from ray florets of Chrysanthemum. Biol Plant 49(1):29–33

Mantiri FR, Kurdyukov S, Lohar DP, Sharopova N, Saeed NA, Wang X-D, VandenBosch KA, Rose RJ (2008) The transcription factor MtSERF1 of the ERF subfamily identified by transcriptional profiling is required for somatic embryogenesis induced by auxin plus cytokinin in *Medicago truncatula*. Plant Physiol 146:1622–1636

Maruyama E, Yoshihisa H, Katsuaki I (2007) Somatic embryogenesis and plant regeneration in yakutane-goyou, *Pinus armandii* Franch. var. amamiana (Koidz.) Hatusima, an endemic and endangered species in Japan. In Vitro Cell Dev Biol Plant 43:28–34

Massey LM (1928) Dry rot of gladiolus corms. Phytopathology 18(6):519–529

Memon N (2012) In vitro propagation of Gladiolus plantlets and cormels. J Hortic Sci Ornam Plants 4(3):280–291

Mujib A, Samaj J (2006) Somatic embryogenesis. Springer, Berlin/Heidelberg/New York

Mujib A, Bandyopadhyay S, Jana BK (1991) *In vitro* propagation of *Caladium bicolor* cv 'Bleeding Heart'. Exp Genet 7:68–71

Mujib A, Maity I, Jana BK (1996) Rapid *in vitro* multiplication of *Caladium sagitifolium*. Adv Plant Sci 9:47–50

Mujib A, Banerjee S, Ghosh PD (2006) Origin, development and structure of somatic embryo in selected bulbous ornamentals: BAP as inducer. In: Mujib A, Samaj J (eds) Somatic embryogenesis, vol 2, Plant cell monograph. Springer, Berlin, pp 15–24

Mujib A, Banerjee S, Ghosh PD (2007) Callus induction, somatic embryogenesis and chromosomal instability in tissue culture-raised Hippeastrum (*Hippeastrum hybridum* cv. United Nations). Prop Ornam Plant 7(4):169–174

Mujib A, Banerjee S, Fatima S, Ghosh PD (2008) Regenerated plant populations from rhizome-calli showed morphological and chromosomal changes in *Caladium bicolor* (Ait.) Vent. cv. 'Bleeding Heart'. Prop Ornam Plant 8(3):138–143

Nagaraju V, Parthasarthy VA (1995) Effect of growth regulators on *in vitro* shoots of *Gladiolus*. Folia Hortic 7:93–100

Nasir IA, Jamal A, Tabassum B, Husnain T (2013) Regeneration response of old cell suspensions of *Gladiolus*. Exp Agric Hortic 2(2):1–11

Nawy T, Lukowitz W, Bayer M (2008) Talk global, act local – patterning the *Arabidopsis* embryo. Curr Opin Plant Biol 11:28–33

Nhut DT, Teixeira da Silva JA, Huyen PX, Paek KY (2004) The importance of explant source on regeneration and micropropagation of *Gladiolus* by liquid shake culture. Sci Hortic 102:407–414

Ogaki M, Furuichi Y, Kuroda K, Chin DP, Ogawa Y, Mii M (2008) Importance of co-cultivation medium pH for successful Agrobacterium-mediated transformation of Lilium x formolongi. Plant Cell Rep 27:699–705

Ohri D, Khoshoo TN (1986) Plant DNA contents and systematics. In: Dutta SK (ed) DNA systematics. Vol II: Plants. CRC Press, Baton Rouge, pp 2–19

Podwyszynska M, Niedoba K, Korbin M, Marasek A (2006) Somaclonal variation in micropropagated tulips determined by phenotype and DNA markers. Acta Hortic 714:211–220

Podwyszynska M, Kuras A, Korbin M (2010) Somaclonal variation in micropropagated tulips as a source of novel genotypes. Acta Hortic 855:225–232

Racosy-Tican E, Bors B, Szatmari A-M (2012) In vitro culture and medium term conservation of the rare wild species *Gladiolus imbricatus*. Afr J Biotechnol 11(81):14703–14712

Rai MK, Asthana P, Singh SK, Jaiswal VS, Jaiswal U (2009) The encapsulation technology in fruit plants-a review. Biotechnol Adv 27:671–679

Remotti PC (1995) Primary and secondary embryogenesis from cell suspension cultures of *Gladiolus*. Plant Sci 107:205–214

Robichaud RL, LesserVC MSA (2004) Treatments affecting maturation and germination of American chestnut somatic embryos. J Plant Physiol 161:957–969

Sakpere AMA, Adebona AC (2007) Tissue culture derived plantlet variation in *Caladium humboldtii* Schott. J Sci Technol 27(1):28–34

Santos-Mendoza M, Dubreucq B, Baud S, Parcy F, Caboche M, Lepiniec L (2008) Deciphering gene regulatory networks that control seed development and maturation in *Arabidopsis*. Plant J 54:608–620

Sarkar S (2010) Indian floriculture industry: the way forward. Floriculture Today pp 40–44

Sarmah DK, Borthakur M, Borua PK (2010) Artificial seed production from encapsulated PLBs regenerated from leaf base of *Vanda coerulea* Grifft. ex. Lindl. an endangered orchid. Curr Sci 98:686–690

Sen J, Sen S (1995) Two-step bud culture technique for a high frequency regeneration of *Gladiolus* corms. Sci Hortic 64:133–138

Singh HP (2009) Floriculture industry in India: the bright future ahead. Indian Hortic 54(1):3–8

Singh HP (2011) Paradigm in marketing of horticultural produce. Indian Hortic 56(3):3–8

Sinha P, Roy SK (2002) Plant regeneration through in vitro cormel formation from callus culture of *Gladiolus primulinus* Baker. Plant Tiss Cult 12(2):139–145

Song Y (2013) Insight into the mode of action of 2,4-Dichlorophenoxyacetic acid (2,4-D) as an herbicide. J Integr Plant Biol 56:106–113

Stefaniak B (1994) Somatic embryogenesis and plant regeneration of Gladiolus (*Gladiolus hort.*). Plant Cell Rep 13:386–389

Su YH, Zhao XY, Liu YB, Zhang CL, O'Neill SD, Zhang XS (2009) Auxin-induced *WUS* expression is essential for embryonic stem cell renewal during somatic embryogenesis in Arabidopsis. Plant J 59:448–460

Su YH, Su YX, Liu YG, Zhang XS (2012) Abscisic acid is required for somatic embryo initiation through mediating spatial auxin response in *Arabidopsis*. J Plant Growth Regul 69:167–176

Takejiro T, Ikuo M, Eisuke M (1995) Somatic embryogenesis of *Cyclamen Persicum* Mill. 'Anneke' from aseptic seedlings. Plant Cell Rep 15(1–2):22–25

Tognetti VB, Mühlenbock P, Van Breusegem F (2012) Stress homeostasis-the redox and auxin perspective. Plant Cell Environ 35:321–333

Vergne P, Maene M, Gabant G, Chauvet A, Debener T, Bendahmane M (2010) Somatic embryogenesis and transformation of the diploid *Rosa chinensis* cv Old Blush. Plant Cell Tiss Org Cult 100(1):73–81

Vieitez A (1994) Somatic embryogenesis in *Camellia* spp. In: Jain SM, Gupta PK, Newton R. (eds.), Somatic embryogenesis in Woody Plants, Kluwer Academic Publishers: The Netherlands, pp 235–276

Wilfret GJ (1986) Inheritance of leaf shape and color patterns in *Caladium* (Araceae). HortSci 21:750 (abstract)

Wilfret GJ (1993) Caladium. In: de Hertogh A, le Nard M (eds) The physiology of flower bulbs. Elsevier, New York, pp 239–247

Wu J, Liu C, Seng S, Khan MA, Sui J, Gong B, Liu C, Wu C, Zhong X, He J, Yi M (2015) Somatic embryogenesis and Agrobacterium-mediated transformation of *Gladiolus hybridus* cv. 'Advance Red'. Plant Cell Tiss Org Cult 120:717–728

Xin WJ, Xu B, Wang GD, Guo WM, Wen FD, Jin JP (2006) Somatic embryogenesis and plant regeneration of *Anthurium andraeanum*. Acta Hortic Sin 33(6):1281–1286

Xu Z, Hao J, He X, Yi M (2009) Callus induction and plant regeneration of *Gladiolus hybridus* Hort. Plant Physiol Commun 45(5):473–478

Zhao J, Li Z, Cui J, Henny R, Gray D, Xie J, Chen J (2013) Efficient somatic embryogenesis and Agrobacterium-mediated transformation of pothos (*Epipremnum aureum)* 'Jade'. Plant Cell Tiss Org Cult 114(2):237–247

Zhu C, Perry SE (2005) Control of expression and autoregulation of AGL15, a member of the MADS-box family. Plant J 41:583–594

Zhu GF, Lu FB, Chen ML, Wang BQ (2004) Somatic embryogenesis and plantlet regeneration of *Euphorbia pulcherrima*. Subtrop Plant Sci 33(4):37–38

Somatic Embryogenesis and Genetic Transformation in Cupressaceae Trees

13

Ken-ichi Konagaya and Toru Taniguchi

Abstract

Coniferous trees belonging to the family Cupressaceae are distributed worldwide. Several species of these conifers possess high economic value and have been a source of important forest products and raw materials for the forest industry. Somatic embryogenesis is a potential tool for clonal propagation of superior lineages within these conifers and for molecular breeding by genetic transformation. In this chapter, the technologies developed to date for somatic embryogenesis in trees of the Cupressaceae family are introduced, and various factors affecting somatic embryogenesis are discussed. In addition, methods for improving the efficiency of stable transformation via somatic embryogenesis are described.

Keywords

Chamaecyparis obtusa • *Cryptomeria japonica* • Genetic transformation • Regeneration

13.1 Introduction

The tree family Cupressaceae sensu lato comprises 32 genera, with more than 130 species, and is the only family of conifers with a cosmopolitan distribution on all continents except Antarctica (Farjón 2005; Adams et al. 2009; Debreczy et al. 2009; Yang et al. 2012). In addition, it is the most diverse of all conifers and includes important ornamental trees and shrubs used in landscaping of gardens and parks, such as the genera *Chamaecyparis*, *Cupressus*, *Juniperus*, and *Thuja*. In particular, these genera as well as other genera, including *Calocedrus*, *Cryptomeria*, *Cunninghamia*, *Sequoia*, and *Taxodium*, collectively form the leading source of timber globally. Juniper (genus *Juniperus*) is widely used as ground cover in landscaping as well as for bonsai trees. Lawson's cypress [*Chamaecyparis lawsoniana* (A. Murray bis) Parl.] and Hinoki cypress (*Chamaecyparis obtusa* Sieb. et Zucc.) also provide a large number of dwarf and color cultivars for horticulture (Farjón 2008).

K.-i. Konagaya (✉) • T. Taniguchi
Forest Bio-Research Center, Forestry and Forest Products Research Institute,
3809-1 Ishi, Juo, Hitachi, Ibaraki 319-1301, Japan
e-mail: konagaya@affrc.go.jp

A. Mujib (ed.), *Somatic Embryogenesis in Ornamentals and Its Applications*,
DOI 10.1007/978-81-322-2683-3_13

Somatic embryogenesis is an effective procedure for not only mass propagation of breeding strains showing preferred traits but also for providing target tissue for genetic transformation. Konar and Oberoi (1965) reported the first embryo-like structures in conifers, termed "embryoids," which were obtained from the cotyledons of the Chinese arborvitae (*Thuja orientalis* L.). Somatic embryos capable of differentiating into normal plants from embryogenic tissues initiated from immature zygotic embryos were first reported in the Norway spruce (Chalupa 1985; Hakman and von Arnold 1985); since then, plant regeneration via somatic embryogenesis has been reported in many conifers (reviewed by Gupta and Grob 1995; Stasolla and Yeung 2003). In the Cupressaceae, somatic embryogenesis has additionally been reported in six genera, namely, *Chamaecyparis*, *Cryptomeria*, *Cunninghamia*, *Cupressus*, *Juniperus*, and *Sequoia* (Table 13.1). Although reports on stable and effective procedures for somatic embryogenesis and a plant regeneration system within Cupressaceae trees are limited, Hinoki cypress, Sawara cypress (*Chamaecyparis pisifera* Sieb. et Zucc.), and Sugi (*Cryptomeria japonica* D. Don) have been studied intensively.

Hinoki cypress, Sawara cypress, and Sugi are widely distributed in Japan (Farjón 2010). Hinoki cypress and Sugi are the two most important commercial forest species in Japan, collectively covering approximately 65 % of the artificial forest area. The wood quality of Sawara cypress is considered to be inferior to that of Hinoki cypress; however, Sawara cypress grows faster and is much more tolerant of cold conditions (Fukuhara 1978; Maruyama et al. 2002). Breeding projects for Hinoki cypress and Sugi, such as plus-tree selection for growth, yield, stem straightness, and resistance to disease and insects, have been conducted in the past for an extended time. However, allergic reactions to pollen of these species have become a severe public health concern in Japan. Therefore, the production of Hinoki cypress and Sugi with a *no-pollen* trait is an important breeding objective.

Genetic transformation is expected to be a powerful method for shortening the breeding period of trees that require many years for the production of progeny. Simply by the introduction of specific genes, genetic transformation can induce desirable traits without unnecessary genetic transitions. Stable transformation via somatic embryogenesis of conifers has been reported in several genera, including *Picea*, *Larix*, and *Pinus* (reviewed by Malabadi and Nataraja 2007). However, stable transformation of Cupressaceae has only been reported for Hinoki cypress and Sugi (Table 13.2).

In this chapter, we will review current approaches for somatic embryogenesis in Cupressaceae trees, focusing primarily on Hinoki cypress and Sugi. We will also review the application of somatic embryogenesis in genetic transformation.

13.2 Culture Initiation

The selection of the explant is a factor that determines success in establishing an embryogenic system. Most examples of somatic embryogenesis shown in Table 13.1 used immature zygotic embryos as explants. Although somatic embryos have not been successfully obtained from immature zygotic embryos of redwood (*Sequoia sempervirens*), they have been obtained from mature embryos, cotyledons, hypocotyls, and needles (Bourgkard and Favre 1989; Favre et al. 1995; Liu et al. 2006). To date, in the Cupressaceae family, somatic embryogenesis using somatic tissues other than zygotic embryos has been reported several times for Hinoki cypress (Ishii et al. 2003), China fir (*Cunninghamia lanceolata*; Xi and Shi 2005), prickly juniper (*Juniperus oxycedrus*; Gomez and Segura 1996), and redwood (Favre et al. 1995; Liu et al. 2006). The efficiency of somatic embryogenesis reported for the aforementioned studies was very low: less than five somatic embryos per callus. Moreover, only one report on the somatic embryogenesis and organogenesis from the in vitro needle leaves derived from mature trees has been published for redwood (Liu et al. 2006). However, utilizing needle explants to initiate embryogenic systems is effective in inducing somatic embryogenesis from the

Table 13.1 Somatic embryogenesis of Cupressaceae trees

Species	Initial explants	Culture response[a]	Established in soil[b]	References
Chamaecyparis obtusa	Seedling shoots	SE, PL	Yes	Ishii et al. (2003)
	Immature zygotic embryos	EMT, SE, PL	Yes	Taniguchi et al. (2004a)
				Maruyama et al. (2005)
Chamaecyparis pisifera	Immature zygotic embryos	EMT, SE, PL	Yes	Maruyama et al. (2002)
Cryptomeria japonica	Immature, mature zygotic embryos	EMT, SE	–	Ogita et al. (1999)
	Immature zygotic embryos	EMT, SE, PL	–	Igasaki et al. (2003a)
				Igasaki et al. (2003b)
				Nakagawa et al. (2006)
	Immature zygotic embryos	EMT, SE, PL	Yes	Maruyama et al. (2000)
				Igasaki et al. (2006)
				Maruyama and Hosoi (2007)
Cunninghamia lanceolata	Cotyledons, hypocotyls	SE, PL	–	Xi and Shi (2005)
	Mature zygotic embryos	SE, PL	–	Xi and Shi (2006)
Cupressus sempervirens	Immature zygotic embryos	EMT	–	Lambardi et al. (1995)
	Immature zygotic embryos	EMT, SE, PL	–	Sallandrouze et al. (1999)
				Lambardi (2000)
Juniperus communis	Immature zygotic embryos	EMT, SE, PL	–	von Arnold and Helmersson (2009)
Juniperus oxycedrus	Needles	EMT, SE	–	Gomez and Segura (1996)
Sequoia sempervirens	Mature zygotic embryos, cotyledons, hypocotyls	EMT, SE, PL	–	Bourgkard and Favre (1988)
				Bourgkard and Favre (1989)
				Favre et al. (1995)
	Needles	EMT, SE, PL	–	Liu et al. (2006)

[a]Abbreviation: *EMT* embryogenic tissues, *SE* somatic embryos, *PL* plantlets
[b]–, no data

clonal genotype that has revealed traits as a plus tree. In addition, when using needles from in vitro shoot cultures, the explants can be supplied all year round.

Genotype affects the ability to induce embryogenic tissues. In Hinoki cypress, the induction frequency of embryogenic tissues from open-pollinated immature zygotic embryos varied from 8 % to 48 % among ten plus-tree clones (Taniguchi et al. 2004a). The induction and maintenance rates in Sugi differed considerably among 20 plus-tree clones (Taniguchi and Kondo 2000), varying from 7.5 % to 78.5 % and from 1.2 % to 27.8 %, respectively. In addition, there was a positive correlation between the initiation and maintenance rates among these clones, thus indicating that clones that initiate embryogenic tissues at high frequency tend to maintain at a high frequency during subculture. Lambardi et al. (1994, 2000) showed that embryos from

Table 13.2 Genetic transformation of Cupressaceae trees

Species	Explants	Transformation method	Transgene[a]	Stable transformation[b]	References
Chamaecyparis nootkatensis	Pollens	Particle bombardment	*GUS*, *NPTII*	–	Hay et al. (1994)
Chamaecyparis obtusa	Shoot primordia	Particle bombardment	*GUS*, *BAR*	Yes	Ishii (2002)
	Shoot primordia	Particle bombardment	*GFP*, *NTPII*	–	Taniguchi et al. (2004b)
	Embryogenic tissues	Agrobacterium	*GFP*, *NTPII*	Yes	Taniguchi et al. (2005)
Cryptomeria japonica	Embryogenic tissues	Particle bombardment	*GUS*, *NPTII*	Yes	Maruyama et al. (2000)
	Zygotic embryos	Particle bombardment	*LUC*	–	Mohri et al. (2000)
	Zygotic embryos	Particle bombardment	*GFP*, *NTPII*	–	Taniguchi et al. (2004b)
	Embryogenic tissues	Agrobacterium	*GFP*, *HPT*, *NTPII*	Yes	Taniguchi et al. (2008)
	Embryogenic tissues	Agrobacterium	*GUS*	Yes	Kurita et al. (2013)
	Embryogenic tissues	Agrobacterium	*GFP*, *GUS*, *HPT*, *NTPII*	Yes	Konagaya et al. (2013)
Cupressus sempervirens	Embryogenic tissues	Particle bombardment	*GFP*, *GUS*, *NTPII*	–	Lambardi et al. (1998)
Sequoia sempervirens	Micropropagated shoots	Agrobacterium	pRi	Yes[c]	Mihaljević et al. (1999)

[a]Abbreviation: *BAR* bialaphos resistance, *GFP* green fluorescent protein, *GUS* β-glucuronidase, *HPT* hygromycin phosphotransferase, *LUC* luciferase, *NPTII* neomycin phosphotransferase II, *pRi* T-DNA in *Agrobacterium rhizogenes* pRi plasmid

[b]–, no data

[c]Gene transfer in adventitious roots only

only five of the 24 different clones of Italian cypress were able to initiate embryogenic tissue. Moreover, the optimal seed collection date to efficiently induce embryogenic tissues was different among the clones. Similar results have been reported for Sugi (Maruyama et al. 2000). These results suggest different timing of the developmental stages of embryo among the clones.

Several reports have suggested that seed collection dates are a critical factor for efficiently inducing embryogenic tissues from zygotic embryos (Ogita et al. 1999; Lambardi 2000; Maruyama et al. 2000; Igasaki et al. 2003a; Taniguchi et al. 2004a). In the Italian cypress (*Cupressus sempervirens*), embryogenic tissues have been reported to originate from embryos that were morphologically at the same stage of maturity, i.e., an early cotyledonary stage characterized by the two cotyledons that have just differentiated and are still tightly joined (Lambardi 2000). Taniguchi et al. (2004a) reported that embryogenic tissues of Hinoki cypress were induced from the immature seeds collected at the beginning of July containing pre-cotyledonary embryos (stages 6–11 as defined by Nagmani et al. 1995). During the middle of July, zygotic embryos had cotyledonary primordia (stage 12) or elongated cotyledons (stage 13), but megagametophytes containing such embryos could not induce embryogenic tissue. Similarly, in Sugi, the pre-cotyledonary embryo stage was also optimal (Igasaki et al. 2003a). Therefore, it is important to determine the optimal development stages in each tree species as well as genotypes.

13.2.1 Embryogenic Tissue Induction and Proliferation in Hinoki Cypress

At the beginning of July, we collected immature seeds of Hinoki cypress that contained precotyledonary embryos. Immature seeds were extracted from the sterilized cone, and the seeds were again surface sterilized with 70 % ethanol for 1 min followed by 6 % H_2O_2 for 5 min and then rinsed three times with sterilized distilled water. Explants (megagametophytes that contained intact immature zygotic embryos) were removed from the seeds using tweezers and a scalpel under a stereomicroscope. The explants were placed on Smith standard embryonic tissue capture medium (SM1; Smith 1996) containing sucrose (sugar source) and activated charcoal, but no plant growth regulator. Within 2–3 weeks of the explant being cultured on SM1, embryogenic tissue was induced from the micropylar end of the megagametophyte (Fig. 13.1a). After 4 weeks of culturing on SM1, embryogenic tissues that had been induced from the explants were transferred to Smith standard embryogenesis medium (SM2; Smith 1996) containing amino acids, but no activated charcoal or plant growth regulator, and cultured for 4 weeks. Following this 4-week culture period, the embryogenic tissues were transferred to Smith embryo development medium (SM3; Smith 1996), which had the same composition as SM2, but contained higher concentrations of amino acids and was solidified by Gelrite. The tissues were then subcultured at 2–3-week intervals on this medium. After 10 weeks of culture, the clumps of embryogenic tissue of each cell line varied greatly in size, with the largest being 2 cm in diameter. The embryogenic tissue was able to proliferate on SM3 (Fig. 13.1b), with an approximately threefold growth rate every 3 weeks (Taniguchi et al. 2004a).

13.2.2 Embryogenic Tissue Induction and Proliferation in Sugi

In Sugi, we collected immature seeds early during July, and explants were extracted from the seeds in the same manner as with Hinoki cypress. The explants were placed on initiation medium (IM), which consisted of half-strength MS medium (Murashige and Skoog 1962) supplemented with 10 μM 2,4-dichlorophenoxyacetic acid (2,4-D), 30 g/l sucrose, and 8 g/l agar (Taniguchi and Kondo 2000). Within 2–3 weeks of the explant being cultured on IM, embryogenic tissue was induced from the micropylar end of the megagametophyte (Fig. 13.2a, b). After 4 weeks of culture on IM, embryogenic tissues were subcultured at 2-week intervals on maintenance medium (MM), which had the same composition as IM, except that it contained 1 g/l casein hydrolysate, 0.5 g/l L-glutamine, 2 μM 2, 4-D, and 50 % of the inorganic nitrate salt composition of IM and was gelled using 4.5 g/l Gelrite (Fig. 13.2c; Taniguchi and Kondo 2000; Taniguchi et al. 2008).

13.3 Somatic Embryo Maturation and Germination

In general, somatic embryo maturation is accelerated by adding abscisic acid (ABA) or a penetrating osmoticant (e.g., sugars, sugar alcohols) or nonpenetrating osmoticant [e.g., polyethylene glycol (PEG)] to the culture medium, which induces water stress (Lipavská and Konrádová 2004). Mature somatic embryos in Hinoki cypress were efficiently formed on the medium containing a combination of 100 μM ABA, 30 g/l maltose as sugars, and 150 g/l PEG (Taniguchi et al. 2004a). These results are consistent with those for other trees of Cupressaceae, such as Sawara cypress (Maruyama et al. 2002) and Sugi (Taniguchi 2007). Furthermore, the effect of various concentrations of Gelrite as a support medium on somatic embryo yield was investigated in the Hinoki cypress, which indicated that the most suitable concentration was 3.0 g/l (Taniguchi et al. 2004a). Although Maruyama et al. (2002) induced the somatic embryos of Sawara cypress on medium containing 5.0 g/l Gelrite, they did not test a lower concentration of Gelrite for somatic embryo induction. In *Pinus taeda*, somatic embryos were induced on a medium

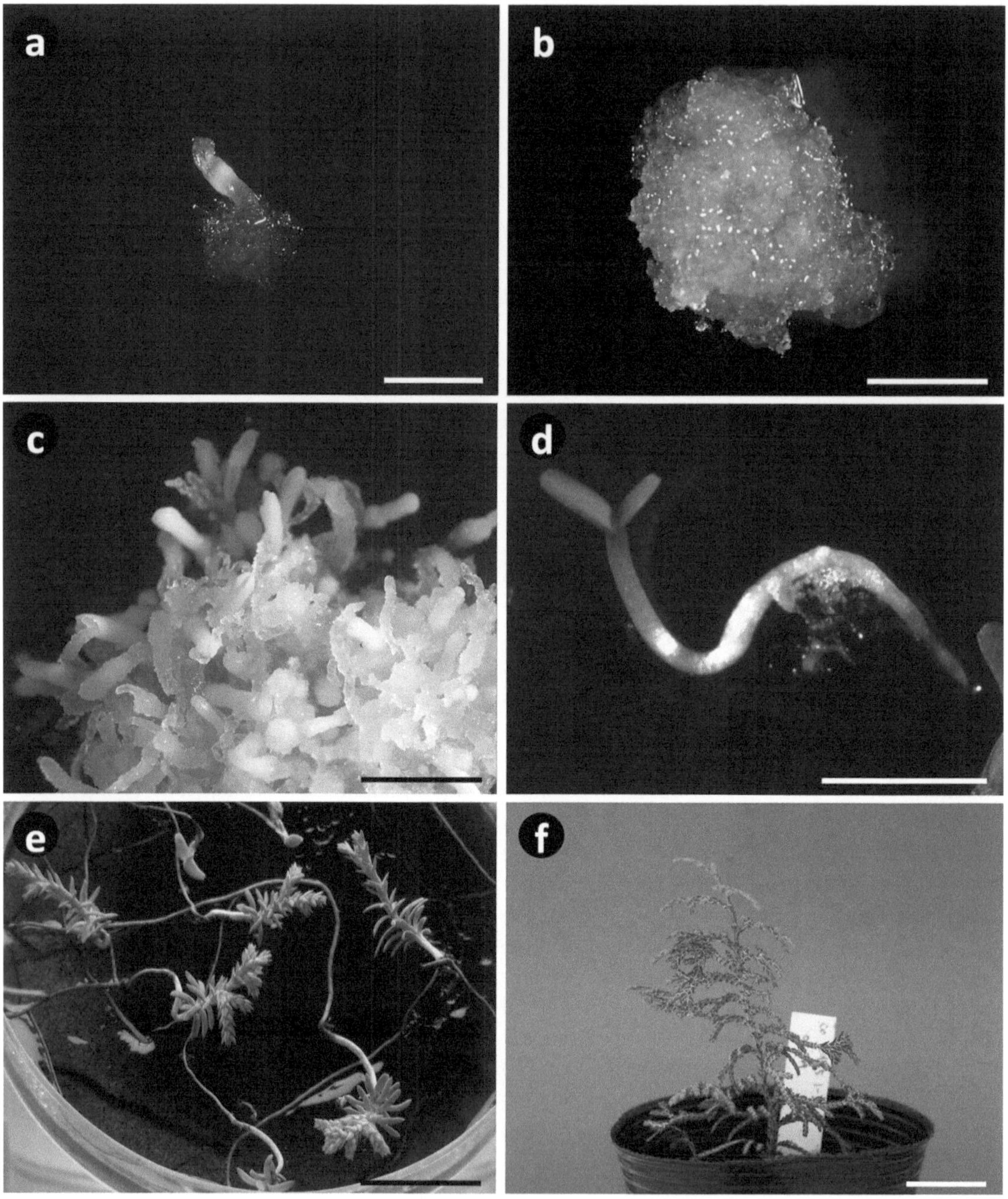

Fig. 13.1 Plant regeneration via somatic embryogenesis from immature embryo cultures in Hinoki cypress (*Chamaecyparis obtusa*; modified after Taniguchi et al. 2004a). (**a**) Embryogenic tissues induced from a megagametophyte explant. (**b**) Embryogenic tissue maintained on SM3 medium. (**c**) Cotyledonary somatic embryos after 8 weeks of culture on maturation medium. (**d**) Germination of a somatic embryo after 1 week of culture on germination medium. (**e**) Plantlets after 2 months of culture on germination medium. (**f**) Regenerated plant following acclimatization. *Bars*: 2 mm (**a**); 4 mm (**b**, **c**, **d**); 2 cm (**e**, **f**)

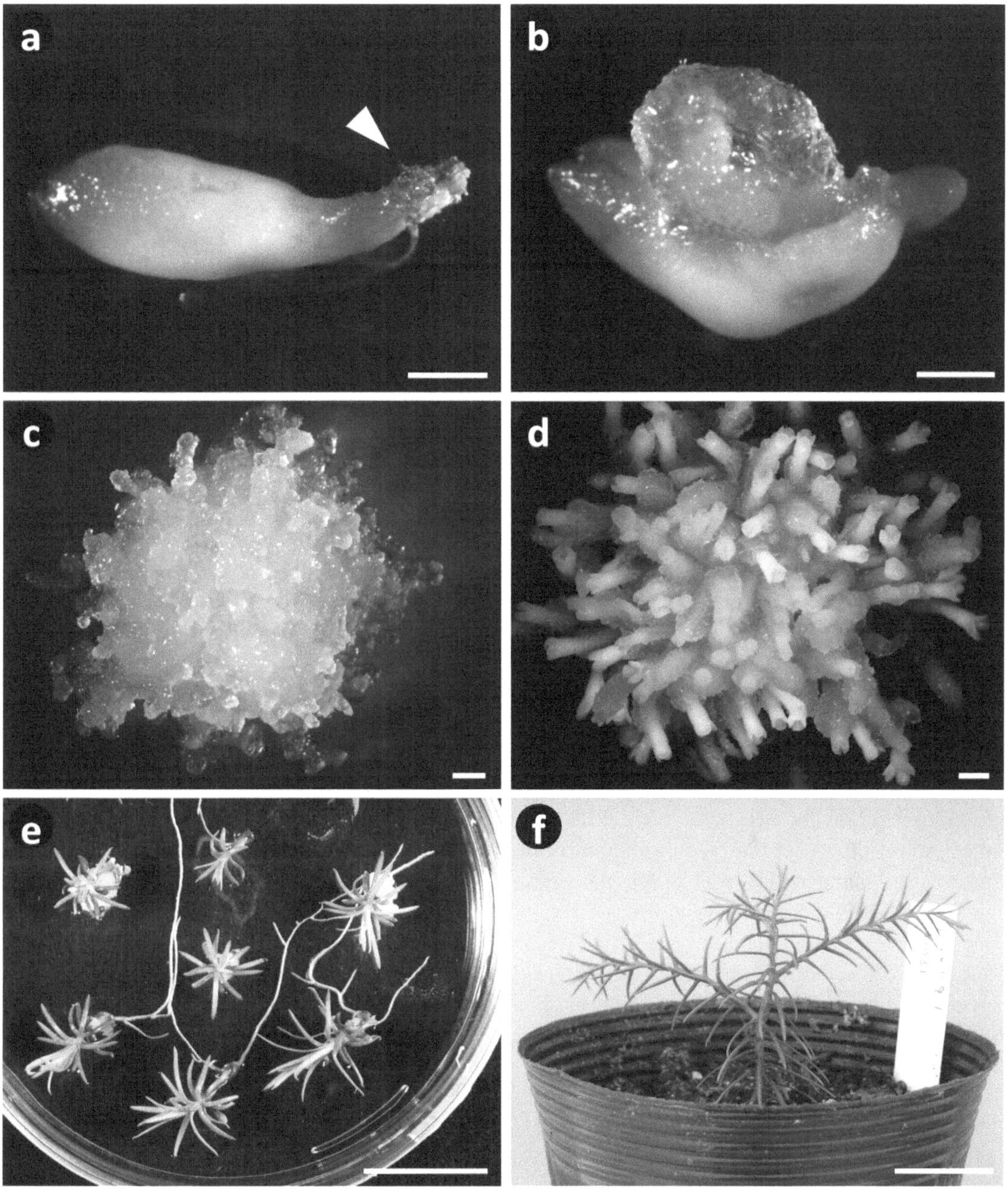

Fig. 13.2 Plant regeneration via somatic embryogenesis from immature embryo cultures in Sugi (*Cryptomeria japonica*). (**a**) Embryogenic tissues induced from the micropylar end of the megagametophyte. (**b**) Embryogenic tissues induced from an incision site in the megagametophyte (**c**) Embryogenic tissue maintained on MM medium. (**d**) Cotyledonary somatic embryos after 6 weeks of culture on maturation medium. (**e**) Plantlets after 2 months of culture on germination medium. (**f**) Regenerated plant following acclimatization. *Bars*: 1 mm (**a, b, c, d**); 2 cm (**e, f**)

containing 2.5 g/l Gelrite (Pullman et al. 2003). A high concentration of Gelrite is thought to be unsuitable for somatic embryogenesis, as it has been reported that Gelrite at high concentrations reduces the bioavailability of magnesium, calcium, zinc, and manganese in the medium (Van Winkle et al. 2003).

Recently, research has determined the peptide hormones that play an important role in plant cell growth and development (Ryan and Pearce 2001). An example of these hormones is phytosulfokine (PSK), which is a five-amino-acid sulfated peptide that is involved in the initial steps of cellular differentiation, proliferation, and redifferentiation (Igasaki et al. 2003b; Matsubayashi et al. 2004). PSK has also been shown to stimulate somatic embryogenesis in carrot, Sugi, Japanese larch, and *Daucus* species (Kobayashi et al. 1999; Igasaki et al. 2003b; Umehara et al. 2005; Mackowska et al. 2014) as well as induce cell division in suspension and protoplast cultures (Matsubayashi and Sakagami 1996; Matsubayashi et al. 1997; Grzebelus et al. 2012). It has also been reported to improve *Agrobacterium*-mediated transformation efficiency (Matsubayashi et al. 2004; Chen et al. 2005). Igasaki et al. (2003b) described how the addition of PSK to the medium results in not only the promotion of the growth and maintenance of embryogenic tissues in Sugi but also a dramatic stimulatory effect on the formation of somatic embryos. These results suggest the possibility of improving the efficiency of the embryogenic and transformation systems in conifers by utilization of PSK.

The ability of embryogenic tissues to produce somatic embryos is also affected by the genotype. We produced somatic embryos for nine clones among ten open-pollinated plus-tree clones of Hinoki cypress, and somatic embryo-forming frequency varied from 4.0 % to 24 % (number of somatic embryo forming cell lines per explants). Somatic embryos of Sugi were produced for all clones among six open-pollinated plus-tree clones, and the somatic embryo-forming frequency varied from 4.0 % to 20 % (Taniguchi 2007). Recently, we investigated the somatic embryo-forming frequency for immature zygotic embryos yielded from artificial crossing among plus trees. As a result, a certain family formed between the specific pollen and seed parent was shown to efficiently produce somatic embryos (unpublished data). This result suggests the possibility that this family harbors the genes to improve somatic embryogenesis.

13.3.1 Somatic Embryo Maturation in Hinoki Cypress and Sugi

To facilitate the maturation of somatic embryos in Hinoki cypress and Sugi, the embryogenic tissues subcultured on SM3 were transferred to a maturation medium (Taniguchi et al. 2004a) consisting of SM3 basal salts, vitamins, amino acids, 2 g/l activated charcoal, 100 μM abscisic acid, 150 g/l polyethylene glycol 4000, 3.0 g/l Gelrite, and 30 g/l maltose. The first cotyledonary embryos appeared at approximately 5 weeks of culture and presented in high numbers by 7–8 weeks of culture (Figs. 13.1c and 13.2d).

13.3.2 Germination in Hinoki Cypress and Sugi

For germination in Hinoki cypress, the mature embryos were cultured on woody plant medium (Lloyd and McCown 1980), which contained 2 g/l activated charcoal, 20 g/l sucrose, and 5 g/l Gelrite. Within 1 week of the mature embryos being placed on germination medium, the embryos rooted and elongated their hypocotyls and cotyledons (Fig. 13.1d). Although the frequency of germination on the medium without activated charcoal was comparable to that on the medium containing charcoal, the germinant grew less vigorously. This result demonstrated that activated charcoal is effective in facilitating the germination of mature somatic embryos in Hinoki cypress (Taniguchi et al. 2004a). After 2 months of culture on germination medium, the germinants with elongated epicotyls (approx. 1 cm long, Fig. 13.1e) were transferred to pots containing soil for acclimatization. All plantlets successfully acclimated and continued growing in the greenhouse (Fig. 13.1f).

The mature embryos of Sugi were cultured on germination medium, which was a modified version of the Gresshoff and Doy medium (Okamura and Kondo 1995) containing 10 g/l sucrose and 5 g/l Gelrite. After 6 weeks of the mature embryos being placed on germination medium, the embryos rooted and elongated their hypocotyls and cotyledons (Fig. 13.2e). The germinants were subcultured every other month on this germination medium. After approximately 3 months, plantlets were transferred to half-strength woody plant medium, which contained 20 g/l sucrose and 2 g/l Gelrite. Elongated plantlets (approx. 10 cm long) were transferred to pots containing soil for acclimatization (Fig. 13.2f).

13.4 Genetic Transformation

Genetic transformation within Cupressaceae was first attempted by particle bombardment using yellow cypress (*Chamaecyparis nootkatensis*) pollen (Hay et al. 1994) and Italian cypress embryogenic tissues (Lambardi et al. 1998). However, only transient expression was observed. By employing *Agrobacterium rhizogenes* to introduce T-DNA derived from Ri plasmid, Mihaljević et al (1999) succeeded in inducing adventitious roots in micropropagated shoot explants of redwood. However, whole plant regeneration of stable transformants has only been reported in Hinoki cypress and Sugi.

Gene delivery to shoot primordia of Hinoki cypress by particle bombardment and the subsequent regeneration of the transformed plantlets were first reported by Ishii (2002). In an earlier investigation, Taniguchi et al. (2004b) similarly attempted the genetic transformation of Hinoki cypress following this method, and although transiently transformed cells were obtained, they failed to produce transformed plants (Taniguchi et al. 2005). With regard to Sugi, there are some reports of particle bombardment, for example, transient expression of the luciferase gene (Mohri et al. 2000) and green fluorescent protein (*GFP*) gene (Taniguchi et al. 2004b) in zygotic embryos. Maruyama et al. (2000) reported bud formation from embryos after introducing the β-glucuronidase gene. Generally, the *Agrobacterium*-mediated method obtains a higher transformation frequency of conifers than the particle bombardment method. Accordingly, we have developed *Agrobacterium tumefaciens*-mediated transformation methods of embryogenic tissue in Hinoki cypress (Taniguchi et al. 2005) and Sugi (Taniguchi et al. 2008, Konagaya et al. 2013). The highest transformation frequency obtained in Hinoki cypress was 22.5 independent transformed lines per dish (250 mg embryogenic tissue). Moreover, we attempted to improve the transformation efficiency in Sugi (Konagaya et al. 2013).

13.4.1 Culture Supports for Cocultivation

Unsuitable cocultivation conditions in *Agrobacterium*-mediated transformation may lead to unfavorable effects, such as bacterial overgrowth and/or tissue necrosis, thereby reducing the transformation efficiency. It has been suggested that an optimal concentration of *Agrobacterium* is required for transformation. Recently, it was reported that the use of filter paper wicks during cocultivation contributed to the efficient transformation of cucumber, kabocha squash, and rice by controlling the growth of *Agrobacterium* (Ozawa 2009; Nanasato et al. 2011, 2013). Therefore, we attempted to determine the optimal cocultivation conditions with filter paper wicks in Sugi. After the embryogenic tissues and *Agrobacterium* were suspended in liquid medium, a filter paper with the embryogenic tissues was placed on a solid cocultured medium or on three sheets of filter paper moistened with 5.5 mL of liquid cocultured medium. The mean number of *Agrobacterium* cells (>17-fold) obtained in the liquid medium conditions was significantly lower than that obtained on the solid medium using the conventional method (Taniguchi et al. 2008).

Compared with cocultivation on the solid medium, culturing in the liquid medium significantly improved the transformation efficiency (>16-fold). Ozawa (2009) showed that liquid medium-moistened filter paper wicks regulated the growth rate of *Agrobacterium* in an effective manner, which improved the cell viability in the transformed callus obtained from rice. Previously, Nanasato et al. (2011, 2013) showed that filter paper wicks increased the *Agrobacterium* infection efficiency in cucumber and kabocha squash. These reports and our experimental results support the utility of filter paper wicks in cocultivation procedures.

13.4.2 Antibiotics to Eliminate *Agrobacterium*

An efficient *Agrobacterium*-mediated transformation system requires the use of antibiotics to eliminate bacteria; the antibiotics should have negligible effects on the growth potential of the transformed cells. Recently, it was reported that meropenem, a novel β-lactam antibiotic, is highly effective against *Agrobacterium*. Also, the transformation efficiencies observed in tobacco, tomato, rice, and apple were higher with meropenem than with other β-lactam antibiotics, such as cefotaxime, carbenicillin, and cefbuperazone (Ogawa and Mii 2004, 2007; Li et al. 2011). In contrast to meropenem, it has been observed that cefotaxime and carbenicillin have negative effects on the growth of embryonic tissue, somatic embryogenesis, and shoot regeneration in woody plants such as Norway spruce, cacao, orange, and pomegranate (Terakami et al. 2007; Malá et al. 2009; Mendes et al. 2009; Silva et al. 2009). However, meropenem had no inhibitory effects on the embryogenic tissue growth in Sugi, even at a higher concentration of 40 mg/l. Similar results with meropenem have been reported using Norway spruce embryogenic tissue (Malá et al. 2009). Moreover, embryogenic tissue growth was significantly higher at lower concentrations (5 mg/l and 10 mg/l) using medium containing meropenem. Based on these results, 10 mg/l was selected as the meropenem concentration in subsequent transformation experiments.

Application of meropenem treatment on Sugi transformation indicated that the transformation efficiency was increased approximately twofold with meropenem than with carbenicillin using the conventional method. In three replicate transformation experiments using our improved method (cocultivation on filter paper wicks and *Agrobacterium* elimination with meropenem), the mean transformation efficiency (the number of GFP-positive colonies per gram of cocultivated embryogenic tissues) was 105.3±9.02. By contrast, the mean transformation efficiency achieved using the conventional method (cocultivation on Gelrite-solidified medium and *Agrobacterium* elimination by carbenicillin) was 3.5±0.71. Thus, the transformation efficiency was increased approximately 30-fold by the improved method than with the conventional method. Moreover, the regrowth of *Agrobacterium* was occasionally observed after cocultivation when using the conventional method. However, no agrobacterial regrowth was observed with the improved method. Similar regrowth control using meropenem was reported after the transformation of *Phalaenopsis* (Sjahril and Mii 2006). These results suggest that the improved method prevents excess *Agrobacterium* growth during cocultivation, and transformed embryogenic tissues are efficiently selected using meropenem, even at low concentrations. High transformation efficiency values (colonies per gram of cocultivated embryogenic tissues) have been reported in conifer plants, such as 42 in *Larix* (Lelu and Pilate 2000), 67.3 in *Pinus* (Trontin et al. 2002), 60–1280 in *Picea* (Klimaszewska et al. 2001), and 90 in Hinoki cypress (Taniguchi et al. 2005). Thus, the transformation efficiency value that we achieved is one of the highest compared with those obtained using previously reported methods.

13.5 Conclusions and Future Prospects

Much research into somatic embryogenesis in Cupressaceae indicates that factors such as explant types and the developmental stage, genotypes, composition, and the supplements of the culture medium play an important role in improving the efficiency of embryogenic tissue induction and somatic embryogenesis. The *Agrobacterium*-

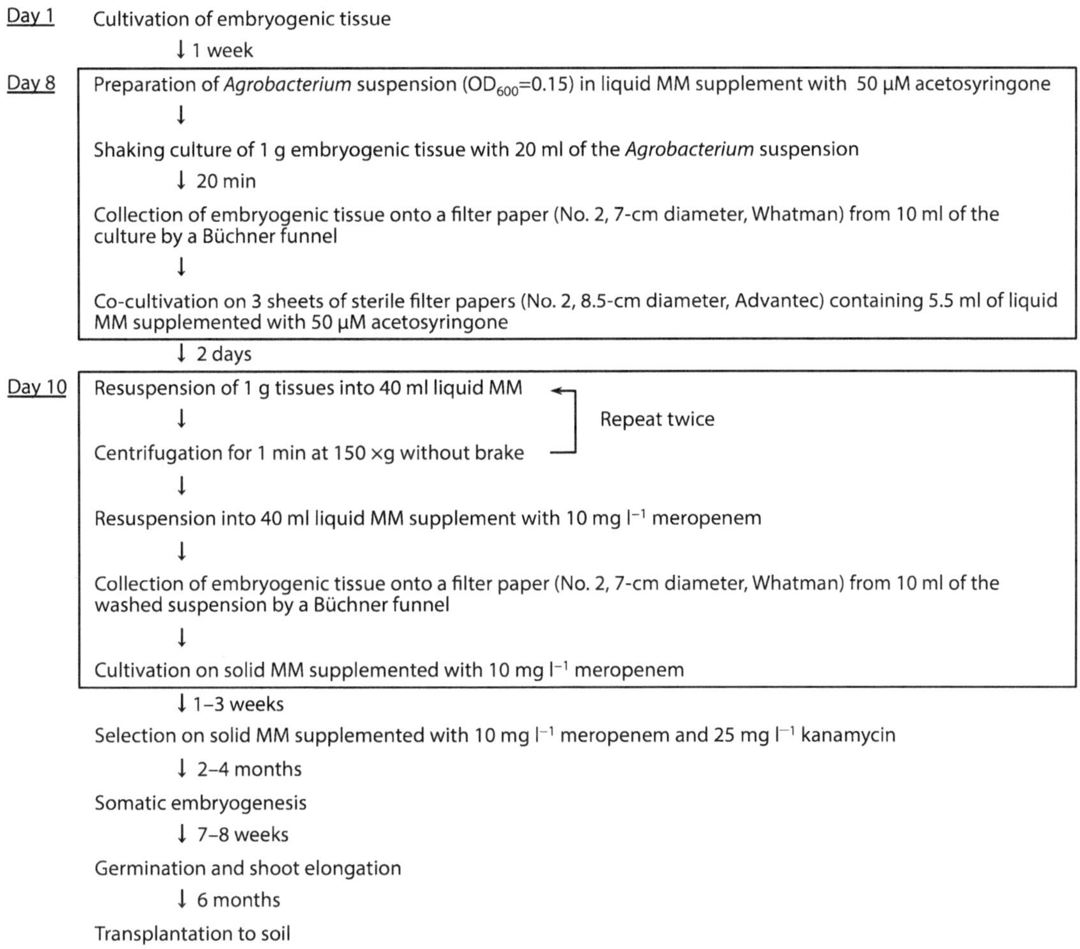

Fig. 13.3 Steps in the transformation of Sugi (Konagaya et al. 2013)

mediated transformation frequency of Sugi significantly increased using the improved method (Fig. 13.3). Currently, we have succeeded in producing the *no-pollen* Sugi by introducing a combination of the male gametophyte-specific promoter (Kurita et al. 2013) and the cytotoxic gene (manuscript in preparation). Although reports of a stable and effective procedure for somatic embryogenesis and genetic transformation of Cupressaceae are limited to *Chamaecyparis* trees and Sugi, these effective methods might be applied to other coniferous trees as well as other Cupressaceae trees. We hope that advances in clonal propagation and molecular breeding technology are accelerated in many coniferous trees.

Acknowledgment This work was partly supported by Grant-in-Aid (Development of Technologies for Control of Pollen Production by Genetic Engineering) from the Forest Agency of Japan.

References

Adams PR, Bartel JA, Price RA (2009) A new genus, *Hesperocyparis*, for the cypresses of the Western Hemisphere (Cupressaceae). Phytologia 91:160–185

Bourgkard F, Favre JM (1988) Somatic embryos from callus of *Sequoia sempervirens*. Plant Cell Rep 7:445–448

Bourgkard F, Favre JM (1989) Somatic embryogenesis in *Sequoia sempervirens*: present possibilities and limits. Annales de Recherches Sylvicoles. Association Foret-Cellulose (AFOCEL), Paris, France, pp 83–95

Chalupa V (1985) Somatic embryogenesis and plantlet regeneration from cultured immature and mature embryos of *Picea abies* (L.). Karst Commun Inst For 14:57–63

Chen D, Xu Z, Ma B, Li S (2005) Effect of phytosulfokine-α on *Agrobacterium*-mediated transformation in rice. Rice Sci 12:255–260

Debreczy Z, Musial K, Price RA, Rácz I (2009) Relationships and nomenclatural status of the nootka cypress (*Callitropsis nootkatensis*, Cupressaceae). Phytologia 91:140–158

Farjón A (2005) A monograph of Cupressaceae and Sciadopitys. Royal Botanic Gardens, Kew, Richmond, Surrey

Farjón A (2008) A natural history of conifers. Timber Press, Portland

Farjón A (2010) A handbook of the world conifers, vol 1. Brill Press, Leiden

Favre JM, Tranvan H, Arnaud Y, Bourgkard F (1995) Present state of somatic embryogenesis m *Sequoia sempervirens* (Lamb.) Endl. In: Jain SM, Gupta PK, Newton RJ (eds) Somatic embryogenesis in woody plants, vol 3., pp 357–370

Fukuhara N (1978) Meiotic observation in the pollen mother cell of interspecific hybrid between *Chamaecyparis obtusa* and *C. pisifera*. J Jpn For Soc 60:437–441

Gomez MP, Segura J (1996) Morphogenesis in leaf and single-cell cultures of mature Juniperus oxycedrus. Tree Physiol 16:681–686

Grzebelus E, Szklarczyk M, Gren J, Sniegowska K, Jopek M, Kacinska I, Mrozek K (2012) Phytosulfokine stimulates cell divisions in sugar beet (*Beta vulgaris* L.) mesophyll protoplast cultures. Plant Growth Regul 67:93–100

Gupta PK, Grob JA (1995) Somatic embryogenesis in conifers. In: Jain SM, Gupta PK, Newton RJ (eds) Somatic embryogenesis in woody plants, vol 1. Kluwer, Dordrecht, pp 81–98

Hakman I, von Arnold S (1985) Plantlet regeneration through somatic embryogenesis in *Picea abies* (Norway spruce). J Plant Physiol 121:149–158

Hay I, Lachance DA, von Aderkas P, Charest PJ (1994) Transient chimeric gene expression in pollen of five conifer species following microparticle bombardment. Can J For Res 24:2417–2423

Igasaki T, Sato T, Akashi T, Mohri T, Maruyama E, Kinoshita I, Walter C, Shinohara K (2003a) Somatic embryogenesis and plant regeneration from immature zygotic embryos of *Cryptomeria japonica* D. Don. Plant Cell Rep 22:239–243

Igasaki T, Akashi T, Ujino-Ihara T, Matsubayashi M, Sakagami Y, Shinohara K (2003b) Phytosulfokine stimulates somatic embryogenesis in *Cryptomeria japonica*. Plant Cell Physiol 44:1412–1416

Igasaki T, Akashi T, Shinohara K (2006) Somatic embryogenesis in *Cryptomeria japonica* D. Don: gene for phytosulfokine (PSK) precursor. Somatic Embryogenesis. Plant Cell Monogr 2:201–213

Ishii K (2002) Liquid culture and transformation of Hinoki cypress (*Chamaecyparis obtusa* Sieb. et Zucc.). J For Res 7:99–104

Ishii K, Maruyama E, Hosoi Y (2003) Plant regeneration by somatic embryogenesis from *in vitro*–cultured shoots of Hinoki cypress (*Chamaecyparis obtusa* Sieb. et Zucc). Propag Ornam Plant 3:19–22

Klimaszewska K, Lachance D, Pelletier G, Lelu M-A, Séguin A (2001) Regeneration of transgenic *Picea glauca*, *P. mariana*, and *P. abies* after cocultivation of embryogenic tissue with *Agrobacterium tumefaciens*. In Vitro Cell Dev Biol Plant 37:748–755

Kobayashi T, Eun CH, Hanai H, Matsubayashi Y, Sakagami Y, Kamada H (1999) Phytosulphokine-α, a peptidyl plant growth factor, stimulates somatic embryogenesis in carrot. J Exp Bot 50:1123–1128

Konagaya K, Kurita M, Taniguchi T (2013) High-efficiency *Agrobacterium*-mediated transformation of *Cryptomeria japonica* D. Don by co-cultivation on filter paper wicks followed by meropenem treatment to eliminate *Agrobacterium*. Plant Biotechnol 30:523–528

Konar RN, Oberoi YP (1965) *In vitro* development of embryoids on the cotyledons of *Biota orientalis*. Phytomorph 15:137–140

Kurita M, Konagaya K, Watanabe A, Kondo T, Ishii K, Taniguchi T (2013) The promoter of an *A9* homolog from the conifer *Cryptomeria japonica* imparts male strobilus-dominant expression in transgenic trees. Plant Cell Rep 32:319–328

Lambardi M (2000) Somatic embryogenesis in cypress (*Cupressus sempervirens* L.). In: Jain SM, Gupta PK, Newton RJ (eds) Somatic embryogenesis in woody plants, vol 6. Kluwer, Dordrecht, pp 553–567

Lambardi M, Menabeni D, Wilson SM (1994) Sviluppo di embrioni somatici ed isolamento di protoplasti da tessuto sospensoriale embriogenico di cipresso (*Cupressus sempervirens* L.). Monti Boschi 45:53–59

Lambardi M, Harry IS, Menabeni D, Thorpe TA (1995) Organogenesis and somatic embryogenesis in *Cupressus sempervirens*. Plant Cell Tiss Org Cult 40:179–182

Lambardi M, Lachance D, Séguin A, Charest PJ (1998) Evaluation of microprojectile-mediated DNA delivery and reporter genes for genetic transformation of the Mediterranean cypress (*Cupressus sempervirens* L.). Plant Cell Rep 18:198–202

Lelu MA, Pilate G (2000) Transgenic in *Larix*. In: Jain SM, Minocha SC (eds) Molecular biology of woody plants, vol 2. Kluwer, Dordrecht, pp 119–134

Li J, Komori S, Sasaki K, Mimida N, Matsumoto S, Wada M, Soejima J, Ito Y, Masuda T, Tanaka N, Shigeta N (2011) Pre-culture before *Agrobacterium* infection to leaf segments and meropenem improves the transformation efficiency of apple (*Malus × domestica* Borkh.). J Jpn Soc Hortic Sci 80:244–254

Lipavská H, Konrádová H (2004) Somatic embryogenesis in conifers: the role of carbohydrate metabolism. In Vitro Cell Dev Biol Plant 40:23–30

Liu C, Xia X, Yin W, Huang L, Zhou J (2006) Shoot regeneration and somatic embryogenesis from needles of redwood (*Sequoia sempervirens* (D.Don.) Endl.). Plant Cell Rep 256:21–628

Lloyd G, McCown B (1980) Commercially feasible micropropagation of mountain laurel *Kalmia latifolia*, by use of shoot tip culture. Proc Int Plant Propag Soc 30:421–427

Mackowska K, Jarosz A, Grzebelus E (2014) Plant regeneration from leaf-derived protoplasts within the *Daucus* genus: effect of different conditions in alginate embedding and phytosulfokine application. Plant Cell Tiss Org Cult 117:241–252

Malá J, Pavingerová D, Cvrčková H, Bříza J, Dostál J, Šíma P (2009) Tolerance of Norway spruce (*Picea abies* [L.] Karst.) embryogenic tissue to penicillin, carbapenem and aminoglycoside antibiotics. J For Sci 55:156–161

Malabadi RB, Nataraja K (2007) Genetic transformation of conifers: applications in and impacts on commercially forestry. Transgenic Plant J 1:289–313

Maruyama E, Hosoi Y (2007) Polyethylene glycol enhance somatic embryo production in Japanese cedar (*Cryptomeria japonica* d. Don). Propag Ornam Plant 7:57–61

Maruyama E, Tanaka T, Hosoi Y, Ishii K, Morohoshi N (2000) Embryogenic cell culture, protoplast regeneration, cryopreservation, biolistic gene transfer and plant regeneration in Japanese Cedar (*Cryptomeria japonica* D. Don). Plant Biotechnol 17:281–296

Maruyama E, Hosoi Y, Ishii K (2002) Somatic embryogenesis in Sawara Cypress (*Chamaecyparis pisifera* Sieb. et Zucc.) for stable and efficient plant regeneration, propagation and protoplast culture. J For Res 7:23–34

Maruyama E, Ishii K, Hosoi Y (2005) Efficient plant regeneration of Hinoki cypress (*Chamaecyparis obtusa*) via somatic embryogenesis. J For Res 10:73–77

Matsubayashi Y, Sakagami Y (1996) Phytosulfokine, sulfated peptides that induce the proliferation of single mesophyll cells of *Asparagus officinalis* L. Proc Natl Acad Sci U S A 93:7623–7627

Matsubayashi Y, Takagi L, Sakagami Y (1997) Phytosulfokine-α, a sulfated pentapeptide, stimulates the proliferation of rice cells by means of specific high-and low-affinity binding sites. Proc Natl Acad Sci U S A 94:13357–13362

Matsubayashi Y, Goto T, Sakagami Y (2004) Chemical nursing: phytosulfokine improves genetic transformation efficiency by promoting the proliferation of surviving cells on selective media. Plant Cell Rep 23:155–158

Mendes AF da S, Cidade LC, Oliveira MLP de, Otoni WC, Soares-Filho W dos S, Costa MGC (2009) Evaluation of novel beta-lactam antibiotics in comparison to cefotaxime on plant regeneration of *Citrus sinensis* L. Osb. Plant Cell Tiss Org Cult 97:331–336

Mihaljević S, Katavić V, Jelaska S (1999) Root formation in micropropagated shoots of *Sequoia sempervirens* using *Agrobacterium*. Plant Sci 141:73–80

Mohri T, Igasaki T, Sato T, Shinohara K (2000) Expression of genes for β-glucuronidase and luciferase in three species of Japanese conifer (*Pinus thunbergii*, *P. densiflora* and *Cryptomeria japonica*) after transfer of DNA by microprojectile bombardment. Plant Biotechnol 17:49–54

Murashige T, Skoog F (1962) A revised medium for rapid growth and bioassays with tobacco tissue cultures. Physiol Plant 15:473–497

Nagmani R, Diner AM, Carton S, Zipf AE (1995) Anatomical comparison of somatic and zygotic embryogeny in conifers. In: Jain SM, Gupta PK, Newton RJ (eds) Somatic embryogenesis in woody plants, vol 1. Kluwer, Dordrecht, pp 23–48

Nakagawa R, Ogita S, Kubo T, Funada R (2006) Effect of polyamines and L-ornithine on the development of proembryogenic masses of *Cryptomeria japonica*. Plant Cell Tiss Org Cult 85:229–234

Nanasato Y, Konagaya K, Okuzaki A, Tsuda M, Tabei Y (2011) *Agrobacterium*-mediated transformation of kabocha squash (*Cucurbita moschata* Duch) induced by wounding with aluminum borate whiskers. Plant Cell Rep 30:1455–1464

Nanasato Y, Konagaya K, Okuzaki A, Tsuda M, Tabei Y (2013) Improvement of *Agrobacterium*-mediated transformation of cucumber (*Cucumis sativus* L.) by combination of vacuum infiltration and co-cultivation on filter paper wicks. Plant Biotechnol Rep 7:267–276

Ogawa Y, Mii M (2004) Screening for highly active β-lactam antibiotics against *Agrobacterium tumefaciens*. Arch Microbiol 181:331–336

Ogawa Y, Mii M (2007) Meropenem and moxalactam: novel β-lactam antibiotics for efficient *Agrobacterium*-mediated transformation. Plant Sci 172:564–572

Ogita S, Ishikawa H, Kubo T, Sasamoto H (1999) Somatic embryogenesis from immature and mature zygotic embryos of *Cryptomeria japonica* I: embryogenic cell induction and its morphological characteristics. J Wood Sci 45:87–91

Okamura M, Kondo T (1995) Manual for tissue culture of pine. Bull For Tree Breed Center 13:139–143

Ozawa K (2009) Establishment of a high efficiency *Agrobacterium*-mediated transformation system of rice (*Oryza sativa* L.). Plant Sci 176:522–527

Pullman GS, Johnson S, Peter G, Cairney J, Xu N (2003) Improving loblolly pine somatic embryo maturation: comparison of somatic and zygotic embryo morphology, germination, and gene expression. Plant Cell Rep 21:747–758

Ryan CA, Pearce G (2001) Polypeptide hormones. Plant Physiol 125:65–68

Sallandrouze A, Faurobert M, Maataoui MEI, Espagnac H (1999) Two-dimensional electrophoretic analysis of proteins associated with somatic embryogenesis development in *Cupressus sempervirens* L. Electrophoresis 20:1109–1119

Silva TER, Cidade LC, Alvim FC, Cascardo JCM, Costa MGC (2009) Studies on genetic transformation of *Theobroma cacao* L.: evaluation of different polyamines and antibiotics on somatic embryogenesis and the efficiency of *uidA* gene transfer by *Agrobacterium tumefaciens*. Plant Cell Tiss Org Cult 99:287–298

Sjahril R, Mii M (2006) High-efficiency *Agrobacterium*-mediated transformation of *Phalaenopsis* using meropenem, a novel antibiotic to eliminate *Agrobacterium*. J Hortic Sci Biotechnol 81:458–464

Smith DR (1996) Growth medium. US Patent 5,565,355

Stasolla C, Yeung EC (2003) Recent advances in conifer somatic embryogenesis: improving somatic embryo quality. Plant Cell Tiss Org Cult 74:15–35

Taniguchi T (2007) Somatic embryogenesis and genetic transformation in forest tree species. Bull For Tree Breed Center 23:63–119

Taniguchi T, Kondo T (2000) Difference in the ability of initiation and maintenance of embryogenic cultures among Sugi (*Cryptomeria japonica* D. Don) seed families. Plant Biotechnol 17:159–162

Taniguchi T, Kurita M, Itahana N, Kondo T (2004a) Somatic embryogenesis and plant regeneration from immature zygotic embryos of Hinoki cypress (*Chamaecyparis obtusa* Sieb. et Zucc.). Plant Cell Rep 23:26–31

Taniguchi T, Ohmiya Y, Kurita M, Kondo T (2004b) Transient expression of the green fluorescent protein gene in *Cryptomeria japonica*, *Chamaecyparis obtusa*, *Pinus densiflora*, *Pinus thunbergii*, *Larix kaempferi* and *Acanthopanax sciadophylloides* following particle bombardment. Bull For Tree Breed Center 20:1–8

Taniguchi T, Kurita M, Ohmiya Y, Kondo T (2005) *Agrobacterium tumefaciens*-mediated transformation of embryogenic tissue and transgenic plant regeneration in *Chamaecyparis obtusa* Sieb. et Zucc. Plant Cell Rep 23:796–802

Taniguchi T, Ohmiya Y, Kurita M, Tsubomura M, Kondo T (2008) Regeneration of transgenic *Cryptomeria japonica* D. Don after *Agrobacterium tumefaciens*-mediated transformation of embryogenic tissue. Plant Cell Rep 27:1461–1466

Terakami S, Matsuta N, Yamamoto T, Sugaya S, Gemma H, Soejima J (2007) Agrobacterium-mediated transformation of the dwarf pomegranate (*Punica granatum* L. var. *nana*). Plant Cell Rep 26:1243–1251

Trontin JF, Harvengt L, Garin E, Lopez-Vernaza M, Arancio L, Hoebeke J, Canlet F, Pâques M (2002) Towards genetic engineering of maritime pine (*Pinus pinaster* Ait.). Ann For Sci 59:687–697

Umehara M, Ogita S, Sasamoto H, Eun CH, Matsubayashi Y, Sakagami Y, Kamada H (2005) Two stimulatory effects of the peptidyl growth factor phytosulfokine during somatic embryogenesis in Japanese larch (*Larix leptolepis* Gordon). Plant Sci 169:901–907

Van Winkle SC, Johnson S, Pullman GS (2003) The impact of gelrite and activated carbon on the elemental composition of two conifer embryogenic tissue initiation media. Plant Cell Rep 21:1175–1182

von Arnold S, Helmersson A (2009) Embryogenic cell lines of *Juniperus communis*; easy establishment and embryo maturation, limited germination. Plant Cell Tiss Org Cult 96:211–217

Xi M, Shi J (2005) Organogenesis and somatic embryogenesis from cotyledons and hypocotyls of *Cunninghamia lanceolata* Hook. Mol Plant Breed 3:846–852

Xi M, Shi J (2006) Organogenesis and somatic embryogenesis from mature zygotic embryos of *Cunninghamia lanceolata*. Scientia Silvae Sinicae 42:29–35

Yang ZY, Ran JH, Wang XQ (2012) Three genome-based phylogeny of Cupressaceae s.l.: further evidence for the evolution of gymnosperms and Southern Hemisphere biogeography. Mol Phylogenet Evol 64:452–470

Somatic Embryogenesis: The Molecular Network Regulating Embryo Formation

14

Mohamed Elhiti and Claudio Stasolla

Abstract

Somatic embryogenesis in plants is a process by which embryos can be produced from somatic cells cultured under specific conditions. A key initial step is represented by the ability of some cells within the explants to dedifferentiate, i.e., reacquire a "young" or immature state, and then redirect their fate into an embryogenic pathway, demarked by precise changes in gene expression. While the initial morphological patterns of somatic embryo formation can be quite different and difficult to categorize, developing somatic embryos can be assigned similar stages ascribed to zygotic embryos. These similarities allow the utilization of somatic embryogenesis as a model system to investigate physiological and molecular events governing zygotic embryogenesis. The aim of this chapter is to provide a general overview of somatic embryogenesis, by describing and analyzing several in vitro embryogenic systems, and to decipher the molecular network responsible for the generation of somatic embryos.

Keywords

Auxin • Embryogenesis • Microspore-derived embryos • Somatic embryos • Root apical meristem • Shoot apical meristem

14.1 Introduction

In flowering plants, embryogenesis is demarked by the fusion of the haploid gametes, i.e., egg and sperm cells, which through a double fertilization process form a single-celled zygote and an endosperm cell. The subsequent development of the zygote is referred to as embryogenesis, during which the zygote forms a fully developed embryo through very precise apical-basal and radial cell

M. Elhiti
Department of Botany, Faculty of Science, Tanta University, Tanta, Egypt, 31527

C. Stasolla (✉)
Department of Plant Science, University of Manitoba, Winnipeg, MB R3T 2N2, Canada
e-mail: stasolla@ms.umanitoba.ca

A. Mujib (ed.), *Somatic Embryogenesis in Ornamentals and Its Applications*,
DOI 10.1007/978-81-322-2683-3_14

division and differentiation patterns (De Smet et al. 2010). A fully developed embryo generally consists of one or more cotyledons surrounding a shoot apical meristem (SAM), an embryonic axis, and a root apical meristem (RAM).

Plant embryogenesis is characterized by three unique features that have immense implications on the elaboration of the different embryonic tissues and organs. Firstly, unlike animal cells, plant cells do not migrate during morphogenesis. Therefore, the final shape of the organism is the mere result of cell division and expansion. Secondly, the plant embryo is not a miniature plant, as it lacks many tissues and organs which are formed during post-embryonic development. Thirdly, the final stage of embryogenesis is characterized by an imposed desiccation period required for the termination of the embryogenic program and the initiation of germination. The time and modality of the desiccation process is species specific and results in a drastic reprogramming of gene expression (Elhiti and Stasolla 2013).

Recapitulation of embryogenesis can also occur in the absence of fertilization through the generation of asexual embryos. Through this process, referred as apomixis, embryos can develop from unfertilized egg cells or cells of the maternal tissue (Nogler 1984). Formation of asexual embryos can also be achieved via in vitro culture through gametophytic or somatic embryogenesis. Somatic embryogenesis can be theoretically initiated from all cells within the sporophyte, except gametic cells, while gametophytic embryogenesis involves the formation of haploid embryos from either the male or female gametophyte (Bhojwani and Razdan 1996; Raghavan 2000).

As hinted above, somatic embryogenesis has acquired relevance in the study of plant embryogenesis for several reasons. Firstly, it allows the synchronous development of embryos which are exposed and easily accessible. This is in contrast to zygotic embryogenesis, where the embryos are encased in the maternal tissue and often impossible to excise. This characteristic becomes problematic especially for collecting a suitable number of zygotic embryos for physiological and/or molecular studies. In addition somatic embryos are similar to their zygotic counterparts, and therefore knowledge acquired in vivo can be transferred in vitro (Yeung and Meinke 1993). As a consequence, several studies dealing with somatic embryogenesis at cellular, tissue, and molecular levels are currently available (Willemsen and Scheres 2004). Finally, generation of embryos in culture allows the targeted manipulations of environmental and/or culture conditions which would be difficult, if not impossible to perform in vivo. The selective addition or removal of specific chemicals in the medium is often used as a strategy not only to optimize culture conditions but also especially to understand the nature of the environment inductive for the proper development of the embryos.

Despite the existence of many similarities between somatic and zygotic embryogenesis, it must be noted that the two processes are also characterized by substantial differences which must be considered in comparative studies. Unlike zygotic embryogenesis, formation of somatic embryos is dependent upon the competence that some somatic cells have to change their developmental fate. This change involves an extensive and poorly understood reprogramming of gene expression which is unique of in vitro systems (Feher et al. 2003). Another relevant consideration is the fact that in vitro conditions are not fully optimized and therefore the "embryonic environment" created in vitro is different from the seed environment. As such, differences in embryo physiology and storage product depositions are often observed between the two systems. Finally, there are instances of in vitro-produced embryos able to germinate without a dormancy period which is often needed in vivo (Elhiti and Stasolla 2013).

Taken together, these considerations suggest that in vitro embryogenesis can indeed be used as a model system to study plant embryogenesis, but with the due care of acknowledging potential differences with zygotic embryogenesis.

14.2 Plant In Vitro Embryogenesis Systems

Over the past years, in vitro embryogenic systems have been developed for many plant species, including *Arabidopsis* and *Brassica napus*. While not a relevant crop, *Arabidopsis* has been used quite extensively for in vitro studies due to the available genetic information which facilitates molecular and genetic analyses. Knowledge on *Arabidopsis* can also be transferred to *Brassica napus*, as the two species are related. The in vitro systems for the two species are very different, as somatic embryogenesis is used for *Arabidopsis*, while microspore-derived embryogenesis (androgenesis) is used for *Brassica*. It must also be mentioned that, unlike canola embryos which develop directly from immature microspores, somatic embryos in *Arabidopsis* arise from a callus derived from the explant. As such, this system is often referred as indirect somatic embryogenesis.

14.2.1 *Arabidopsis* Somatic Embryogenesis System

Although reports of *Arabidopsis* somatic embryogenesis from mature tissues are available, somatic embryos are more easily produced from immature explants, such as zygotic embryos (Mordhorst et al. 1998). Dissected zygotic embryos, preferably at the bent cotyledon stage of development, are cultured in a medium containing the auxin 2,4-dichlorophenoxyacetic acid (2,4-D), considered the inductive signal required for the dedifferentiation process of the somatic cells within the explants. Under these conditions, embryogenic callus is generated by the apical regions of the zygotic embryos and in particular from the adaxial side of the cotyledons. Removal of the auxin induces the formation of somatic embryos.

14.2.2 *Brassica napus* Microspore-Derived Embryogenesis System

A key event during androgenesis in *Brassica napus* is the redifferentiation step in which the genetic program of the immature microspores is redirected toward the embryonic pathway. This redirection is triggered by several treatments including elevated temperatures (usually 32 °C; Keller and Armstrong 1979), colchicines (Zhao et al. 1996), gamma irradiation (Pechan and Keller 1989), ethanol (Pechan and Keller 1989), low temperatures (Kasha et al. 1995), change in pH (Barinova et al. 2004), and sucrose starvation (Touraev et al. 1996). The first sign of dedifferentiation of the microspores, as reviewed by Telmer et al. 1992, involves changes in cytoskeletal organization.

Simmonds and Keller (1999) observed that the pre-prophase band which is composed by arrays of microtubules tends to localize in the middle region of induced microspores. This positioning ensures the symmetric cell division of the microspore, which demarks the completion of the inductive events and the acquisition of the embryogenic competence (Yeung 2002). This is in contrast to the gametophytic developmental pathway which is initiated with an asymmetric mitotic division of the microspore. Of the two daughter cells originating from the microspore, one is committed to form the suspensor of the embryo, while the other, referred to as the pro-embryogenic cell, will generate the embryo proper. The whole process is accomplished within 3 days. Through a series of anticlinal and periclinal divisions, the pro-embryogenic cell gives rise to a cluster of cells demarking the globular stage of embryogenesis after 5 days in culture. Within 7–9 days in culture, a globular embryo is produced characterized by a well-developed protoderm, the precursor of the epidermis. During the following days, the growth of the embryo is characterized by the formation of two cotyledons and a morphologically visible SAM and RAM. A detailed description of the histodifferentiation events occurring during microspore-derived embryogenesis is available (Yeung et al. 1996).

14.3 Genetic Components of In Vitro Embryogenesis

The mechanisms by which plant somatic embryogenesis is accomplished are quite complex but somehow conserved among plant species (Elhiti et al. 2013b). Simplified "molecular" steps of the in vitro embryogenic process have been reviewed by Elhiti et al. (2013a), and they are referred to as embryonic induction and development.

The embryonic induction leads to the formation of embryogenic tissue and is further subdivided in dedifferentiation, acquisition of totipotency, and commitment. During the dedifferentiation step, cells within the explants must lose their pre-acquired fate; this is accompanied up by the acquisition of totipotency which enables the cells with the potential to differentiate into any cell type. The concept of totipotency is often associated to that of "stemness" as stem cells are indeed totipotent. The new developmental fate acquired by the totipotent cells is regulated by extrinsic factors, which in culture are often determined by the presence of plant growth regulators. During somatic embryogenesis, the fate of the totipotent cells is redirected or "committed" toward the embryogenic pathway.

The different phases of embryogenesis are accompanied by major "molecular" reprogramming. As described by Elhiti et al. (2013a), somatic embryogenesis encompasses two developmental stages: (1) embryonic induction and (2) development. The embryogenic induction stage is further subdivided into three main phases: (a) dedifferentiation, (b) expression of totipotency, and (c) commitment of induction phase. Hereafter we will describe the genetic networking during each stage of somatic embryogenesis. These initial phases are followed up by the "development" step which is characterized by the growth of the embryos which is often achieved in the absence of plant growth regulators (Sugiyama 1999; Elhiti 2010). The following sections will provide an updated description of the molecular events underlying the induction and development phases.

14.4 Genetic Network of Early Embryogenesis

14.4.1 Genetic Networking Regulating the Induction Phase

During this stage, the genetic program of the somatic cells under culture condition is reprogrammed by either applications of exogenous hormones or stresses (Feher et al. 2003). The induction stage of somatic embryogenesis is very difficult to study at molecular levels because of the lack of clear cytological markers permitting the identification of those clusters of somatic cells undergoing reprogramming in gene expression leading to the acquisition of the embryogenic fate. As such, gene network modeling and bioinformatic analyses are the only means to identify candidate genes required during the three different phases (dedifferentiation, acquisition of totipotency, and commitment) of the induction step (Elhiti et al. 2013a).

14.4.1.1 Dedifferentiation

The dedifferentiation of somatic cells, which in culture is often promoted by auxins, possibly involves a major reprogramming in gene expression. Microarray analyses in *Arabidopsis* have identified *LATERAL ORGAN BOUNDARIES DOMAIN 29* (*LBD29*) as a key developmental gene controlling cell dedifferentiation processes both in vitro and in vivo (Liu et al. 2010). *LBD29* has been identified as a downstream target of the auxin response factors ARF7 and ARF19 (Feng et al. 2012), and lbd29 cells show a reduced sensitivity to auxin and are unable to dedifferentiate. These observations suggest that the native function of this gene is necessary for dedifferentiation and reinforce the notion that auxin acts as the inductive signal (reviewed by Elhiti et al. 2013a). Other possible candidate genes participating in the dedifferentiation step are *KRYPTONITE* (*KYP*)/*SUVH4*, a gene encoding H3 lysine 9 methyltransferase, which, if mutated, reduces the formation of embryogenic tissue, and POLYCOMB REPRESSIVE COMPLEX 1 (PRC1) which has a repressive effect on the ability of cells to dedifferentiate upon the imposition of inductive signals (Bratzel et al. 2010).

14.4.1.2 Totipotency

A key characteristic of plant cells is their inherent ability to retain all the genetic information required to alter their development fate even once fully differentiated (Birnbaum and Alvarado 2008). If expressed by appropriate environmental conditions, this ability, referred to as totipotency, allows the regeneration of the whole organism, as exemplified during somatic embryogenesis

(Verdeil et al. 2007). Despite extensive efforts to identify key elements required for the expression of totipotency, our knowledge on the molecular regulation of this process is very scarce. Independent studies suggest that epigenetic changes play an important role in totipotency (Costa and Shaw 2007; Birnbaum and Alvarado 2008). Furthermore, *Arabidopsis* mutant analysis showed that the concomitant knockout of *CURLY LEAF* (*CLF*) and *SWINGER* (*SWN*), genes encoding two polycomb repressor protein 2 (PRC2) proteins, results in the spontaneous production of embryogenic callus in culture in the absence of plant growth regulators which are normally required for callus formation. Based on these observations, the involvement of PCR2 proteins in the manifestation of totipotency cannot be excluded (Chanvivattana et al. 2004). Two other genes possibly implicated with the manifestation of totipotency are *PICKLE (PKL)* and *SOMATIC EMBRYOGENESIS RECEPTOR KINASE 1 (SERK1)*. It has been suggested that the function of *PICKLE* is to repress totipotency since embryogenic tissue and ultimately somatic embryos are produced spontaneously from *Arabidopsis pkl* roots in the absence of the inductive signals (Aichinger et al. 2009). As *PKL* encodes a putative CHD3 (chromatin helicase DNA binding protein 3), the authors suggest a possible implication of chromatin remodeling processes in totipotency. *SERK1*, which encodes a leucine-repeat receptor protein kinase, is highly expressed during early embryogenesis (Hecht et al. 2001). Ectopic expression of this gene favors the formation of embryogenic tissue and encourages somatic embryo production indicating an involvement in embryogenic competence (Hecht et al. 2001). Using gain-of-function screening approach Zuo et al. (2002) revealed that a shoot apical meristem-related gene, *WUSHEL*, is also expressed in *Arabidopsis* explants during the early inductive phases of somatic embryogenesis in specific domains giving rise to embryogenic cells. In the same study it was observed that overexpression of *WUS* in *Arabidopsis* roots, leaf petioles, stems, or leaves is sufficient to induce somatic embryogenesis. It must be noted, however, that the ability to produce somatic embryos is retained in *WUS* tissue (Zuo et al. 2002), thus suggesting the existence of complex and possible multiple pathways regulating embryogenesis in vitro.

Generation of *Brassica* microspore-derived embryos is dependent upon the ability of immature microspores to lose their gametophytic fate and acquire an embryogenic fate. Transcription studies have identified *LEAFY COTYLEDON1* and *LEAFY COTYLEDON2* (*LEC1*, *LEC2*) as potential candidates mediating this developmental switch and molecular markers for embryogenicity (Malik et al. 2007), a function which appears to be retained across species. A significant repression in *Arabidopsis* somatic embryo production was indeed observed in *lec* mutants (Harada 2001). It must be noted that during somatic embryogenesis, *WUS*, *LEC1*, and *LEC2* share similar expression profiles (Elhiti and Stasolla 2011). Taken together the authors speculated that *WUS* and *LEC* genes may be involved in the acquisition of totipotency possibly through parallel mechanisms. Genetic studies, including the analyses of *wus/lec1/lec2* triple mutants, might be needed to unravel the function of these genes in early embryogeny.

It is well known that *LEC* genes are required to promote the expression of *AGAMOUS-LIKE 15* (*AGL15*), encoding a MADS-domain protein (Zheng et al. 2009). Induction of *AGL15* strongly activates the gibberellin 2-oxidase *GA2ox6* which represses gibberellic acid synthesis. Therefore it cannot be excluded that *LECs* operate through the inhibition of gibberellins, which have been shown to act in an antagonistic fashion to auxin, the signal promoting the dedifferentiation of somatic cells.

The participation of auxin during the early embryogenic phases was also demonstrated in *Brassica napus* using *BABY BOOM*1 (*BBM1*), an AP2/ERF transcription factor (Boutilier et al. 2002). Ectopic expression of *BBM1* in *Brassica* seedlings results in the production of somatic embryos from leaf margins, while its overexpression in *Arabidopsis* produces cotyledon-like structures (Boutilier et al. 2002). Overall, the overexpression of *BBM1* was associated to other changes in leaf and flower morphology as well as neoplastic growth. Furthermore, *BBM1* overexpressing explants were able to regenerate through organogenesis and embryogenesis without applications of exogenous plant hormones, an observation suggesting that *BBM1* may interfere with auxin sensitivity. The requirement of auxin signaling during early embryogeny was also demonstrated during *Arabidopsis* somatic embryogenesis. Elhiti et al. (2013b) demonstrated that the increased number of somatic embryos obtained by suppressing *GLB2*, a type 2 nonsymbiotic hemoglobin, was the result of elevated levels of auxins which accumulate at the sites of the explants where embryogenic tissue forms. The authors developed a model in which suppression of *GLB2* results in an increase in nitric oxide which represses the transcription factor MYC2, a repressor of auxin synthesis. Collectively, these studies demonstrated a solid link between the acquisition and manifestation of totipotency to auxin.

14.4.1.3 Commitment

It is believed that once somatic cells express their totipotency, specific signal cascades must be activated to promote cell division and encourage the acquisition of meristematic identity. Both events are crucial for the proper development of the embryos. Overall, the genes involved in this phase of somatic embryo induction may be divided into three main categories: genes participating in cell cycle, genes required for meristematic cell formation and regulation, and genes involved in several signal transduction cascades.

Genes Participating in Cell Cycle

Cell division in plants is controlled by complicated mechanisms which are governed by the expression of cyclin-dependent kinases (*CDKs*). *CDK*s are proteins influencing the entry time into the different phases of the cell cycle (Elhiti et al. 2013a). According to their internal motives, *CDK*s are classified into eight groups, *CDKA* through *CDKG* and *CDKL* (Zhang et al. 2012). Functional genetic analyses revealed that only *CDKA1* (also referred to as *CDC2A*) is involved in embryogenesis and its expression is induced by the plant growth regulators auxins and cytokinins (Nowack et al. 2006). In *Arabidopsis*, overexpression of *CDC2A* represses somatic embryogenesis, while a downregulation of the same gene enhances the number of somatic embryos produced (Hemerly et al. 2000). Another CDK possibly participating in embryogenesis is *CDKA*, the transcript levels of which increase during the early phases of somatic embryogenesis prior to declining as the embryos develop (Cortes et al. 2010).

It has been reported that PROPORZ1 (PRZ1), a putative *Arabidopsis* transcriptional adaptor, mediates cell proliferation through auxin and cytokinin signaling (Sieberer et al. 2003). Compared to WT tissue, in which ectopic cell proliferation is observed in the presence of both auxin and cytokinins, *prz1* tissue is able to produce callus when cultured with either auxin or cytokinin (Sieberer et al. 2003). Based on these observations, the authors suggested that PRZ1 mediates cell proliferation and differentiation by affecting the behavior of cell cycle regulators. Another possible component of the mitotic machinery with a possible involvement on somatic embryogenesis is histone H3-11, a mitosis-specific phosphorylation protein. Hendzel et al. (1997) suggested that histone H3-11 is particularly required during the inductive phases of embryogenesis, an observation consistent with the high levels of *histone H3-11* transcripts measured in alfalfa tissue subjected to 2,4-D treatments which stimulate embryogenic tissue formation (Kapros et al. 1992).

Genes Involved in Meristematic Cell Formation

The competent cells formed in culture on the explants respond to an inductive signal, usually provided by specific culture addenda such as growth regulators, and become meristematic cells. Elhiti et al. (2010) proposed that meristematic cell formation in culture is regulated by similar mechanisms involved in the formation and maintenance of the meristematic cells within the SAM in vivo. Proper SAM homeostasis relies on two classes of genes: those promoting cell division and those favoring cell differentiation. Members of the former class are *SHOOTMERISTEMLESS* (*STM*) and

WUSCHEL (*WUS*), while promoters of cell differentiation are *CLAVATA1* and *CLAVATA2* (*CLV1* and *CLV2*). Using *Brassica* and *Arabidopsis*, it was demonstrated that while the constitutive expression of *STM* induces embryo formation in culture, overexpression of *CLV1* represses the production of embryos (Elhiti et al. 2010). This contrasting behavior in vitro is analogous to that observed during the maintenance of the SAM in vivo.

A lot of attention has also been directed toward the interaction of *WUS* and *CLV1*, the function of which has been well documented. In the SAM the role of *CLV1*, a transmembrane receptor serine/threonine kinase with leucine-rich repeat (Clark et al. 1997), is to promote the differentiation of meristematic cells by repressing *WUS* expression through a complex signaling model involving other CLV proteins (Dodsworth 2009). In this model *CLV3* produced by the apical cells of the SAM binds to CLV1/CLV2 receptor kinase complexes located in the subapical cells and through the activation of downstream signaling components downregulates *WUS* which is expressed in the "organizing center" (Dodsworth 2009). In *Arabidopsis* somatic embryogenesis system, the expression of *WUS*, induced by auxin, is first visible in those domains of the explants giving rise to the embryogenic tissue (Su and Zhang 2009). Chen et al. (2009) also demonstrated a cytokinin-mediated activation of *WUS*. The WUS-CLV interaction was shown to occur during *M. truncatula* somatic embryogenesis where the two genes competitively modulate the formation of embryogenic tissue formation (Chen et al. 2009). It must be noted, however, that all the SAM-related genes described above are not necessary for somatic embryo formation, as their respective *Arabidopsis* mutants are still able to produce somatic embryos in culture (Mordhorst et al. 1998).

It is well known that *WUS* acts as a transcription factor repressing A-type *Arabidopsis* response regulators, thereby activating cytokinin responses contributing to meristem maintenance (Leibfried et al. 2005). Several studies suggest that *WUS* activity in vivo requires the expression of the ARGONAUTE (AGO) protein ZWILLE/AGO10 (Tucker et al. 2008). AGO proteins are central elements of the RNA interference (RNAi) pathway and mediate the repression of target mRNA through mRNA degradation or translational inhibition (Mallory and Vaucheret 2010). Specifically, ZLL/AGO10 blocks the accumulation of microRNA165/microRNA166 in the stem cell niche of the SAM by sequestration mechanisms preventing the degradation of microRNA165/microRNA166 targets' transcripts of *HD-ZIPIII* transcription factor (Knauer et al. 2013). Based on these observations, it might therefore be interesting to ascertain the participation of AGO proteins in the initial phases of embryogenesis.

Genes Involved in Signal Transduction Cascade

The CLV signaling described in the first section is modulated by downstream components, the function of which, although not tested during in vitro embryogenesis, might participate during in vitro embryogenesis. Two intermediary modulators of CLV signaling are a kinase-associated protein phosphatase (KAPP) and a rho-like GTPase (Rop) (Song et al. 2006). These two proteins interact directly with *CLV1* forming a 450 kDa active signaling complex. KAPP functions in vivo as a negative regulator of the CLV signaling through direct dephosphorylation of *CLV1*, while Rop is assumed to transduce the CLV signal into the nucleus (reviewed by Elhiti et al. 2010). Future studies assessing the involvement of these two proteins during in vitro embryogenesis might further validate the notion that the formation of meristematic cells in vitro uses signaling systems governing SAM homeostasis.

Besides KAPP and Rop, SHEPHERD, a HSP90-like protein predicted to be required for correct folding of CLV complex (Ishiguro et al. 2002), and POLTERGEIST (*POL*), a nuclear-localized protein phosphatase 2C (PP2C) which acts downstream within the CLV transduction (Carles and Fletcher 2003), can be additional candidates to be tested during in vitro embryogenesis. These proposed studies would verify the proposed notion that meristematic cell formation in vitro relies on similar mechanisms governing SAM formation and maintenance in vivo.

14.5 Genes/Gene Homologues Influence Embryo Development

14.5.1 Genetic Networking Controlling Somatic Embryo Development

The developmental phase of in vitro embryogenesis culminates with the formation of fully developed embryos, the growth of which occurs along two axes: an apical-basal axis and a radial axis. While the apical-basal growth ensures the proper positioning of the cotyledons surrounding the SAM, a hypocotyl, and a RAM, the radial growth specifies concentric layers of tissues: the stele, cortex, and epidermis. Precise coordination of these events is paramount for the accurate establishment of the embryo body. The development of many *Arabidopsis* mutants, as well as high-resolution molecular techniques, has aided our understanding on the molecular networks coordinating apical-basal and radial growth.

14.5.1.1 Establishment of the Apical-Basal Body Plan

The formation of apical-basal axis of a somatic embryo is responsible for the proper positioning of the SAM and RAM at the opposite regions of the hypocotyl. Among the processes ensuring this axis pattern are asymmetric cell division and preferential elongation along the desired axis (De Jong et al. 1993; Emons 1994). While asymmetric cell divisions are promoted by plant hormones that alter cell polarity by interfering with pH gradients or the electrical fields across membranes (Smith and Kirkorian 1990), cell expansion is associated with the composition of polysaccharides within the cell wall and specific hydrolytic enzymes (De Jong et al. 1993; Emons 1994; Fry 1995). The participation of asymmetric cell divisions and elongation for the establishment of the apical-basal axis during in vivo embryogenesis are manifested at the zygotic stage when the zygote elongates and undergoes an asymmetric division leading to the formation of small apical cells, precursors of the embryo proper, and larger basal cells forming the suspensor cell. The contribution of these two events is also crucial for the later stages of embryogenesis (Zhang and Laux 2011). Although the early phases of embryogenesis in vitro follow less precise patterns, the roles of asymmetric cell divisions and elongations are still apparent in some systems, including *Brassica* microspore-derived embryogenesis where the type of division observed in the microspore, i.e., symmetric or asymmetric, marks its developmental fate. Molecular analyses during the earliest phases of *Brassica* microspore-derived embryogenesis identified some potential genes possibly involved in this fate acquisition, including *FUSCA3*, *LEAFY COTYLEDON1* (*LEC1*), *LEC2*, *BABY BOOM* (*BBM*), *PINFORMED7* (*PIN7*), two *WUSCHEL*-related homeobox (*WOX*) genes, *WOX2*, *WOX8* and *WOX9*, and *ABSCISIC ACID INSENSITIVE3* (Joosen et al. 2007; Malik et al. 2007; Tsuwamoto et al. 2007). While the involvement of these genes in asymmetric cell division is known, more information is available for *PIN7* and *WOX2*. During the asymmetric cell division of the zygote, *PIN7* is preferentially localized in the basal cell, while expression of *WOX2* is restricted in the apical cell. A mutation in either of the two genes compromises the ability of the zygote to divide properly (reviewed by Elhiti and Stasolla 2013).

The participation of auxin for the execution of asymmetric cell divisions is well established, and a precise distribution of this growth regulator is also required for the specification of somatic cells embarking in the embryogenic pathway. According to Su and Zhang (2009), the formation of an auxin gradient within the *Arabidopsis* embryogenic tissue is crucial for inducing the stem cell formation through the regulation of *PIN1*. This regulation would also mediate the expression of *WUS* and other *WOX* genes required for the establishment of the apical-basal axis. Of note, the observation that *WOX8* and *WOX9* are also expressed during conifer embryogenesis possibly through auxin-mediated mechanisms (Palovaara and Hakman 2009) raises the possibility of a more general involvement of these groups of genes in embryo patterning.

The apical domain of a fully developed embryo consists of cotyledons and a SAM. Independent studies have shown that the establishment of the apical embryonic domains in vivo is specified by *GURKE* and *TOPLESS*. *GURKE* encodes an acetyl-CoA carboxylase and, if mutated, precludes the formation of cotyledons and the SAM (Baud et al. 2003). Knockout of *TOPLESS* results in the formation of a root in the apical pole, thus indicating that the function of this gene is to abolish the manifestation of the basal patterning in the apical domains (Szemenyei et al. 2008).

The central embryonic domain consists of a hypocotyl, the specification of which is regulated by *FACKEL*, *HYDRA1*, and *CEPHALOPOD* (Willemsen and Scheres 2004). With mutations in these genes, embryos form without hypocotyls in which the apical domain is directly connected to the embryonic root (Lindsey et al. 2003). These genes participate in the biosynthesis of sterols, suggesting an involvement of these compounds in hypocotyl formation. Not surprisingly, auxin is also required for the development of a functional hypocotyl. Mutations of *MONOPTEROS*, a gene encoding an auxin responsive factor, produce embryos lacking a hypocotyl (Schruff et al. 2006). In the same study it is speculated that the specific function of these genes might be related to the formation of a functional stele, as this is the most affected tissue in the mutants.

Other genes interfering with auxin signaling: *AUXIN-RESISTANT6* and *BONDELOS* are also required for the proper establishment of the central embryonic domain (Park and Harada 2008).

The embryonic basal domain includes the RAM which is composed of quiescent cells surrounded by the stem cells. During early phases of embryogenesis, expression of *PINFORMED1*, *4*, and *7*, all encoding auxin efflux carriers, are required for the formation of an auxin maximum at the basal domain, which is essential for the specification of the RAM (Willemsen and Scheres 2004). Mutations in auxin downstream components, such as *PLETHORA*, which is expressed in quiescent cells and encodes AP2 domain transcription factor, cause the mis-specification of quiescent cells and consequently the improper formation of the embryonic root (Aida et al. 2004). Analyses of these mutants showed that the effects of *PLETHORA* in the formation of embryonic root are mediated through interaction with *SCARECROW* and *SHORTROOT* (Aida et al. 2004). Furthermore, *HOBBIT*, a homologue of a subunit of the anaphase-promoting complex, is also required for proper localization of quiescent cells in embryonic root (Willemsen et al. 1998). Collectively, these studies show that the apical, central, and basal embryonic domains appear to be controlled by independent genetic mechanisms which are coordinated by a proper flow of auxin. The majority of these studies, however, have been conducted in vivo, and it is not clear whether similar mechanisms also operate in vitro, where tissue patterning is less organized and predictable.

14.5.1.2 Establishment of Embryonic Shoot Apical Meristem (SAM)

The establishment of the SAM is considered a key event during embryogenesis and encompasses three phases: the specification of apical domain (discussed in the previous sections), the formation of the stem cell niche, and the separation of the central and peripheral domains. The transcription factor *WUS* defines the organizing center of the meristem and is considered the initial marker for the specification of the stem cell niche. Localization studies in *Arabidopsis* demonstrate that *WUS* transcripts appear very early during somatic embryogenesis (Su and Zhang 2009). The main function of *WUS* is to maintain the stem cells in an undifferentiated state, thereby ensuring the proper maintenance of the apical region. As previously described, *WUS* is regulated by *CLV* feedback mechanisms through the interaction of *CLV1-3*. Another marker of the initial formation of the SAM is the homeodomain transcription factor *STM* which is also expressed in somatic embryos starting from the globular stage of development (Elhiti 2010). Downregulation of *STM* results in fusion in the embryonic cotyledons resulting in the production of trumpet-shaped embryos (Elhiti et al. 2010). The demarcation between the central and peripheral domains of the SAM is necessary for the proper positioning of the cotyledons relative to the SAM (reviewed by Elhiti and Stasolla 2013). This process is mediated by *CUP-SHAPED COTYLEDON* (*CUC1*, *2*, *3*), expressed at the boundary between the cotyledons and the SAM (Aida et al. 1999). Knockout of *CUC* phenocopies the *stm* phenotype (trumpet-shaped embryo), suggesting that *STM* and *CUC* may share the same pathway. It has been observed that accumulation of *CUC* transcripts is regulated by microRNA164 (Zhang et al. 2006).

14.5.1.3 Establishment of the Embryonic Radial Pattern

Radial patterning results in the proper specification of the epidermis, cortical tissue, and vascular tissues. The first hint of radial pattern formation during in vivo and in vitro embryogenesis corresponds with the separation of the protoderm from the inner cells (Elhiti and Stasolla 2013). Expression analyses in *Arabidopsis* indicate that *ARABIDOPSIS THALIANA MERISTEM LAYER1* and *PROTODERMAL FACTOR2*, encoding transcription factors containing the START domain, are implicated in the radial specification of the protodermal layer (Abe et al. 2003). Other genes involved in radial patterning are *KEULE* and *KNOLLE*, as a radial axis is never initiated in the two mutants. While their function is not fully clear, it has been shown that *KNOLLE* encodes a syntaxin-like protein involved in secretary processes (Song et al. 2000). A mutation in this gene results in abnormal cytokinesis due to incomplete formation of the cell wall separating the two daughter cells (Song et al. 2000).

SHORT ROOT (*SHR*) and *SCARECROW* (*SCW*), encoding transcription factors of the GRAS family, are required for the proper specification of endodermal and cortical layers. Knockout of *SHR* results in absence of the endodermis, while *scw* mutants have a single file of cells in place of cortex and endodermis (Di Laurenzio et al. 1996). Localization studies indicated that *SHR* is expressed in the vascular tissue and translocated into the endodermal layer where *SCW* is expressed (Di Laurenzio et al. 1996).

14.6 Conclusions

Embryo formation in vivo is initiated with the fusion of the gametes, i.e., sperm and egg, resulting in the formation of the zygote. Through precise cell division and differentiation processes, the zygote produces a fully develop embryo, composed of an apical, a central, and a basal domain. Recapitulation of embryogenesis can also occur in vitro through somatic and gametophytic embryogenesis. Formation of in vitro embryos relies on similar genetic mechanisms operating during in vivo embryogenesis although the culture conditions are less stable and often not optimized. As a result, the molecular events controlling in vitro embryogenesis are less defined. Overall, somatic embryogenesis can be divided in two distinct phases: induction and development. The first phase requires the dedifferentiation of the somatic cells, the acquisition of totipotency, and the commitment to embark an embryogenic fate. These events, critical for the overall embryogenesis, do not occur in vivo and are therefore specific to culture systems. Independent studies have demonstrated the relevance of auxin for the inductive step and the participation of genes regulating SAM formation and maintenance. Removal of plant regulators is often required to initiate the development of the somatic embryos, and during this event, the embryo body is elaborated. Like the in vivo system, the tissue patterning of in vitro-produced embryos occurs through an apical-basal and a radial axis. Growth along the two axes is mediated by distinct genetic networks, although auxin seems to be implicated with both. As the developmental phases of in vitro embryogenesis are very similar to those observed in zygotic embryos, knowledge on the molecular mechanisms operating in the latter system are often transferred to the former.

References

Abe M, Katsumata H, Komeda Y, Takahashi T (2003) Regulation of shoot epidermal cell differentiation by a pair of homeodomain proteins in *Arabidopsis*. Development 130:635–643

Aichinger E, Villar CBR, Farrona S, Reyes JC, Hennig L, Kohle C (2009) CHD3 proteins and polycomb group proteins antagonistically determine cell identity in *Arabidopsis*. PLoS Genet 5, e1000605

Aida M, Beis D, Heidstra R, Willemsen V, Blilou I, Galinha C, Nussaume L, Noh YS, Amasino R, Scheres

B (2004) The PLETHORA genes mediate patterning of the *Arabidopsis* root stem cell niche. Cell 119:109–120

Aida M, Ishida T, Tasaka M (1999) Shoot apical meristem and cotyledon formation during *Arabidopsis* embryogenesis: interaction among the CUP-SHAPED COTYLEDON and SHOOT MERISTEMLESS genes. Development 126:1563–1570

Barinova I, Clement C, Martiny L, Baillieul F, Soukupova H, Heberle-Bors E, Touraev A (2004) Regulation of developmental pathways in cultured microspores of tobacco and snapdragon by medium pH. Planta 219:141–146

Baud S, Guyon V, Kronenberger J, Wuillème S, Miquel M, Caboche M, Lepiniec L, Rochat C (2003) Multifunctional acetyl-CoA carboxylase 1 is essential for very long chain fatty acid elongation and embryo development in *Arabidopsis*. Plant J 33:75–86

Bhojwani S, Razdan M (1996) Plant tissue culture: theory and practice. Elsevier, Amsterdam, pp 125–166

Birnbaum KD, Alvarado AS (2008) Slicing across kingdoms: regeneration in plants and animals. Cell 132:697–710

Boutilier K, Offringa R, Sharma V, Kieft H, Ouellet T, Zhang L, Hattori J, Lui C, van Lammeren A, Miki B (2002) Ectopic expression of BABY BOOM triggers a conversion from vegetative to embryonic growth. Plant Cell 14:1737

Bratzel F, Lopez-Torrejon G, Koch M, Del Pozo JC, Calonje M (2010) Keeping cell identity in *Arabidopsis* requires PRC1 RING-finger homologs that catalyze H2A monoubiquitination. Curr Biol 20:1853–1859

Carles CC, Fletcher JC (2003) Shoot apical meristem maintenance: the art of a dynamic balance. Trends Plant Sci 8:394–401

Chanvivattana Y, Bishopp A, Schubert A, Stock C, Moon YH, Sung ZR, Goodrich J (2004) Interaction of Polycomb-group proteins controlling flowering in *Arabidopsis*. Development 131:5263–5276

Chen SK, Kurdyukov S, Kereszt A, Wang XD, Gresshoff PM, Rose RJ (2009) The association of homeobox gene expression with stem cell formation and morphogenesis in cultured *Medicago truncatula*. Planta 230:827–840

Clark S, Williams R, Meyerowitz E (1997) The *CLAVATA1* gene encodes a putative receptor kinase that controls shoot and floral meristem size in *Arabidopsis*. Cell 89:575–585

Cortes MM, Paredes FR, Burgeff C, Nunez TP, Cordova I, Oropeza C, Verdeil JL, Saenz L (2010) Characterisation of a cyclin-dependent kinase (*CDKA*) gene expressed during somatic embryogenesis of coconut palm. Plant Cell Tiss Org Cult 102:251–258

Costa S, Shaw P (2007) 'Open minded' cells: how cells can change fate. Trends Cell Biol 17:101–106

De Jong AJ, Heidstra R, Spaink HP, Hartog MV, Meijer EA, Hendriks T, Lo Shiavo F, Terzi M, Bisseling T, van Kammen A, de Vries SC (1993) Rhizobium lipo-oligosacharides rescue a *Daucus carota* somatic embryo variant. Plant Cell 5:615–620

De Smet I, Lau S, Mayer U, Jürgens G (2010) Embryogenesis – the humble beginnings of plant life. Plant J 61:959–970

Di Laurenzio L, Wysocka-Diller J, Malamy JE, Pysh L, Helariutta Y, Freshour G, Hahn MG, Feldmann KA, Benfey FN (1996) The *SCARECROW* gene regulates an asymmetric cell division that is essential for generating the radial organization of the *Arabidopsis* root. Cell 8:423–433

Dodsworth S (2009) A diverse and intricate signalling network regulates stem cell fate in the shoot apical meristem. Dev Biol 336:1–9

Elhiti M, Hebelstrup KH, Wang A, Li C, Cui Y, Hill RD, Stasolla C (2013a) Function of the type-2 *Arabidopsis* hemoglobin in the auxin-mediated formation of embryogenic cells during morphogenesis.The. Plant J 74:946–958

Elhiti M, Stasolla C (2011) Ectopic expression of the Brassica SHOOTMERISTEMLESS attenuates the deleterious effects of the auxin transport inhibitor TIBA on somatic embryo number and morphology. Plant Sci 180:383–390

Elhiti M, Stasolla C, Wang A (2013b) Molecular regulation of plant somatic embryogenesis. In Vitro Cell Dev Biol Plant 49:631–642

Elhiti M, Stasolla S (2013) Genetic of embryogenesis. Brenner's encyclopedia of genetics (2nd edition) 5:343–345

Elhiti M, Tahir M, Gulden RH, Khamiss K, Stasolla C (2010) Modulation of embryo-forming capacity in culture through the expression of *Brassica* genes involved in the regulation of the shoot apical meristem. J Exp Bot 61:4069–4085

Elhiti MA (2010) Molecular characterization of several brassica shoot apical meristem genes and the effect of their altered expression during in vitro morphogenesis. Ph.D. thesis, Faculty of Graduate Studies, University of Manitoba

Emons AMC (1994) Somatic embryogenesis: cell biological aspects. Acta Bot Neerl 43:1–14

Feher A, Pasternak TP, Dudits D (2003) Transition of somatic plant cells to an embryogenic state. Plant Cell Tiss Org Cult 74:201–228

Feng Z, Sun X, Wang G, Liu H, Zhu J (2012) LBD29 regulates the cell cycle progression in response to auxin during lateral root formation in *Arabidopsis thaliana*. Ann Bot 160:2–10

Fry SC (1995) Polysaccharide-modifying enzymes in the plant-cell wall. Annu Rev Plant Physiol Plant Mol Biol 46:497–520

Harada JJ (2001) Role of *Arabidopsis LEAFY COTYLEDON* genes in seed development. J Plant Physiol 158:405–409

Hecht V, Vielle-Calzada JP, Hartog MV, Schmidt EDL, Boutilier K, Grossniklaus U, De Vries SC (2001) The *Arabidopsis SOMATIC EMBRYOGENESIS RECEPTOR KINASE 1* gene is expressed in developing ovules and embryos and enhances embryogenic competence in culture. J Plant Physiol 127:803–816

Hemerly AS, Ferreira PC, Van Montagu M, Engler G, Inze D (2000) Cell division events are essential for embryo patterning and morphogenesis: studies on dominant negative cdc2aAt mutants of *Arabidopsis*. Plant J 23:123–130

Hendzel MJ, Wei Y, Mancini MA, Van Hooser A, Ranalli T, Brinkley BR, Bazett-Jones DP, Allis CD (1997) Mitosis-specific phosphorylation of histone H3 initiates primarily within pericentromeric heterochromatin during G2 and spreads in an ordered fashion coincident with mitotic chromosome condensation. Chromosoma 106:348–360

Ishiguro S, Watanabe Y, Ito N, Nonaka H, Takeda N, Sakai T, Kanaya H, Okada K (2002) SHEPHERD is the *Arabidopsis* GRP94 responsible for the formation of functional CLAVATA proteins. EMBO J 21:898–908

Joosen R, Cordewener J, Supena EDJ, Vorst O, Lammers M, Maliepaard C, Zeilmaker T, Miki B, America T, Custers J, Boutilier K (2007) Combined transcriptome and proteome analysis identifies pathways and markers associated with the establishment of rapeseed microspore-derived embryo development. Plant Physiol 144:155–172

Kapros T, Bogre L, Nemeth K, LaszBako L, Gyorgyey J, Wu SC, Dudits D (1992) Differential expression of histone H3 gene variants during cell cycle and somatic embryogenesis in alfalfa. Plant Physiol 98:621–625

Kasha K, Yao Q, Simon E, Oro R, Hu T (1995) Production and application of double haploids in crops. University of Vienna Pusblisher, IAEA

Keller W, Armstrong K (1979) Stimulation of embryogenesis and haploid production in *Brassica campestris* anther culture by elevated temperature treatments. Theor Appl Genet 55:65–67

Knauer S, Holt AL, Rubi-Somoza I, Tucker EJ, Hinze A, Pisch M, Javelle M, Timmermans MC, Tucker M, Laux T (2013) A protodermal miR394 signal defines a region of stem cell competence in the *Arabidopsis* shoot meristem. Dev Cell 24:125–132

Leibfried A, To J, Busch W, Stehling S, Kehle A, Demar M, Kieber J, Lohmann J (2005) WUSCHEL controls meristem function by direct regulation of cytokinin inducible response regulators. Nature 438:1172–1175

Lindsey K, Pullen ML, Topping JF (2003) Importance of plant sterols in pattern formation and hormone signalling. Trends Plant Sci 8:521–525

Liu HI, Wang GC, Feng Z, Zhu J (2010) Screening of genes associated with dedifferentiation and effect of LBD29 on pericycle cells in *Arabidopsis* thaliana. Plant Growth Regul 62:127–136

Malik M, Wang F, Dirpaul J, Zhou N, Polowick P, Ferrie A, Krochko J (2007) Transcript profiling and identification of molecular markers for early microspore embryogenesis in *Brassica napus*. Plant Physiol 144:134–154

Mallory A, Vaucheret H (2010) Form, function, and regulation of ARGONAUTE proteins. Plant Cell 22:3879–3889

Mordhorst AP, Voerman KJ, Hartog MV, Meijer EA, van Went J, Koornneef M, de Vries SC (1998) Somatic embryogenesis in *Arabidopsis* thaliana is facilitated by mutations in genes repressing meristematic cell divisions. Genetics 149:549–563

Nogler G (1984) Gametophytic apomixis. In: Johri BM (ed) Embryology of angiosperms. Springer, Berlin, pp 475–518

Nowack MK, Grini PE, Jakoby MJ, Lafos M, Koncz C, Schnittger A (2006) A positive signal from the fertilization of the egg cell sets off endosperm proliferation in angiosperm embryogenesis. Nat Genet 38:63–67

Palovaara J, Hakman I (2009) WOX2 and polar auxin transport during spruce embryo pattern formation. Plant Signal Behav 4:153–155

Park S, Harada JJ (2008) Arabidopsis embryogenesis. In: Plant embryogenesis, Humana Press, Totowa, pp 3–16

Pechan P, Keller W (1989) Induction of microspore embryogenesis in *Brassica napus* L. by gamma irradiation and ethanol stress. In vitro Cell Dev Biol Plant 25:1073–1074

Raghavan V (2000) Developmental biology of flowering plants. Springer, New York

Schruff MC, Spielman M, Tiwari S, Adams S, Fenby N, Scott RJ (2006) The AUXIN RESPONSE FACTOR 2 gene of *Arabidopsis* links auxin signalling, cell division, and the size of seeds and other organs. Development 133:251–261

Sieberer T, Hauser MT, Seifert GJ, Luschnig C (2003) PROPORZ1, a putative *Arabidopsis* transcriptional adaptor protein, mediates auxin and cytokinin signals in the control of cell proliferation. Curr Biol 13:837–842

Simmonds D, Keller W (1999) Significance of preprophase bands of microtubules in the induction of microspore embryogenesis of *Brassica napus*. Planta 208:383–391

Smith DL, Kirkorian AD (1990) Somatic proembryo production from excised, wounded zygotic carrot embryos on hormone-free medium: evaluation of the effects of pH, ethylene and activated charcoal. Plant Cell Rep 9:468–470

Song JY, Leung T, Ehler LK, Wang C, Liu Z (2000) Regulation of meristem organization and cell division by TSO1, an *Arabidopsis* gene with cysteine-rich repeats. Development 127:2207–2217

Song S, Lee M, Clark S (2006) POL and PLL1 phosphatases are CLAVATA1 signaling intermediates required for *Arabidopsis* shoot and floral stem cells. Development 133:4691–4698

Su YH, Zhang XS (2009) Auxin gradients trigger de novo formation of stem cells during somatic embryogenesis. Plant Signal Behav 4:574–576

Sugiyama M (1999) Organogenesis in vitro. Curr Opin Plant Biol 2:61–64

Szemenyei H, Hannon M, Long JA (2008) TOPLESS mediates auxin-dependent transcriptional repression during *Arabidopsis* embryogenesis. Science 319:1384–1386

Telmer C, Simmond D, Newcomb W (1992) Determination of developmental stage to obtain high frequencies of embryogenic microspores in *Brassica napus*. Physiol Plant 84:417–424

Touraev A, Pfosser M, Vicente O, Heberle-Bors E (1996) Stress as the major signal controlling the developmental fate of tobacco microspores: towards a unified model of induction of microspore/pollen embryogenesis. Planta 200:144–152

Tsuwamoto R, Fukuoka H, Takahata Y (2007) Identification and characterization of genes expressed in early embryogenesis from microspores of *Brassica napus*. Planta 225:641–652

Tucker M, Hinze A, Tucker E, Takada S, Jurgens G, Laux T (2008) Vascular signaling mediated by ZWILLE potentiates WUSCHEL function during shoot meristem stem cell development in the *Arabidopsis* embryo. Development 135:28–39

Verdeil JL, Alemanno L, NiemenakN TTJ (2007) Pluripotent versus totipotent plant stem cells: dependence versus autonomy? Trends Plant Sci 12:245–252

Willemsen V, Scheres B (2004) Mechanisms of pattern formation in plant embryogenesis. Annu Rev Genet 38:587–614

Willemsen V, Wolkenfelt H, de Vrieze G, Weisbeek P, Scheres B (1998) The HOBBIT gene is required for formation of the root meristem in the *Arabidopsis* embryo. Development 125:521–531

Yeung E (2002) The canola microspore-derived embryo as a model system to study developmental process in plants. J Plant Biol 45:119–133

Yeung E, Rahman M, Thorpe T (1996) Comparative development of zygotic and microspore-derived embryos in *Brassica napus* L. cv Topas. I. Histodifferentiation. Int J Plant Sci 157:27–39

Yeung EC, Meinke DW (1993) Embryogenesis in angiosperms: development of the suspensor. Plant Cell 5:1371–1381

Zhang B, Pan X, Cannon CH, Cobb GP, Anderson TA (2006) Conservation and divergence of plant microRNA genes. Plant J 46:243–259

Zhang G, Song C, Zhao MM, Li B, Guo SX (2012) Characterization of an A-type cyclin-dependent kinase gene from *Dendrobium candidum*. Biologia 67:360–368

Zhang Z, Laux T (2011) The asymmetric division of the *Arabidopsis* zygote: from cell polarity to an embryo axis. Sex Plant Reprod 24:161–169

Zhao J, Simmonds D, Newcomb W (1996) Induction of embryogenesis with colchicines instead of heat in microspores of *Brassica napus* L.c. Topaz. Planta 198:433–439

Zheng Y, Ren N, Wang H, Stromberg AJ, Perrya SE (2009) Global identification of targets of the *Arabidopsis* MADS domain protein AGAMOUS-Like15. Plant Cell 21:2563–2577

Zuo J, Niu QW, Frugis G, Chua NH (2002) The *WUSCHEL* gene promotes vegetative-to-embryonic transition in *Arabidopsis*. Plant J 30:349–359

15

Advances in Somatic Embryogenesis of Palm Trees (Arecaceae): Fundamentals and Review of Protocols

Emília Ordones Lemos Saleh
and Jonny Everson Scherwinski-Pereira

Abstract

Micropropagation via somatic embryogenesis (SE) is one of the forms of propagation of plants and the most commonly recommended method for clonal propagation of palms. Despite the existence of a vast number of works in the literature, SE is still far from being completely mastered and understood, although it is known that callus induction and the use of auxins are fundamental prerequisites for obtaining satisfactory results. In fact, in vitro development of cells and tissues via SE is dependent on different factors, such as plant genotype and type, age, and the physiological and developmental state of the explant and donor plant. External factors include the composition of the medium and the physical cultivation conditions. The interaction between all these factors leads to the induction and expression of a specific mode of cell differentiation and development, which in turn leads to embryogenesis. The present review discusses the main factors involved during somatic embryogenesis of palms. At the same time, a significant number of protocols developed in recent years are compiled to provide readers about the current state of the art on the subject.

Keywords

Arecaceae • Micropropagation of ornamental palms • Somatic embryogenesis • Morphogenesis • Cell differentiation • Clonal propagation

E.O.L. Saleh
Center of Natural Sciences, State University of Piauí (UESPI), Campus Poeta Torquato Neto, Rua João Cabral, 2231, Teresina, PI 64002-150, Brazil

J.E. Scherwinski-Pereira (✉)
Embrapa Genetic Resources and Biotechnology, Av. W5 Norte (final), Brasília, DF 70770-917, Brazil
e-mail: jonny.pereira@embrapa.br

A. Mujib (ed.), *Somatic Embryogenesis in Ornamentals and Its Applications*,
DOI 10.1007/978-81-322-2683-3_15

15.1 Introduction

The family Arecaceae comprises plants known as palms, native to tropical and subtropical regions; the plants are key species for frugivore communities and offer resources to pollinators (Dransfield 2008; Lorenzi 2010). The Arecaceae are the third most economically important family as the plants provide material used in construction; as food, fuel, and medications; in decoration; and also for subsistence farming (Eiserhardt et al. 2011; Meerow et al. 2012). Palms are important landscaping components in regions with tropical, Mediterranean, and subtropical climates and are used for landscaping in temperate zones. The plants are also found in greenhouses and indoor gardens in shopping centers, office buildings, and other public areas, where their function is to create a tropical feel to the environment (Broschat et al. 2014) (Fig. 15.1).

Arecoideae is the largest and most diverse subfamily within the family Arecaceae, with around 50 % of the species; it has an exceptional degree of endemism, notably in the Americas and in the Indo-Pacific region. Some of the most important palms, from an economic point of view, are Arecoideae, such as the oil palm (*Elaeis guineensis* Jacq.), coconut palm (*Cocos nucifera* L.), peach palm (*Bactris gasipaes* Kunth), and many other species of importance to both local and global trade (Baker et al. 2011). Important species used in landscaping include the macaw palm [*Acrocomia aculeata* (Jacq) Lood. ex Mart.], the betel palm (*Areca catechu* L.), and the date palm (*Phoenix dactylifera* L.) (Broschat et al. 2014).

Palms have various characteristics that make the propagation difficult, such as the presence of a single meristem per shoot and the absence of secondary growth in stems and roots. A few palms produce offshoots that can easily be propagated (Tomlinson 2006; Broschat et al. 2014). These characteristics, combined with their perennial life cycle, the recalcitrance of their seeds, and the fact that it takes a long time to be able to determine the value of the progeny, limit the genetic improvement of various palms if conducted using conventional methods (Pérez-Núñez et al. 2006). Besides, most of the species are allogamous, and great heterogeneity is observed when these are planted from seeds (Rajanaidu and Ainul 2013).

In vitro propagation is an important tool used in plants and for which there are no known techniques for vegetative propagation available (Costa and Aloufa 2007; Rajanaidu and Ainul 2013). In vitro systems have enabled the regeneration of plants of more than 1000 different species through two morphogenic alternatives: organogenesis (direct differentiation into new tissues) and embryogenesis (Gaj 2004). Somatic embryogenesis (SE) is an inductive process in which a competent cell or group of cells undergo a series of biochemical and molecular changes that result in the formation of a bipolar somatic embryo (Rai et al. 2011). The induction of SE in plants is only possible if the somatic cells recover their totipotency and acquire the competence necessary to respond to the embryogenic signals and initiate embryogenesis (Pasternak et al. 2002), i.e., the induction of embryogenic competence in cells that are not naturally embryogenic (Dodeman et al. 1997). SE has numerous applications including clonal propagation, the regeneration of genetically modified plants, the formation of somatic hybrids, and the induction and selection of mutants (Gaj 2004). Indeed, progress in the development of in vitro systems has enabled the induction of SE in many economically important plants (Santos et al. 1997; Zelena 2000; Aoshima 2005; Te-chato and Hilae 2007). An important example is the vegetative propagation of the oil palm (*Elaeis guineensis*), which has enabled the production of uniform material for planting programs in the field and genetic improvement (Durand-Gasselin et al. 1990; Low et al. 2008; Konan et al. 2010; Beulé et al. 2011).

Besides being a tool for vegetative propagation, somatic embryos can also be used for studies on the regulation of plant embryo development (Bhojwani and Razdan 1992; Von Arnold et al. 2002; Karami et al. 2009; Zhang and Ogas 2009). SE includes the ways in which cells develop into structures similar to zygotic embryos, through an ordered series of characteristic embryologic stages, without gamete fusion having occurred (Jiménez 2005). Somatic embryos display many

Fig. 15.1 Ornamental palms: A) *Acrocomia aculeata*, B) *Areca* sp., C) *Cocos nucifera*, D) *Elaeis guineensis*, E) *Euterpe* sp., F) *Phoenix* sp. (Source: EOL Saleh)

characteristics that are similar to those of the development stages of zygotic embryos: somatic embryos are bipolar and have shoot and root meristems with a closed vascular system separate from the maternal tissue (Luis and Scherwinski-Pereira 2014). Somatic embryos are often formed from a single cell and produce specific proteins (Jayasankar et al. 1999).

In vitro development of cells and tissues via SE is dependent on different factors, such as plant genotype and type, age, and the physiological and developmental state of the explant and donor plant. External factors include the composition of the medium and the physical cultivation conditions (Vinãs and Jiménez 2011). The interaction between all these factors leads to the induction and expression of a specific mode of cell differentiation and development, which in turn leads to embryogenesis (Gaj 2004). Usually, a distinction is made between direct and indirect SE. In the first case, the embryo is induced directly from the cells of the original explant (without a dedifferentiation stage), while in the second, the embryo is induced indirectly (going through dedifferentiation) from the formation of a callus. Indirect SE is the most common form of embryogenesis (Gaj 2004; Fehér 2008; Zavattieri et al. 2010). Indeed, callogenesis is a prerequisite for the formation of somatic embryos in various palms such as the date palm (Sané et al. 2006). Othmani et al. (2009) confirmed the formation of somatic embryos in date palm leaves by the direct route on 5 mg L^{-1} of 2,4-dichlorophenoxyacetic acid (2,4-D) added medium. When higher concentrations of 2,4-D were used (10 mg L^{-1}), indirect somatic embryogenesis occurred, indicating that there are different modes of action of the phytoregulator on the explant or species studied (Othmani et al. 2009). Compared with indirect somatic embryogenesis, the induction of embryos directly from the original explant reduces the somatic embryo formation time. Furthermore, this method presents fewer problems related to the accumulation of phenolic compounds and has lower contamination rates (less manipulation of the explants and shorter cultivation time) (Kumar et al. 2006). Additionally, the genetic variability of the material obtained by the direct route is relatively lower (Cuenca et al. 1999). Although it is not common, some studies report the propagation of palms via direct embryogenesis in saw palmetto (Gallo-Meagher and Green 2002), via organogenesis in date palm (Asemota et al. 2007; Kriaa et al. 2012; Mazri and Meziani 2013) and in peach palm (Almeida and Almeida 2006).

Indirect somatic embryogenesis is usually a process of regeneration involving multiple steps, beginning with the formation of primary, embryogenic calli that evolve to form proembryogenic masses. The development of a somatic embryo then occurs, followed by its maturation and finally the regeneration of the plant (Von Arnold et al. 2002, Von Arnold 2008). In the SE induction phase, the differentiated somatic cells acquire embryogenic competence, either directly or indirectly (Zavattieri et al. 2010). For this, the somatic cells undergo a reorganization of the cellular state, including physiological, metabolic, and gene expression (Jiménez et al. 2001). Although the mechanisms involved in the transition from the cell's destination, from somatic to embryogenic, are unclear, in general, it is a process of dedifferentiation and activation of cell division (Fehér 2003), which usually culminates in the formation of a proembryogenic callus (Vinãs and Jiménez 2011). Following induction, the somatic embryos go through the typical stages of zygotic embryogenesis, i.e., globular, scutellar, and coleoptilar, in monocotyledons (Toonen and De Vries 1996; Jiménez 2005). Finally, various reserve proteins, specific late embryogenesis abundant (LEA) protein, fatty acids, and sugars, all required for germination, are synthesized (Von Arnold et al. 2002; Rai et al. 2011). This accumulation of reserves and the desiccation process enables the somatic embryos to prepare for germination and develop into normal plants (Jiménez et al. 2001; Rai et al. 2011).

To summarize, the steps normally considered in indirect somatic SE in palms are (Jiménez 2005; Fehér 2008; Von Arnold 2008) (Fig. 15.2):

(I) Callogenesis: dedifferentiation of the somatic tissue, which becomes responsive, evolving to form proembryogenic masses.
(II) Induction of embryogenic structures: the multiplication/proliferation of the callus and organized embryogenic development take place.
(III) Maturation of the embryos: the initial structures accumulate reserve substances, passing through the desiccation process and completing embryonic development, evolving from the globular to the coleoptilar stage.
(IV) Conversion: the establishment and regeneration of complete plantlets (with the aerial part and the root) from the somatic embryos.

Although the four steps can be defined, not all SE protocols for palms differentiate between them or specify different treatments for each of these steps. SE consists, therefore, of replacing the pattern of gene expression in the tissue of the explant with a new embryogenic program (Zeng et al. 2007). Thus, the more the pattern of gene expression of the somatic embryo resembles or corresponds to that of zygotic embryos, the greater the chance of obtaining highly efficient regeneration systems (Merkle et al. 1995), and this is only possible if the cells are competent and receive appropriate inductive stimuli (Zavattieri et al. 2010).

15.2 Plant Material and Embryogenic Competence

Somatic cells with embryogenic capacity cultivated in vitro need stimuli to initiate embryogenesis, unlike embryogenic cells formed in vivo,

Fig. 15.2 Phases of somatic embryogenesis in palms using zygotic embryos as primordial explants in three Arecaceae species. Scale bar: *Acrocomia aculeata*, 1 cm; *Elaeis guineensis*, 0.5 cm; *Euterpe oleracea*, 0.1 cm (Authorized reproduction: *A. aculeata* (ZG Luis), *E. guineensis* (TA Balzon), *E. oleracea* (JE Scherwinski-Pereira))

which do not need external stimuli (Zimmerman 1993). Each tissue has embryogenic potential translated by the specific sensitivity into an external hormonal signal (Gaj 2004). After a change in one or more growing conditions, competent cells can reach the expression stage, in which the cell can differentiate into proembryos and develop into somatic embryos (Jiménez 2005). Thus, two categories of inductive conditions are recognized that allow differentiated cells to develop into competent dedifferentiated cells: growth regulators (internal and/or external cellular levels) and stress factors (Zavattieri et al. 2010). Competent cells have specific characteristics, such as early activation of the division cycle, more alkaline vacuolar pH, altered auxin metabolism, and the presence of leucoplasts (non-differentiated chloroplasts) (Pasternak et al. 2002).

During the induction of SE, a temporary increase in the endogenous levels of indoleacetic acid (IAA) seems to be a common characteristic in various plants and tissues (Jiménez 2005; Gueye et al. 2009). Thus, auxin synthesis induced by embryogenic conditions should be one of the key signals that determine the embryogenic destiny of a cultured cell (Gaj 2004). However, the interaction in this process between endogenous hormones and the growth regulators should be taken into account (Jiménez 2005). Gueye et al. (2009) reported that the capacity of date palm leaf segments to form calli depends on the state of cellular differentiation and that the most callogenic segments are found within the leaf lengthening zone. Thus, the establishment of a successful SE system is dependent on choosing plant material that has competent cells, and physical and chemical factors trigger/stimulate the embryogenic development paths of these cells (Zavattieri et al. 2010). According to Gaj (2004), immature zygotic embryos represent an important source of propagules in SE programs because the explants are used in more than a fifth of the protocols established between 1995 and 2004, with mature embryos being the second most used source.

In palms, the use of embryonic tissue has been successfully tested in vitro for various species: mature zygotic embryos (ZE) of areca palm, peach palm, macaw palm, and oil palm (Wang et al. 2002; Steinmacher et al. 2007a, 2007b; Moura et al. 2009; Thawaro and Te-Chato 2009; Balzon et al. 2013; Luis and Scherwinski-Pereira 2014), isolated coconut plumules (Fernando et al. 2003; Pérez-Núñez et al. 2006; Sáenz et al. 2006) and the immature ZE of coconut and açaí palm (Karunaratne and Periyapperuma 1989; Fernando and Gamage 2000; Saldanha and Martins-Corder 2012; Scherwinski-Pereira et al. 2012) (Table 15.1). When cloning is performed using zygotic embryos, these are usually genetically different from the mother plant due to the occurrence of cross-pollination, common among palms (Perera et al. 2007).

When considering somatic tissues, different types of explants are tested, but the response is usually limited to certain species. A few successful examples include date palm, oil palm, and peach palm leaves (Othmani et al. 2009; Konan et al. 2010; Santos et al. 2012) and roots of *Areca* (Wang et al. 2006). The reproductive structures are also used, successful examples are female date palm flowers (Kriaa et al. 2012), immature ovaries (Perera et al. 2007), and anthers (Perera et al. 2009a) of the coconut palm (Table 15.1). The latter are very promising, enabling clones to be obtained that are true to the matrix (Gueye et al. 2009).

Somatic tissues of young plants are good sources of explants. The caulinar apexes were successfully used in date palm (Sané et al. 2006) and oil palm (Thawaro and Te-Chato 2009). Using thin cell layer technique (Tran Thanh Van and Bui 2000) in explants of germinating seedlings, Steinmacher et al. (2007a) and Scherwinski-Pereira et al. (2010) obtained somatic embryos of peach palm and oil palm, respectively. The thin cell layer technique is based on the principle that smaller explants are more responsive because these have proportionally more surface area in contact with the growing medium, making them subject to higher stress, capable of increasing the cellular metabolism and the embryogenic response (Fehér et al. 2003; Othmani 2009). In small explants, the synthesis occurs on new cell wall components, such as oligosaccharides, that act as a signal to the cell to reinitiate the cellular cycle (Tran Thanh Van and Bui 2000). Othmani et al. (2009) tested date palm leaf explants of different sizes and confirmed that the smaller ones (5–10 mm) had a higher frequency of embryogenic callus formation. The authors suggest that the larger explants (15–20 mm), which did not form embryogenic calluses, allow the normal interactions between the cells of the tissue to be

Table 15.1 Protocols for indirect somatic embryogenesis of Arecaceae species (subfamily Arecoideae)

Tribe	Specie	Plant material and developmental stage	Embryogenesis phase	Growth regulator	Other components	Time	Light and temperature conditions	Activated charcoal (conc.)	Medium	Gelling agent	Carbohydrate source	Reference
Tribe Cocoseae subtribe *Attaleinae*	*Cocos nucifera*	Unfertilized ovaries from immature female flowers	C	100 μM 2,4-D	–	3 months Without subculturing	Dark, 28 °C	0.1 %	CRI 72 (Karunaratne and Periyapperuma 1989)	0.8 % (w/v) agar	4 % sucrose	Perera et al. (2007)
			I	5 μM ABA	10 μM AgNO3	5 weeks						
			M	–	–	4 weeks						
			R	5 μM BAP	–	N. m.	16-h photoperiod (17 μmol $m^{-2} s^{-1}$)	0.25 % (w/v)	Y3 (Eeuwens 1976, (Fernando and Gamage 2000)			
	Cocos nucifera	Anthers (microspore)	C	100 μM NAA 100 μM 2,4-D 10 μM kinetin	–	8 months, subcultured every month	Dark, 28 °C	0.1 %	Y3 (Karunaratne and Periyapperuma 1989)	–	9 % sucrose	Perera et al. (2009a)
			I	66 μM 2,4-D	–	Subcultured every month						
			M	–	–	N. m.						
			R	0.35 μM 2,4-D 5 μM BA 0.35 μM GA3	–	N. m.	16-h photoperiod (25 μmol $m^{-2} s^{-1}$)					
	Cocos nucifera	Unfertilized ovaries from immature female flowers	C	100 μM 2,4-D 9 μM TDZ	–	10 weeks; not subcultured	Dark, 28 °C	0.1 %	CRI 72	0.55 % (w/v) agar	4 % sucrose	Perera et al. (2009b)
			I	66 μM 2,4-D	–	4 weeks						
			M	–	–	Subcultured every month			Y3 (Eeuwens 1976)			
			R	5 μM 6-BA 0.1 μM 2,4-D 5 μM 2-iP 0.35 μM GA_3	–	Subcultured every month	16-h photoperiod 28 °C					

(continued)

Table 15.1 (continued)

Tribe	Specie	Plant material and developmental stage	Embryogenesis phase	Growth regulator	Other components	Time	Light and temperature conditions	Activated charcoal (conc.)	Medium	Gelling agent	Carbohydrate source	Reference
	Cocos nucifera	Plumule from the zygotic embryo	C	0.65 mM 2,4-D	–	3 months	Dark, 27 ± 2 ° C	2.5 g.L^{-1}	Y3	3 g L^{-1} Gelrite	N. m.	Sáenz et al. (2006)
			I	–	–	–	–					
			M	6 μM 2,4-D 300 μM 6-BA	–	Subcultured every 2 months	16-h photoperiod 27 ± 2 ° C					
			R	–	–	N. m.						
Tribe Cocoseae Subtribe *Bactridinae*	*Acrocomia aculeata*	Zygotic embryos	C 1	9 μM picloram 1 μM TDZ	1.0 g L^{-1} hydrolyzed casein 100 mg L^{-1} myo-inositol	60 days	Dark, 25 ± 2 °C	–	Y3	2.5 g L^{-1} Gelrite	68.46 g L^{-1} sucrose	Moura et al. (2009)
			C 2	9 μM picloram 1 μM TDZ		60 days		3 g.L^{-1}				
			I / M	–		120, days, subcultured every 60 days	16-h photoperiod (30 μmol m^{-2} s^{-1}), 25 ± 2 °C					
			R	–		60 days						
Tribe Cocoseae subtribe *Bactridinae*	*Acrocomia aculeata*	Zygotic embryos	C	1.5 mg L^{-1} 2,4-D	500 mg L^{-1} glutamine	240 days, subcult. every mo.	Dark, 25 ± 2 °C	–	Y3	2.5 g L^{-1} Phytagel	3 % sucrose	Luis and Scherwinski-Pereira (2014)
			I	1.0 mg L^{-1} 2,4-D		90 days, subcult. every 45 days						
			M	–		60 days					4.5 % sucrose	
			R	–		>35 days	16-h photoperiod (38 μmol m^{-2} s^{-1}), 25 ± 2 °C	2.5 g L^{-1}			3 % sucrose	

	Bactris gasipaes	Plantlet TCL	C	300 μM picloram	500 glutamine mg/L^{-1}	Without subculturing	Dark, 26 ± 1 °C	1.5 g L^{-1}	MS (Murashige and Skoog 1962) + Morel and Wetmore (1951)	2.5 g L Phytagel	3.0 % sucrose	Steinmacher et al. (2007a)
			I	40 μM 2,4-D 10 μM 2-iP	1 g L^{-1} glutamine 0.5 g L^{-1} hydrolyzed casein	Subcultured at 4-week intervals				7.0 g L^{-1} agar		
			M	24.6 mM 2-iP 0.44 mM NAA	–	30 days	16-h photoperiod (50–60 μmol m^{-2} s^{-1}) 26 ± 1 °C	–				
			R	–	–	N. m.		1.5 g L^{-1}				
	Bactris gasipaes	Mature zygotic embryos	C	20 μM picloram 0 or 5 μM 2-iP	–	12 weeks	Dark, 25 ± 2 °C	–	MS + Morel & Wetmore (1951)	0.7 % agar	3 % sucrose	Steinmacher et al. (2007b)
			I	10 μM of picloram 0 or 5 μM 2-iP	1 μM AgNO3	20–24 weeks		–		2.5 g L^{-1} Phytagel		
			M	40 μM 2,4-D 10 μM 2-iP	1 g L^{-1} glutamine 0.5 g L^{-1} hydrolyzed casein	4 weeks		1.5 g L^{-1}		0.7 % agar		
			R	0.44 μM NAA 24.6 μM 2-iP	–	4 weeks	16-h photoperiod (50–60 μmol m^{-2} s^{-1})	–				
			Plantlet growth	–	–	16 weeks		1.5 g L^{-1}				

(continued)

Table 15.1 (continued)

Tribe	Specie	Plant material and developmental stage	Embryogenesis phase	Growth regulator	Other components	Time	Light and temperature conditions	Activated charcoal (conc.)	Medium	Gelling agent	Carbohydrate source	Reference
Tribe Cocoseae subtribe *Elaeidinae*	*Elaeis guineensis*	Young leaves	C/I/M	1 mg L^{-1} dicamba	200 mg L^{-1} ascorbic acid	Subcultured every 4 weeks for 3–6 months	14-h photoperiod (1,300 lux), 26 ± 4 °C	–	MS	N. m.	3 % sucrose	Hilae and Te-Chato (2005)
			R shoot	–	–	3 months		–			0.2 M sorbitol	
			R rooting	–	–	4 weeks		–	1/2 MS		0.2 M sacarose	
	Elaeis guineensis	Mature zygotic embryos	C	450 μM picloram	–	Subcultured every 4 weeks for 150 days	Dark, 25 ± 2 °C	2.5 g L^{-1}	MS	2.5 g L^{-1} Phytagel	30 g L^{-1} sucrose	Balzon et al. (2013)
			I	40 μM picloram 10 μM 2-iP	–	12 weeks		–				
			M	0.54 μM NAA 12.3 μM 2-iP	–	12 weeks		–				
			R	–	–	From 8 to 24 weeks	16-h photoperiod (38 μmol m^{-2} s^{-1}) 25 ± 2 °C	2.5 g L^{-1}	MS 1/2			
Tribe Cocoseae subtribe *Elaeidinae*	*Elaeis guineensis*	Immature leaf	C	0.2 mg L^{-1} 2,4-D	0.2 mg L^{-1} TCPP (tris 2-chloro methyl phosphate)	60–100 days	Dark, 27 ± 1 °C	–	MS	N. m.	30 g L^{-1} sucrose	Konan et al. (2010)
			I	100 mg L^{-1} 2,4-D 1 mg L^{-1} BAP (benzylaminopurine)	30 mg L^{-1} adenine sulfate	3–60 months	12-h photoperiod (40 μ Em^{-2} s^{-2}), 27 ± 1 °C	4.5 g L^{-1}			20 g L^{-1} sucrose	
			M	–	500 mg L^{-1} casein hydrolysate	Subcultured every month for over 20 years		–			30 g L^{-1} sucrose	
			R	–	–	8–12 weeks		–			45 g L^{-1} sucrose	
			R rooting	NAA 1 mg L^{-1}	–	N. m.		–			60 g L^{-1} sucrose	

	Elaeis guineensis	Zygotic embryo	C	450 μM picloram		Subcultured every 4 weeks	Dark, 25 ± 2 °C	2.5 g L^{-1}	MS	2.5 g L^{-1}	30 g L^{-1} sucrose	Gomes et al. (2015)
			I	40 μM picloram 10 μM 2iP	–	12 weeks	Dark, 25 ± 2 °C	–				
			M	NAA 0.54 μM 12.3 μM 2iP	–	N. m.						
			R	–	–	N. m.	16-h photoperiod (50 μmol m^{-2} s^{-1}) 25 ± 2 °C	2.5 g L^{-1}	½ MS			
			R rooting	53.7 μM IBA	–	N. m.		–		DP System *		
Tribe Euterpeae	*Euterpe edulis*	Immature zygotic embryos	C / I	3 mg L^{-1} 2iP 100 mg L^{-1} 2,4-D	0.5 g L^{-1} glutamine	150 d	Dark, 25 ± 2 °C	1.5 g L^{-1}	MS + Morel & Wetmore (1951)	7 g L^{-1} agar	30 g L^{-1} sucrose	Saldanha and Martins-Corder (2012)
			M	3 mg L^{-1} 2iP 50 mg L^{-1} 2,4-D	0.5 g L^{-1} glutamine	180 d						
			R	–	–	60 d	16-h photoperiod 25 ± 2 °C					
	Euterpe oleracea	Immature zygotic embryos	C	225 μM picloram	2 mg L^{-1} myo-inositol	20 weeks	Dark, 25 ± 2 °C	1.5 g L^{-1}	MS MS 1/2	6 g L^{-1} agar	3 % sucrose	Scherwinski-Pereira et al. (2012)
			I / M	0.537 μM NAA 12.3 μM 2iP	2 mg L^{-1} myo-inositol	12 weeks, subcult. every 4 weeks		–				
			R	–	2 mg L^{-1} myo-inositol	12–18 weeks	16-h photoperiod (35–40 lmol m^{-2} s^{-1} 25 ± 2 °C)	2.5 g L^{-1}				
Tribe Phoenicaceae	*Phoenix dactylifera*	Young leaves from seedlings	C 1	2 mg L^{-1} 2,4-D	0.01 mg L^{-1} biotin Sodium ascorbate 100 mg L^{-1} 100 mg L^{-1} myo-inositol	2 + 1 months	N. m.	–	MS + Morel & Wetmore (1951)	8 g L^{-1} agar	30 g L^{-1} sucrose	Sané et al. (2006)
			C 2	2 mg L^{-1} 2,4-D		1 month Shaker 90 rpm				Liquid		
			I	–		N. m.					20 g L^{-1} glucose	
			M	2 mg L^{-1} BA (benzyladenine)		1 week + 5 weeks without BA, subcultured weekly				8 g L^{-1} agar + filter paper	20 g L^{-1} glucose	
			R	NAA 1 mg L^{-1}	–				MS	8 g L^{-1} agar	N.m	

(continued)

Table 15.1 (continued)

Tribe	Specie	Plant material and developmental stage	Embryogenesis phase	Growth regulator	Other components	Time	Light and temperature conditions	Activated charcoal (conc.)	Medium	Gelling agent	Carbohydrate source	Reference
Tribe Phoenicaceae	*Phoenix dactylifera*	Juvenile leaves from offshoots	C	10 mg L^{-1} 2,4-D	–	6–7 months, subcultured monthly	Dark, 28 ± 2 °C	0.3 g.L^{-1}	MS	0.7 % agar	30 g L^{-1} sucrose	Othmani et al. (2009)
			I	0.1 mg L^{-1} 2,4-D	–	Subcult. every 4–5 weeks		–				
			M 1	1 mg L^{-1} ABA	–	2 months, subcult. monthly	16/8-h photoperiod (80 μmol m^{-2}. s^{-1}) 28 ± 2 °C	0.1 mg L^{-1}				
			Extra desiccation and chopping	–	Fine chopping	12-h desiccation	Dark, 27 ± 2 °C	–	No medium used	–	–	
				1 mg L^{-1} ABA	–	N. m.	16/8-h photoperiod (80 μmol m^{-2}. s^{-1}) 28 ± 2 °C	0.1 mg L^{-1}	MS	0.7 % agar	30 g L^{-1} sucrose	
			M 2	1 mg L^{-1} ABA	–	2 months, subcult. monthly		–				
			R	1 mg L^{-1} NAA	–	N. m.		0.1 g L^{-1}				

	Phoenix dactylifera	Mature female flowers	C	1 mg L^{-1} 2,4-D	120 mg L^{-1} H_2PO_4 100 mg L^{-1} myo-inositol 2 mg L^{-1} glycine 100 mg L^{-1} L-glu 25 mg L^{-1} adenine 0.5 mg L^{-1} nicotinic acid 0.5 mg L^{-1} HCL-pyridoxine	8–12 months	Dark, 28 °C	–	MS	8 g L^{-1} agar	50 g L^{-1} sucrose	Kriaa et al. (2012)
			I / M	0.1 mg L^{-1} 2,4-D	–	Rotary shaker (120 rpm) subcultured every 3 months	16-h photoperiod 26 ± 2 °C	0.3 g L^{-1}	½ MS	–	N.m	
			R	0.5 mg L^{-1} BAP 0.5 mg L^{-1} NAA	–	N. m.	16-hour daylight condition 26 ± 2 °C	0.3 g L^{-1}	MS		N.m	
			R rooting	2 mg L^{-1} IBA	–	N. m.		–			N.m	

(continued)

Table 15.1 (continued)

Tribe	Specie	Plant material and developmental stage	Embryogenesis phase	Growth regulator	Other components	Time	Light and temperature conditions	Activated charcoal (conc.)	Medium	Gelling agent	Carbohydrate source	Reference
	Phoenix dactylifera	Immature inflorescences	C	10 mg L^{-1} 2,4-D	100 mg L^{-1} myo-inositol 2 mg L^{-1} glycine 100 mg L^{-1} L-glu 120 mg L^{-1} H_2PO_4 30 mg L^{-1} adenine	4–5 months, subcultured after 2 months	Dark, 28 ± 2 °C	–	MS	8 g L^{-1} agar	50 g L^{-1}	Fki et al. (2003)
		Juvenile leaves from offshoots	C	0.5 mg L^{-1} 2,4-D	–							
		Immature inflorescences and juvenile leaves from offshoots	I 1	1 mg L^{-1} 2,4-D	–	1 month Rotary shaker (120 rpm)	16/8-h photoperiod (28 μE m^{-2} s^{-1}) 28 ± 2 °C	300 mg L^{-1}	½ MS	–	30 g L^{-1}	
			I 2	10 mg L^{-1} 2,4-D	–	20 days	Dark, 28 ± 2 °C	500 mg L	MS	8 g L^{-1} agar	50 g L^{-1}	
			M	1 mg L^{-1} 2,4-D	–	N. m.	16/8-h photoperiod (28 μE m^{-2} s^{-1}) 28 ± 2 °C	300 mg L^{-1}	½ MS	–	30 g L^{-1}	
			R	–	–	N. m.	Dark, 28 ± 2 °C	–	MS	8 g L^{-1} agar	50 g L^{-1}	
			R rooting	1 mg L^{-1} NAA 1 mg L^{-1} BAP	–	N. m.	16/8-h photoperiod (28 μE m^{-2} s^{-1}) 28 ± 2 °C				–	

Tribe Phoenicaceae	*Phoenix dactylifera*	Shoot tips	C	5 mg L^{-1} 2,4-D 5 mg L^{-1} BAP	–	6 months	Dark, 28 ± 2 °C	0.15 g L^{-1}	MS	6.8 g L^{-1} carrageenan	30 g L^{-1}	Zouine and El Hadrami (2007)
			I	0.5 mg L^{-1} 2,4-D 0.1 mg L^{-1} BAP	–	Subcultured every 5 weeks						
			M	0.1 mg L^{-1} 2,4-D 0.5 mg L^{-1} BAP	6.7×10^{-4} M glutamine	2 months, rotary shaker (100 rpm)		0.25 g L^{-1}	½ MS	–		
			R	0.1 mg L^{-1} NAA 0.1 mg L^{-1} IBA 0.05 mg L^{-1} BAP	–	N. m.	N. m.	0.25 g L^{-1}		6.8 g L^{-1} carrageenan		
Subtribe Arecineae	*Areca catechu*	Inflorescence tissues Juvenile leaves of seedlings	C	200 µM picloram Reducing by half every 2 weeks	–	16 weeks	Dark, 27 ± 1 °C	0.1 %	MS	0.6 % agar	3 % sucrose	Karun et al. 2004
			I	5 µM picloram	–	6 weeks	16-h photoperiod ($40\ \mu E\ m^{-2}\ s^{-1}$) 27 ± 1 °C	0.1 %				
			M	–	–	N. m.		0.2 %				
			R	20 µM BA	–			0.1 %				
			R rooting	5 µM BA		6 weeks		–	½ MS	Liquid + filter paper		

(continued)

Table 15.1 (continued)

Tribe	Specie	Plant material and developmental stage	Embryogenesis phase	Growth regulator	Other components	Time	Light and temperature conditions	Activated charcoal (conc.)	Medium	Gelling agent	Carbohydrate source	Reference
	Areca catechu	Leaf, roots, and stem from young seedlings	C	2 mg L^{-1} dicamba	100 mg L^{-1} myo-inositol 0.5 mg L^{-1} niacin 0.5 mg L^{-1} pyridoxine HCl 0.1 mg L^{-1} thiamine HCl 2.0 mg L^{-1} glycine 1.0 mg L^{-1} peptone 170 mg L^{-1} NaH_2PO_4	60 days	Dark, 26 ± 1 °C	–	MS	2.2 g L^{-1} Gelrite	30 g L^{-1}	Wang et al. (2006; 2010)
			I	2 mg L^{-1} dicamba		Subcultured every 8 weeks						
			M	–		10 weeks	16-h photoperiod (28–36 µmol $m^{-2}s^{-1}$)					
			R	–		3 months						

Indirect embryogenesis stages: C, callogenesis; I, induction of somatic embryogenesis and proliferation; M, maturation; R, regeneration; N. m., not mentioned

*DP system: double-phase system (semisolid medium with a layer of liquid medium on the top)

L-glu: L-glutamine

maintained and that these interactions must inhibit cellular division by maintaining the symplastic domains (Othmani et al. 2009). The thin cell layer technique has been successfully used for the micropropagation of various ornamental plants including monocotyledons (Silva and Dobránszki 2013).

15.3 Endogenous Hormones

The development of techniques sensitive enough to analyze molecules like plant hormones, present in very small tissue samples, made it possible to investigate the role of these compounds in directing SE and to link their endogenous concentration to the morphogenesis and development (Gaj 2004). In leaf explants, the embryogenic tissues have higher concentrations of IAA (3-indolacetic acid) and lower concentrations of endogenous cytokinins than non-embryogenic tissues, although many results do not indicate a direct relationship between the concentration of these hormones and the embryogenic competence of the tissue (Jiménez 2005). Other evidence that high levels of endogenous auxins are correlated with embryogenic competence comes from studies in which a decrease in embryogenic capacity coincides with a decrease in endogenous IAA, practically to the levels that exist in non-embryogenic lineages (Jiménez 2005). Auxin synthesis and the polar transport of auxin are key events in the formation of meristem that leads to the development of the embryo (Nawy et al. 2008). In date palm (*Phoenix dactylifera*), the leaf segments that have greater competence in callus formation had a higher level of free endogenous IAA, which suggests that auxin homeostasis is an important factor for the cells to be able to dedifferentiate and be induced by 2,4-D (Gueye et al. 2009).

As regards to the other classes of regulators, several studies point to the important role of endogenous abscisic acid (ABA) during the SE induction phase (Jiménez 2005). On the other hand, it is believed that ethylene has a negative effect on the induction of SE, as observed in studies using ethylene action inhibitors, such as silver nitrate, to obtain somatic embryos of soy and date palm (Santos et al. 1997; Al-Khayri and Al-Bahrany 2001). With respect to the gibberellins (GA_3), the situation is less clear because few studies have analyzed the endogenous content of these hormones in embryogenic and non-embryogenic cultures and those that have ambiguous data (Jiménez 2005).

15.3.1 Growth Regulators Added to the Medium

The concentration and relative composition of growth regulators (GRs) added to the medium determine both the explant response ability and the type of morphogenic reaction (formation of embryos or axillary buds). GRs are often added to the culture medium in trial and error experiments; different substances, concentrations, and application times are being evaluated for their ability to induce the desired development pattern (Jiménez 2005). For the induction of somatic embryos, auxins are used alone or in combination with cytokines in 80 % of the protocols analyzed by Gaj (2004), and 2,4-D (2,4-dichlorophenoxyacetic acid) is the auxin used in 65 % of the protocols. Of the 20 SE protocols for palms analyzed from articles published in the period from 2000 to 2015, 14 (70 %) used 2,4-D in the formation of callus and/or in the induction of embryogenesis (Table 15.1). 2,4-D appears to act not only as an analog of auxin but also as a stress factor that triggers embryogenic development in cells in cultivation, especially when used in high concentrations (Fehér 2003). The suppression of inductor auxin is essential for embryogenesis to be complete, as continuous growth in medium containing 2,4-D does not permit a decrease in the endogenous concentrations of auxin and results in the inhibition of embryo development (embryo expression and differentiation), even in induced cells (Nissen and Minocha 1993).

Auxins and cytokines are the most used phytoregulators because these compounds are able to regulate the cellular cycle and trigger cell division (Francis and Sorrell 2001; Jiménez 2005). In palms, the attempt to use cytokinin coupled with auxin during callogenesis occurred in only one study (Steinmacher et al. 2007b). However, during the phases of induction and maturation of embryogenic structures, 8 out of 20 studies sampled used a cytokinin, usually 2-isopentenyladenine (2iP), together with the auxin (Sáenz et al. 2006; Sané et al. 2006; Steinmacher et al. 2007b; Zouine and El Hadrami 2007; Konan et al. 2010; Saldanha and Martins-Corder 2012; Scherwinski-Pereira et al. 2012; Balzon et al. 2013).

Using 2iP, it was possible to proliferate embryogenic calluses and regenerate plants from somatic embryos of *Elaeis guineensis* (Balzon et al. 2013; Gomes et al. 2015), *Phoenix dactylifera* (Gabr and Tisserat 1985), and *Cocos nucifera* (Perera et al. 2009b). In coconut palm, regeneration of plants using immature inflorescences as explants in the SE was successful with the combined use of 6-benzyladenine (BA), 2iP, and 2,4-D (Perera et al. 2009b).

Abscisic acid (ABA) promotes the transition of somatic embryos from the proliferation phase to the maturation phase. Both the synthesis, the deposit of reserve, and late embryogenesis (LEA) proteins during somatic embryogenesis are regulated by ABA and by the expression of genes induced by stress (Dodeman et al. 1997; Von Arnold et al. 2002, Rai et al. 2011). ABA was used in the induction of somatic embryogenesis in the cultivation of coconut palm plumules (Fernando and Gamage 2000) and in the maturation of somatic embryos of date palm (Othmani et al. 2009) and coconut palm (Perera et al. 2007). However, for the coconut palm, there is one report that states that the use of ABA reduced the formation of shoots in plants regenerated from somatic embryos (Perera et al. 2009b).

Besides GRs, other biologically active molecules like polyamines can induce SE in different plant species (Gaj 2004). Moreover, a new generation of GRs, such as thidiazuron, a cytokinin that belongs to the diphenylureas, has emerged as an alternative for high frequencies of direct regeneration of somatic embryos, even from differentiated tissues (Jiménez 2005). Perera et al. (2009b) obtained good results using thidiazuron (TDZ) in the induction phase of calluses in unfertilized coconut palm ovaries (*Cocos nucifera* L.).

15.3.2 Sugars

Sugars are a critical source of energy and create the osmotic conditions appropriate for in vitro cell growth. The carbon source appears to play an important role in somatic embryogenesis, and sucrose is the most commonly used sugar in SE cultures (Gaj 2004; Yaseen et al. 2013). The carbohydrate requirement of a culture depends on the species and on the stage of development of cultivated tissue (Yaseen et al. 2013). In palms, sucrose at a concentration of 30 g L^{-1} is the most commonly used carbohydrate (Table 15.1). However, this concentration was increased in specific cases, as in the induction of callus and maturation of somatic embryos of date palms (Fki et al. 2003) and in the maturation and rooting of oil palm (Konan et al. 2010). In date palm, Sané et al. (2006) used a sucrose concentration of 20 g L^{-1} in all steps, starting from the induction of embryogenesis.

Sugar alcohols, like sorbitol, mannitol, and glycerol, are also used in tissue culture, and the addition of carbohydrates to the culture medium plays an important role in the osmoregulation of hydric stress (Yaseen et al. 2013). The presence of substances with osmotic potential such as sorbitol in the medium increases hydric stress, which favors the germination of somatic embryos of oil palm, inducing the formation of shoot and roots (Hilae and Te-chato 2005).

15.3.3 Activated Charcoal

Activated charcoal (AC) is commonly used in tissue culture to increase the growth and development of cells and tissues and plays an important role in micropropagation and somatic embryogenesis (Verdeil and Buffard-Morel 1995) (Table 15.1). The promotive effects of AC on morphogenesis may be mainly due to its irreversible absorption of inhibitory compounds in the culture medium and the substantial reduction of toxic metabolics, aromatic compounds including phenols and exudates that cause oxidation and browning in the medium (De Touchet et al. 1991; Teixeira et al. 1993; Thomas 2008; Othmani et al. 2009). Additionally, the substances, naturally present in AC released into the medium promote growth, alter the culture medium, cause it to darken, and absorb various compounds that have stimulatory or inhibitory effects, such as vitamins, metallic ions, CRs, acid, and gaseous ethylene (Thomas 2008). Thus, the use of AC in the culture medium can help with the induction of somatic embryogenesis and with the maturation and regeneration of clones (Thomas 2008; Othmani et al. 2009). Adding AC to the culture medium alters some of its properties, such as conductivity and osmotic concentration, with most significant change occurring at the pH levels, which undergo changes from the day of preparation of medium (Sáenz et al. 2010). There are differences between the different brands of AC in their capacity to absorb 2,4-D, and these differences are reflected in optimum concentration of 2,4-D required for an embryogenic callus formation response. This data is particularly relevant because one of the most important factors affecting the morphogenic response of explants cultivated in vitro is the presence of this auxin in the optimum concentration (Ebert and Taylor 1990; Verdeil and Buffard-Morel 1995; Sáenz et al. 2010). The time elapsed from the preparation of medium influences the final availability of 2,4-D. Thus, when AC is added to the medium, the balance is only reached after about 20 days and depends on the absorption rate of the AC and the rate of diffusion of 2,4-D in medium (Ebert and Taylor 1990). With the use of radioactive markers, an absorption rate of 99.5 % of the 100 μm of 2,4-D added to the liquid medium can be calculated 5 days after the preparation of the medium. The absorption of 2,4-D by the charcoal increases in an acid pH and decreases with the addition of Phytagel or agarose to the culture medium (Ebert and Taylor 1990).

Among the palms, the coconut palm (*Cocos nucifera*) has the highest number of examples of positive use of AC in somatic embryogenesis. For example, by using 2,4-D in concentrations higher than 100 μM and AC in concentrations ranging from 0.1 to 0.25 %, the induction of calluses from anthers (Perera et al. 2009a), immature ovaries (Perera et al. 2007), and plumules was possible (Fernando et al. 2003; Sáenz et al. 2006). Other species respond satisfactorily to the use of AC in SE, such as *Elaeis guineensis*, from mature zygotic embryos (Balzon et al. 2013), with 450 μM of picloram, and *Euterpe edulis* and *Euterpe oleracea*, from immature zygotic embryos, with 100 mg L^{-1} of 2,4-D (Saldanha and Martins-Corder 2012) or with 225 μM of picloram (Scherwinski-Pereira et al. 2012). Besides the induction of calluses in *Elaeis guineensis*, AC has a positive impact on the maintenance of calluses (Teixeira et al. 1993; Eeuwens 2002). In *Bactris gasipaes*, the use of AC was positive in SE, both during the formation of primary callus, in induction, during the maturation of embryos, and the establishment of plantlets (Steinmacher et al. 2007a; Steinmacher et al. 2007b).

Many authors indicate AC for *Phoenix dactylifera* (Gueye et al. 2009). However, there is no consensus as to its use in the induction of primary callus, and it may or may not be used for the same explant. For caulinar apexes excised from plants in vitro, Sané et al. (2006) did not use AC, while Zouine and El Hadrami (2007) added 0.15 g L^{-1} to the medium, using 2,4-D in different concentrations, 2 mg L^{-1} and 5 mg L^{-1}, respectively. Fki et al. (2003) did not use AC in the shoots of young leaves, while Othmani et al. (2009) used 0.3 g L^{-1} of AC in concentrations of 2,4-D of 0.5 mg L^{-1} and 10 mg L^{-1}, respectively. In the induction of calluses in mature female flowers, Kriaa et al. (2012) used no AC in medium, adding 1 mg L^{-1} of 2,4-D, and for *Phoenix canariensis*, the use of AC had a negative effect on the formation of embryogenic callus (Thomas 2008). The use of 2.0 g L^{-1} of AC in medium for induction of direct somatic embryos in endangered bottle palm (*Hyophorbe lagenicaulis*) prevented darkening of the tissues and showed reduced growth (Sarasan et al. 2002).

15.3.4 Other Factors Added to the Basic Medium

15.3.4.1 Complex Mixtures

Complex mixtures include hydrolyzed proteins, yeast extract, coconut water, etc. and are used to complement the culture media when expected results are not achieved (Barrueto Cid and Teixeira 2010). The addition of coconut water to the culture medium yielded positive results for two varieties of date palm at two different concentrations (10 and 15 %), resulting in greater callus growth and an increase in the number of embryos formed (Al-Khayri 2010) (Table 15.1). Hydrolyzed casein extract was used in SE in various species like *Acrocomia aculeata* (Moura et al. 2009), *Bactris gasipaes* (Valverde et al. 1987; Steinmacher et al. 2007a, b), and *Elaeis guineensis* (Konan, et al. 2010).

15.3.4.2 Organic Compounds

Inositol is an organic compound of low molecular weight that has various isomers, normally found in plants (Barrueto Cid and Teixeira 2010), and this myo-inositol is widely used in SE protocols in palms like *Acrocomia aculeata* (Moura et al. 2009), *Bactris gasipaes* (Valverde et al. 1987), and *Euterpe oleracea* (Scherwinski-Pereira et al. 2012).

The main vitamins used are water soluble or components of group B (Barrueto Cid and Teixeira 2010). In palms, the most often used B complex vitamins are niacin, pyridoxine, and thiamine, usually in acid form (Wang et al. 2006, 2010; Kriaa et al. 2012). Ascorbic acid was included in the composition of culture medium for *Elaeis guineensis* (Te-Chato and Hilae 2007; Thawaro and Te-Chato 2009) and sodium ascorbate in the composition of medium for *Phoenix dactylifera* (Sané et al. 2006). Several protocols have developed media using vitamins of Morel and Wetmore (1951), which are the same as those of MS, but in lower concentrations (Sané et al. 2006; Steinmacher et al. 2007a; Saldanha and Martins-Corder 2012).

Many amino acids are also added to the medium, the most common being L-glutamine (Steinmacher et al. 2007a, b; Zouine and El Hadrami 2007; Saldanha and Martins-Corder 2012; Luis and Scherwinski-Pereira 2014). However, L-cysteine, adenine, and glycine are also present in various protocols (Wang et al. 2006, 2010) (Table 15.1).

15.3.5 Inorganic Compounds

Most protocols use MS medium (Murashige and Skoog 1962) as the basic medium, with a total or modified salt concentration depending on the stage of SE (Table 15.1). Y3 medium (Eeuwens 1976) and modified versions of it (Karunaratne and Periyapperuma 1989) were used exclusively with two species, *Cocos nucifera* (Sáenz et al. 2006; Perera et al. 2007; Perera et al. 2009a, 2009b) and *Acrocomia aculeata* (Moura et al. 2009; Luis and Scherwinski-Pereira 2014).

Valverde et al. (1987) replaced NH_4NO_3 with NH_4Cl and increased the concentration of KNO_3 in the SE of *Bactris gasipaes*, although other studies with the same species maintained the initial concentrations of MS medium (Steinmacher et al. 2007a, b). Another study increased the final concentration of KH_2PO_4 (Fki et al. 2003). The objective of this modification was to increase the availability of essential groups such as nitrates and phosphates in the medium, increasing the efficiency of the stages of SE (Barrueto Cid and Teixeira 2010). Some protocols include silver nitrate ($AgNO_3$) in the medium (Steinmacher et al. 2007b; Perera et al. 2007), which acts as an ethylene action inhibitor.

15.3.6 Light Conditions

Systematic studies of the effects of light on the response of explants cultivated in vitro are limited. Photoperiod and conditions of total darkness were required to induce SE in 49 % and 44 % of the protocols, respectively (Gaj 2004). However, embryos in primary stages of development are two times more efficient when cultivated in the dark. The influence of light on morphogenesis in vitro may be related to the stimulating or inhibitory effect of light on different endogenous substances including GRs (Zelena 2000). In palms, only one of the protocol analyzed conditions by keeping the explants under light regime during callogenesis (Hilae and Te-Chato 2005), while all the other protocols (90 %) kept the explants in the dark. During the induction of embryogenesis, 30 % of the protocols transferred the calluses into light conditions, with photoperiod conditions ranging from 12 to 16 h of light. This percentage increases during the maturation phase, extending to 50 % of the protocols. During the regeneration phase of the plant, all protocols transferred the embryos to the light, with a photoperiod of 16 h of light in 85 % of the cases (Table 15.1).

15.4 Conversion of Somatic Embryos into Plants

The efficiency of in vitro multiplication systems depends not only on the formation of embryos but also on the capacity of these embryos to be transformed into plants. The process involves the changes in development and undergoes conversion by the appearances of primary roots, of a caulinar meristem with primordial leaves, of a hypocotyl, and of functional cotyledons (Gaj 2004). To stimulate the conversion of the embryo and to increase regeneration efficiency, gibberellic acid is commonly used in the culture medium and is believed to be especially necessary in cultures of embryos from plant species that normally exhibit dormancy. Other GRs can also be used such as IBA (indolebutyric acid), ABA (abscisic acid), and cytokinins (Table 15.1). The excess or the poor distribution of auxins concentration principally 2,4-D can cause morphogenic abnormalities in the embryos (Jiménez, 2005).

15.5 Concluding Remarks

After the grass, the Arecaceae is the second most important and useful plant family among monocotyledons and is the third among all plant families after Leguminosae. Although economically important, palms are recognized to be a neglected group of plants, and these are considered to be a difficult and time-consuming material in terms of research. In fact, in palms, the application of asexual embryogenesis faces several limitations. The major problems in palm tissue culture are the difficulty in standardizing and acquiring responsive explants, microbial contamination, browning, low explant responses (mainly leaves), slow growth of callus and somatic embryos, and poor embryo germination.

However, a better understanding of initial requirements and of special physiology of explants has been gradually optimized. Recent advances in molecular techniques have also created new opportunities for the study on somatic embryogenesis. Thus, studies of different stages of somatic embryogenesis by using molecular biology tool including gene expression pattern, associated with better understanding of biochemical and histochemical changes occurring in somatic cells to become embryogenic, are probably the next steps to be followed. Although challenging, there still remains a considerable potential for scientific progress in palms.

Acknowledgements The authors thank CNPq (Conselho Nacional de Desenvolvimento Cientifico e Tecnologico) for financial support, and FAPEPI (Fundação de Amparo à Pesquisa do Estado do Piauí) by doctoral scholarship to the first author.

References

Al-Khayri JM (2010) Somatic embryogenesis of date palm (*phoenix dactylifera* L.) improved by coconut water. Biotech 9(4):477–484

Al-Khayri JM, Al-Bahrany AM (2001) Silver nitrate and 2-isopentyladenine promote somatic embryogenesis in date palm (*Phoenix dactylifera* L.). Sci Hortic 89(4):291–298

Almeida M, Almeida CV (2006) Somatic embryogenesis and in vitro plant regeneration from pejibaye adult plant leaf primordia. Pesq Agropec Bras 41(9):1449–1452

Aoshima Y (2005) Efficient embryogenesis in the callus of tea (*Camellia sinensis*) enhanced by the osmotic stress or antibiotics treatment. Plant Biotechnol 22(4):277–280

Asemota O, Eke CR, Odewale JO (2007) Date palm (*Phoenix dactylifera* L.) in vitro morphogenesis in response to growth regulators, sucrose and nitrogen. Afr J Biotechnol 6:2353–2357

Baker WJ, Norup MV, Clarkson JJ, Couvreur TLP, Dowe JL, Lewis CE, Pintaud JC, Savolainen V, Wilmot T,

Chase MW (2011) Phylogenetic relationships among arecoid palms (Arecaceae: Arecoideae). Ann Bot 108:1417–1432

Balzon TA, Luis ZA, Scherwinski-Pereira JE (2013) New approaches to improve the efficiency of somatic embryogenesis in oil palm (*Elaeis guineensis* Jacq.) from mature zygotic embryo. In Vitro Cell Dev Biol – Plant. doi:10.1007/s11627-012-9479-3

Barrueto Cid LP, Teixeira JB (2010) Explante, meio nutritivo, luz e temperatura. In: Barrueto Cid LP (ed.) Cultivo in vitro de plantas. Embrapa, Brasília, pp 15–49

Beulé T, Camps C, Debiesse S, Tranchant C, Dussert S, Sabau X, Jaligot E, Alwee SSRS, Tregear JW (2011) Transcriptome analysis reveals differentially expressed genes associated with the mantled homeotic flowering abnormally in oil palm (*Elaeis guineensis*). Tree Genet Genomes 7:169–182

Bhojwani SS, Razdan MK (1992) Plant tissue culture: theory and practice, 6th edn. Elsevier, Amsterdam

Broschat TK, Hodel DR, Elliott ML (2014) Ornamental palms: biology and horticulture. Hortic Rev 42:1–120

Costa NM de S, Aloufa MAI (2007) Desenvolvimento in vitro de embriões zigóticos de tamareira. Rev Ciênc Agr Fortaleza 38(3):276–279

Cuenca B, San-José MC, Martínez MT, Ballester A, Vieitez AM (1999) Somatic embryogenesis from stem and leaf explants of *Quercus robur* L. Plant Cell Rep 18:538–543

De Touchet B, Duval Y, Pannetier C (1991) Plant regeneration from embryogenic suspension cultures of oil palm (*Elaeis guineensis* Jacq.). Plant Cell Rep 10(10):529–532

Dodeman VL, Ducreux G, Kreis M (1997) Zygotic embryogenesis versus somatic embryogenesis. J Exp Bot 48(7):1493–1509

Dransfield J (2008) Genera Palmarum: the evolution and classification of palms. Kew Publisihing, Richmond

Durand-Gasselin T, Guen V, Konan K, Duval Y (1990) Oil palm (*Elaeis guineensis* Jacq.) plantations in Ivory Coast obtained through *in vitro* culture. First results. Oléagineux 45:1–11

Ebert A, Taylor HF (1990) Assessment of the changes of 2,4-dichlorophenoxyacetic acid concentrations in plant tissue culture media in the presence of activated charcoal. Plant Cell Tiss Org Cult 20:165–172

Eeuwens CJ (1976) Mineral requirements for growth and callus initiation of tissue explants excised from mature coconut palms (*Cocos nucifera*) and cultured in vitro. Physiol Plant 36(1):23–28

Eeuwens CJ, Lord S, Donough CR, Rao V, Vallejo G, Nelson, S (2002) Effects of tissue culture conditions during embryoid multiplication on the incidence of "mantled" flowering in clonally propagated oil palm. Plant Cell Tissue Org Cult 70(3):311–323

Eiserhardt WL, Svenning J-C, Kissling WD, Balslev H (2011) Geographical ecology of the palms (Arecaceae): determinants of diversity and distributions across spatial scales. Ann Bot 108:1391–1416

Fehér A (2003) Transient of somatic plant cells to an embryonic state. Plant Cell Tiss Org Cult 74:202–228

Fehér A (2008) The initiation phase of somatic embryogenesis: what we know and what we don't. Acta Biol Szegediensis 52(1):53–56

Fernando SC, Gamage CKA (2000) Abscisic acid induced somatic embryogenesis in immature embryo explants of coconut (*Cocos nucifera* L.). Plant Sci 151:193–198

Fernando SC, Verdeil JL, Hocher V, Weerakoon LK, Hirimburegama K (2003) Histological analysis of plant regeneration from plumule explants of *Cocos nucifera* L. Plant Cell Tiss Org Cult 72:281–284

Fki L, Masmoudi R, Drira N, Rival A (2003) An optimized protocol for plant regeneration from embryogenic suspension cultures of date palm, *Phoenix dactylifera* L. cv. Deglet Nour. Plant Cell Rep 21:517–524

Francis D, Sorrell DA (2001) The interference between the cell cycle and plant growth regulators: a minireview. Plant Growth Regul 33:1–12

Gabr MF, Tisserat B (1985) Propagating palms in vitro with special emphasis on the date palm (*Phoenix dactylifera* L.). Sci Hortic 25(3):255–262

Gaj MD (2004) Factors influencing somatic embryogenesis induction and plant regeneration with particular reference to *Arabidopsis thaliana* (L.) Heynh. Plant Growth Regul 43:27–47

Gallo-Meagher M, Green J (2002) Somatic embryogenesis and plant regeneration from immature embryos of saw palmetto, an important landscape and medicinal plant. Plant Cell Tiss Org Cult 68:253–256

Gomes HT, Bartos PMC, Scherwinski-Pereira JE (2015) Optimizing rooting and survival of oil palm (*Elaeis guineensis*) plantlets derived from somatic embryos. In Vitro Cell Dev Biol Plant. doi:10.1007/s11627-015-9669-x

Gueye B, Morcillo F, Collin M, Gargani D, Overvoorde P, Aberlenc-Bertossi F, Verdeil JL (2009) Acquisition of callogenic capacity in date palm leaf tissues in response to 2,4-D treatment. Plant Cell Tiss Org Cult 99:35–45

Hilae A, Te-chato S (2005) Effects of carbon sources and strength of MS medium on germination of somatic embryos of oil palm (*Elaeis guineensis* Jacq.) Songklanakarin J Sci Technol 27(Suppl. 3):629–635

Jayasankar S, Gray DJ, Litz RE (1999) High-efficiency somatic embryogenesis and plant regeneration from suspension cultures of grapevine. Plant Cell Rep 18(7–8):533–537

Jiménez VM (2005) Involvement of plant hormones and plant growth regulators on in vitro somatic embryogenesis. Plant Growth Regul 47:91–110

Jiménez VM, Guevara E, Herrera J, Bangerth F (2001) Endogenous hormone levels in habituated nucellar *Citrus* callus during the initial stages of regeneration. Plant Cell Rep 20:92–100

Karami O, Aghavaisi B, Pour AM (2009) Molecular aspects of somatic-to-embryogenic transition in plants. J Chem Biol 2:177–190

Karun A, Siril EA, Radha E, Parthasarathy VA (2004) Somatic embryogenesis and plantlet regeneration from leaf and inflorescence explants of arecanut (*Areca catechu* L.). Curr Sci 86(12):1623–1628

Karunaratne S, Periyapperuma K (1989) Culture of immature embryos of coconut (*Cocos nucifera* L.): callus proliferation and somatic embryogenesis. Plant Sci 62:247–253

Konan EE, Durand-Gasselin T, Kouadio YJ, Flori A, Rival A, Duval Y, Pannetier C (2010) In vitro conservation of oil palm somatic embryo for 20 years on a hormone-free culture medium: characteristics of the embryogenic cultures, derived plantlets and adult palms. Plant Cell Rep 29:1–13

Kriaa W, Sghaier-Hammami B, Masmoudi-Allouche F, Benjemaa-Masmoudi R, Drira N (2012) The date palm (*Phoenix dactylifera* L.) micropropagation using completely mature female flowers. C R Biol 335:194–204

Kumar V, Naidu MM, Ravishankar GA (2006) Developments in coffee biotechnology-in vitro plant propagation and crop improvement. Plant Cell Tiss Org Cult 87(1):49–65

Lorenzi H (2010) Flora Brasileira: Arecaceae. Instituto Plantarum, Nova Odessa

Low ET, Alias H, Boon SH, Shariff EM, Tan CYA, Ooi LCL, Cheah SC, Raha AR, Wan KL, Singh R (2008) Oil palm (*Elaeis guineensis* Jacq.) tissue culture ESTs: identifying genes associated with callogenesis and embryogenesis. BMC Plant Biol 8(1):62

Luis ZG, Scherwinski-Pereira JE (2014) An improved protocol for somatic embryogenesis and plant regeneration in macaw palm (*Acrocomia aculeata*) from mature zygotic embryos. Plant Cell Tiss Org Cult 118:485–496

Mazri MA, Meziani R (2013) An improved method for micropropagation and regeneration of date palm (*Phoenix dactylifera* L.). J Plant Biochem Biotechnol 22(2):176–184. doi:10.1007/s13562-012-0147-9

Meerow AW, Krueger RR, Singh R, Low ET, Ithnin M, Ooi LCL (2012) Coconut, date, and Oil palm genomics. In: Schnell RJ, Priyadarshan PM (eds) Genomics of tree crops. Springer, New York/London. doi:10.1007/978-1-4614-0920-5_10

Merkle SA, Parrott WA, Flinn (1995) Morphogenic aspects of somatic embryogenesis. In: Thorpe TA (ed) In vitro Embryogenesis in Plants. Kluwer Academic Publishers, Dordrecht, pp 155–203

Morel GM, Wetmore RH (1951) Fern callus tissue culture. Am J Bot 38:141–143

Moura EF, Motoike SY, Vendrella MC, Sá AQ, Carvalho M (2009) Somatic embryogenesis in macaw palm (*Acrocomia aculeata*) from zygotic Embryos. Sci Hortic 119:447–454

Murashige T, Skoog F (1962) A revised medium for rapid growth and bioassay with tobacco tissue. Physiol Plant 15:473–497

Nawy T, Lukowitz W, Bayer M (2008) Talk global, act local – patterning the *Arabidopsis* embryo. Curr Opin Plant Biol 11:28–33

Nissen P, Minocha SC (1993) Inhibition by 2,4-D of somatic embryogenesis in carrot as explored by its reversal by difluoromethylornithine. Physiol Plant 89:673–680

Othmani A, Bayoudh C, Drira N, Marrakchi M, Trifi M (2009) Somatic embryogenesis and plant regeneration in date palm *Phoenix dactylifera* L., cv. Boufeggous is significantly improved by fine chopping and partial desiccation of embryogenic callus. Plant Cell Tiss Org Cult 97:71–79

Pasternak TP, Prinsen E, Ayaydin F, Miskolczi P, Potters G, Asard H, Van Onckelen HA, Dudits D, Fehér A (2002) The role of auxin, pH, and stress in the activation of embryogenic cell division in leaf protoplast derived cells of alfalfa. Plant Physiol 129(4):1807–1819

Perera PIP, Hocher V, Verdeil JL, Doulbeau S, Yakandawala DM, Weerakoon LK (2007) Unfertilized ovary: a novel explant for coconut (*Cocos nucifera* L.) somatic embryogenesis. Plant Cell Rep 26:21–28

Perera PIP, Yakandawala DMD, Hocher V, Verdeil JL, Weerakoon LK (2009a) Effect of growth regulators on microspore embryogenesis in coconut anthers. Plant Cell Tissue Org Cult 96(2):171–180

Perera PIP, Vidhanaarachchi VRM, Gunathilake TR, Yakandawala DMD, Hocher V, Verdeil JL, Weerakoon LK (2009b) Effect of plant growth regulators on ovary culture of coconut (*Cocos nucifera* L.). Plant Cell Tiss Org Cult 99:73–81

Pérez-Núñez MT, Chan JL, Sáenz L, González T, Verdeil JL, Oropeza C (2006) Improved somatic embryogenesis from *Cocos nucifera* L. plumule explants. In Vitro Cell Dev Biol Plant 42(1):37–43

Rai MK, Shekhawat NS, Gupta AK, Phulwaria M, Ram K, Jaiswal U (2011) The role of abscisic acid in plant tissue culture: a review of recent progress. Plant Cell Tiss Org Cult 106:179–190. doi:10.1007/s11240-011-9923-9

Rajanaidu N, Ainul MM (2013) Conservation of oil palm and coconut genetic resources. In: Normah MN, Chin HF, Reed BM (eds) Conservation of tropical plant species. Springer, New York

Sáenz L, Azpeitia A, Chuc-Armendariz B, Chan JL, Verdeil JL, Hocher V, Oropeza C (2006) Morphological and histological changes during somatic embryo formation from Coconut plumule explants. In Vitro Cell Dev Biol Plant 42:19–25

Sáenz L, Herrera-Herrera G, Uicab-Ballote F, Chan JL, Oropeza C (2010) Influence of form of activated charcoal on embryogenic callus formation in coconut (*Cocos nucifera*). Plant Cell Tiss Org Cult 100:301–308

Saldanha CW, Martins-Corder MP (2012) In vitro germination and embryogenic competence acquisition of *Euterpe edulis* Martius immature zygotic embryos. Crop Breed Appl Biotechnol 12:171–178

Sané D, Aberlenc-Bertossi F, Gassama-Dia YK, Sagna M, Trouslot MF, Duval Y, Borgel A (2006) Histocytological analysis of callogenesis and somatic embryogenesis from cell suspensions of date palm (*Phoenix dactylifera*). Ann Bot 98:301–308

Santos KGB, Mundstock E, Bodanese-Zanettini MH (1997) Genotype-specific normalization of soybean somatic embryogenesis through the use of an ethylene inhibitor. Plant Cell Rep 16:859–864

Santos MRA, Ferreira MGR, Rocha JF, Correia AO (2012) Estabelecimento in vitro e calogênese em explantes foliares de pupunheira. Rev Cienc Agrar 55(3):197–203

Sarasan V, Ramsay M, Roberts A (2002) In vitro germination and induction of direct somatic embryogenesis in "Bottle Palm" (*Hyophorbe lagenicaulis* (L. Bailey) H. E. Moore), a critically endangered Mauritian palm. Plant Cell Rep 20:1107–1111

Scherwinski-Pereira JE, Guedes RS, Fermino PCP Jr, Silva TL, Costa FHS (2010) Somatic embryogenesis and plant regeneration in oil palm using the thin cell layer technique. Vitro Cell Dev Biol Plant 46(4):378–385

Scherwinski-Pereira JE, Guedes RS, Silva RA, Fermino PCP, Luis ZG, Freitas EO (2012) Somatic embryogenesis and plant regeneration in açaí palm (*Euterpe oleracea*). Plant Cell Tiss Org Cult 109:501–508

Silva JAT, Dobránszki J (2013) Plant Thin Cell Layers: A 40-Year Celebration. J Plant Growth Regul 32:922–943

Steinmacher DA, Krohn NG, Dantas AC, Stefenon VM, Clement CR, Guerra MP (2007a) Somatic embryogenesis in Peach Palm using the Thin Cell Layer Technique: induction, morpho-histological aspects and AFLP analysis of somaclonal variation. Ann Bot 100:699–709

Steinmacher DA, Cangahuala-inocente GC, Clement CR, Guerra MP (2007b) Somatic embryogenesis from peach palm zygotic embryos. In Vitro Cell Dev Biol Plant 43:124–132

Te-chato S, Hilae A (2007) High-frequency plant regeneration through secondary somatic embryogenesis in oil palm (*Elaeis guineensis* Jacq. var. *tenera*). J Agric Technol 3(2):345–357

Teixeira JB, Sondahl MR, Kirby EG (1993) Somatic embryogenesis from immature zygotic embryos of oil palm. Plant Cell Tiss Org Cult 34:227–233

Thawaro S, Te-Chato S (2009) Effect of genotypes and auxins on callus formation from mature zygotic embryos of hybrid oil palms. J Agric Technol 5(1):167–177

Thomas TD (2008) The role of activated charcoal in plant tissue culture. Biotechnol Adv 26:618–631

Tomlinson PB (2006) The uniqueness of palms. Bot J Linn Soc 151:5–14

Toonen MAJ, De Vries SC (1996) Initiation of somatic embryos from single cells. In: Wang TL, Cuming A (eds) Embryogenesis: the generation of a plant. Bios Scientific Publishers, Oxford, pp p173–p189

Tran Than Van K, Bui VL (2000) Current status of thin cell layer method for the induction of organogenesis or somatic embryogenesis. In: Mohan SJ, Gupta PK, Newton RJ (eds) Somatic embryogenesis in woody plants. Kluwer, Dordrecht, pp 51–92

Valverde R, Arias O, Thorpe TA (1987) Picloram-induced somatic embryogenesis in pejibaye palm (*Bactris gasipaes* HBK). Plant Cell Tiss Org Cult 10:149–156

Verdeil JL, Buffard-Morel J (1995) Somatic embryogenesis in coconut (*Cocos nucifera* L.) In: Bajaj YPS (ed) Somatic Embryogenesis and Synthetic Seed-Biotechnology in Agriculture and Forestry 30:299–317

Vinãs M, Jiménez VM (2011) Factores que influyen em la embriogénesis somática in vitro in palmas (Arecaceae). Rev Colomb Biotecnol 13(2):229–242

Von Arnold S (2008) Somatic embryogenesis. In: George EF, Hall MA, De Klerk GJ (eds) Plant propagation by tissue culture, 3rd edn. Springer, Dordrecht, pp p335–p354

Von Arnold S, Sabala I, Bozhkov P, Dyachok J, Filonova J (2002) Developmental pathways of somatic embryogenesis. Plant Cell Tiss Org Cult 69:233–249

Wang HC, Chen JT, Wu SP, LIN MC, Chang WC (2002) Plant regeneration through somatic embryogenesis from zygotic embryo-derived callus of *Areca catechu* L. (Arecaceae). In vitro Cell Dev Biol Plant 39:34–36

Wang HC, Chen JT, Chang WC (2006) Somatic embryogenesis and plant regeneration from leaf, root and stem-derived callus cultures of *Areca catechu*. Biol Plant 50(2):279–282

Wang HC, Chen JT, Chang WC (2010) Morphogenetic routes of long-term embryogenic callus culture of *Areca catechu*. Biol Plant 54(1):1–5

Yaseen M, Ahmad T, Sablok G, Standardi A, Hafiz IA (2013) Review: role of carbon sources for in vitro plant growth and development. Mol Biol Rep 40:2837–2849

Zavattieri MA, Frederico AM, Lima M, Sabino R, Arnholdt-Schmitt B (2010) Induction of somatic embryogenesis as an example of stress-related plant reactions. Electr J Biotechnol 13(1):1–9. doi:10.2225/vol13-issue1-fulltext-4

Zelena E (2000) The effect of light on metabolism of IAA in maize seedlings. Plant Growth Regul 30:23–29

Zeng F, Zhang X, Cheng L, Hu L, Zhu L, Cao J, Guo X (2007) A draft gene regulatory network for cellular totipotency reprogramming during plant somatic embryogenesis. Genomics 90(5):620–628

Zhang H, Ogas J (2009) An epigenetic perspective on developmental regulation of seed. Genes Mol Plant 2(4):610–627

Zimmerman JL (1993) Somatic embryogenesis: a model for early development in higher plants. Plant Cell 5:1411–1423

Zouine J, El Hadrami I (2007) Effect of 2,4-D, glutamine and BAP on embryogenic suspension culture of date palm (*Phoenix dactylifera* L.). Sci Hortic 112:221–226

16 Effects of Plant Growth Regulator, Auxin Polar Transport Inhibitors on Somatic Embryogenesis and *CmSERK* Gene Expression in *Cattleya maxima* (Lindl.)

Augusta Yadira Cueva-Agila, Jhelisa Medina, Lorenzo Concia, and Rino Cella

Abstract

The somatic embryogenesis is commonly used for rapid propagation of species of interest hard to reproduce, like orchids. The induction of somatic embryogenesis requires the establishment of a peculiar genetic expression pattern in the presence of the suitable environmental conditions and a favorable hormonal background, which can be obtained providing synthetic plant growth regulators.

We tested several combinations of plant growth regulators in different illumination conditions on leaf explants of *Cattleya maxima*. The most efficient production of embryos was achieved with exogenous cytokinin (thidiazuron) in dark conditions at the cut end of the leaf. The expression of *CmSERK* gene was higher in the presence of cytokinins. The effects of the different treatments are discussed.

Keywords

Somatic embryogenesis • 2,3,5-Triiodobenzoic acid (TIBA) and naftilpropionic acid (NPA) • Darkness • Somatic embryogenesis receptor-like kinase (SERK)

Abbreviations

NPA	N-1-Naphthylphthalamic acid
TIBA	2,3,5-Triiodobenzoic acid
PGRs	Plant growth regulators
PAT	Polar auxin transport
SE	Somatic embryogenesis
TDZ	Thidiazuron [1-phenyl-3-(1,2,3-thiadiazol-5-yl)-urea]

A.Y. Cueva-Agila (✉) • J. Medina
Departamento de Ciencias Naturales, Universidad Técnica Particular de Loja,
San Cayetano alto s/n, Loja, Ecuador
e-mail: acueva@utpl.edu.ec

L. Concia • R. Cella
Dipartimento di Biologia e Biotecnologie, Università degli Studi di Pavia, Via Ferrata 9, 27100 Pavia, Italy

A. Mujib (ed.), *Somatic Embryogenesis in Ornamentals and Its Applications*,
DOI 10.1007/978-81-322-2683-3_16

2,4-D 2,4-Dichlorophenoxyacetic acid
BA N6-Benzyladenine
NAA Naphthaleneacetic acid

16.1 Introduction

Orchids are a diverse and important family of ornamental plants. One of the most popular, renowned, and colorful genus of Orchids is *Cattleya*, also known as the "Queen of Orchids." Several techniques have been developed for clonal propagation of *Cattleya* species by tissue culture (reviewed by Krapiec et al. 2003; Arditti 2008).

Among the various in vitro systems available, somatic embryogenesis offers opportunities for production of true-to-type plants by clonal propagation, and it is a useful tool for studying totipotency and fundamental processes of plant morphogenesis (Gaj 2004). Somatic embryogenesis (SE) allows competent cells to reprogram toward the embryogenic pathway (Chugh and Khurana 2002). Since its discovery in carrot in 1958 (Reinert 1958; Steward et al. 1958), SE has been extensively studied in higher plants, and protocols have been established for many species. Orchid-specific efficient protocols have been described for some hybrids, such as *Oncidium*, *Cymbidium*, and *Phalaenopsis* (Ishii et al. 1998; Tokuhara and Mii 2001; Chen and Chang 2001; 2006; Huan et al. 2004; Fang-Yi et al. 2006; Su et al. 2006; Chung et al. 2007; Hong et al. 2008), and for few Andean orchids (Cueva and González 2009; Cueva et al. 2013). In the process of SE, somatic cells develop the ability to form embryos and generate organized structures from undifferentiated calli. This process involves the establishment of a specific gene expression pattern (Chugh and Khurana 2002), triggered by a signal transduction mechanism mediated by auxins and cytokinins.

The synthetic growth regulator most frequently used for induction of somatic embryogenesis are thidiazuron [1-phenyl-3-(1,2,3-thiadiazol-5-yl)-urea] (TDZ), N^6-benzyladenine (BA), and 2,4-dichlorophenoxyacetic acid (2,4-D) (Chugh and Khurana 2002), while noncompetitive inhibitor of auxin polar transport like 2,3,5-triiodobenzoic acid (TIBA) and N-1-naphthylphthalamic acid (NPA), known as antiauxins, can prevent it. Schiavone and Cooke (1987) confirmed the role of polar auxin transport in somatic embryogenesis by treating carrot somatic embryos with TIBA and NPA. A concentration of 1 μM was sufficient to disrupt the ability by induced somatic embryos to progress further. The inhibitory action of TIBA and NPA on embryogenesis had also been documented in *Picea morrisonicola* (Liao et al. 2008), Norway spruce (Hakman et al. 2009), and *Pinus sylvestris* (Abrahamsson et al. 2012). In case of orchids, the process had been studied for *Oncidium* (Chen and Chang 2004; Chen and Hong 2012).

Considerable efforts have been made to identify genes controlling somatic embryogenesis. Genes involved in cell differentiation, morphogenesis, desiccation tolerance, and signal transduction are expressed during SE and function as part of this developmental program (Ikeda et al. 2006). Among all these genes, only *SERK* gene was shown to be a specific marker, characteristic of individual carrot embryo-forming masses (Schmidt et al. 1997). However, the latest studies show that *SERK1* gene and other members of the same family are only a part of the complex process of developmental transitions in plants.

In the present work, we studied the process of SE induction in leaf explants using combinations of auxins, cytokinins, auxin transport inhibitors, and light regime during the first month of culture. In addition, the expression of *SERK* gene in different leaf tissues was analyzed.

16.2 Explained Molecular, Histological and Statistical Methods

16.2.1 Induction of Somatic Embryogenesis

Eight-month old *Cattleya maxima* plants, obtained from in vitro germination, were used as donor plants. Apical tip leaves (0.5 cm) were explanted and placed with the adaxial side up on the surface of half-strength semisolid Murashige and Skoog medium (½MS). The same medium was used as control in all the assays with the exception of those involving auxin polar transport inhibitors, in which dimethyl sulfoxide (DMSO) was added to ½MS to solubilize them. In all the assays, we performed seven replicates (Petri dishes) per treatment, with eight-leaf segments per replicate. The observation of embryo formation was made after 30 days under a stereoscope (MOTIC). The effects of three factors were analyzed: plant growth regulators, light regime, and auxin polar transport inhibitors.

16.2.2 Plant Growth Regulators

To evaluate the effect of plant growth regulators (PGRs), we performed a test with eighteen combinations of plant growth regulators (Fig. 16.1). For this test, during the first 30 days of culture, the explants were maintained in darkness.

16.2.3 Auxin Polar Transport Inhibitors

To better understand the influence of auxins in the induction of somatic embryogenesis, we used inhibitory factors of auxin polar transport: 0.4, 2.5, and 5.1 μM TIBA and 0.02, 0.2, and 10 μM NPA. As well as for PGR during the first 30 days of culture, the explants were maintained in darkness.

16.2.4 Light/Darkness during the First Month of Culture

In order to determine the role of darkness in the induction of somatic embryos, we selected some of the growing conditions above mentioned and repeated the assays culturing the explants in the dark or in 16 h light photoperiod.

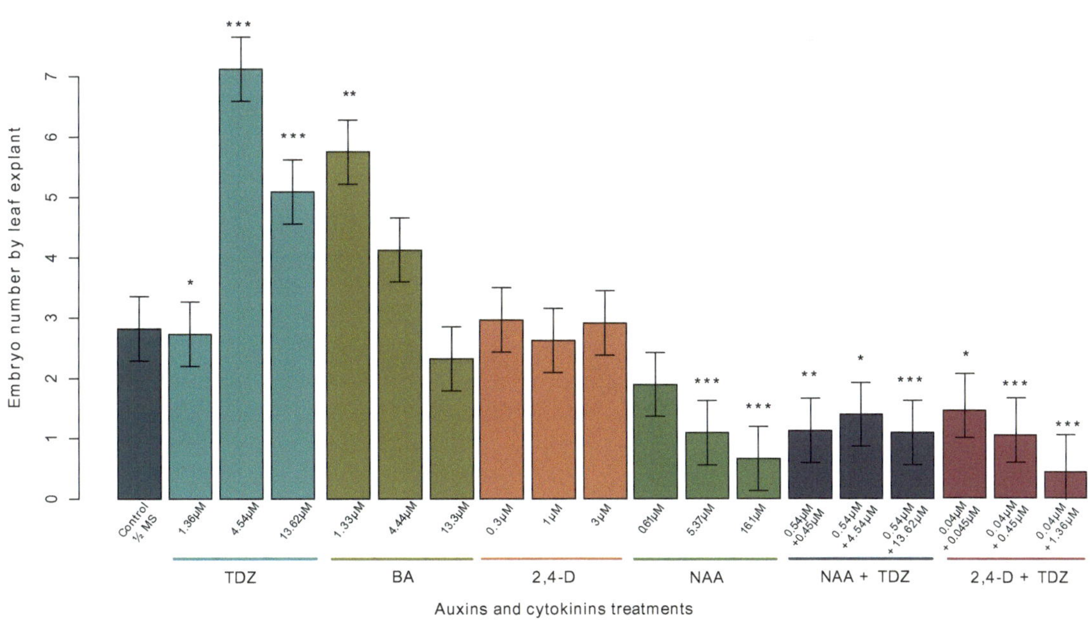

Fig. 16.1 Effect of auxins and cytokinins on embryo production on leaf explants of *Cattleya maxima*. Asterisks show the significances of each treatment after the analysis with general linearized models (GLMs) with negative binomial distribution. $p>0$ '***'; $p>0.001$ '**'; $p>0.01$ '*'

16.2.5 Histological Analysis

For the histological analysis, samples were fixed for 24 h at room temperature in FAA solution (9:1:1 of 50 % of ethanol, 5 % formaldehyde and glacial acetic acid). Fixed samples were dehydrated in a graded series of ethanol-xylol and then embedded in paraffin. Tissues were sectioned at 8 μm thickness using a microtome (American Optical). Samples stained with hematoxylin-eosin were used for the microscopic analysis (OlympusCX32).

16.2.6 *CmSERK* Gene Expression

In a previous project, we isolated and characterized a *CmSERK* gene. The expression level of *CmSERK* gene during somatic embryogenesis induction was analyzed using gene-specific primer: *SERK1_Cm_RT*_Fw: 5′ TTT AAT CCT CCG GTT ACA GTC T 3′ and *SERK1_Cm_*RT_Rv: 5′ AAT GTT CTT GTG GCT TAC GAC G 3′.

The primers were designed so as to overlap the boundary of two adjacent exons in order to prevent an unwanted contribution of genomic DNA (Cueva et al. in preparation). Some *C. maxima* samples treated with TDZ, naphthaleneacetic acid (NAA), and TIBA and those on control ½MS were analyzed after 15 days from induction. We selected two tissues for RNA extraction: cut end of the leaf where embryo formation was present and the tip of the leaf where embryos were not.

RNA was extracted using the total RNA isolation system kit (Promega) following the manufacturer's instructions. Each RNA sample was analyzed by electrophoresis in 1 % agarose to verify its integrity, and concentration was determined using a NanoDropTM 1000 spectrophotometer (Thermo Scientific). RT-qPCR was performed in Fast SYBR SYBR®Green Master Mix (Applied Biosystems) following the manufacturer's instruction with an Applied Biosystems 7500 Fast Real-Time PCR System, using two-step cycling condition (95 °C for 10 min, 1 cycle, 95 °C 10 s, 58 °C 15 s, and 72 °C 30 s, 45 cycles). Reactions were performed pooling two biological replicates for each sample and setup in double. For each sample, we prepared a mix with 10 μl of 2× RT buffer, 1 μl of 20× RT enzyme mix, and 9 μl of RNA, obtaining a final volume of 20 μl.

RT-qPCR expression levels were normalized to α-tubulin, using the following primers: α-tubulin_Onc_Hou_Fw: 5′ GGA TTA GGC TCT CTG CTG TTG G 3′ and tubulin_Onc_Hou_Rv: 5′ GTG TGG ATA AGA CGC TGT TGT ATG 3′ (Hou and Yang 2009). PCR efficiency of each run was calculated using LinRegPCR program http://LinRegPCR.nl (Pfaffl 2001).

16.2.7 Statistical Analysis

For each assay, we scored the number of embryos produced by each leaf after 30 days of culture and the register the site where embryos were produced. The results were analyzed in R program (R Development Core Team 2004) using generalized linear models (GLMs) suitable for biological data that do not have constant variance. We used negative binomial distribution that gives a better description of count data (Crawley 2007).

16.3 Plant Growth Regulators' Influence on Somatic Embryo Induction

Cytokinins were the best inductors of embryo formation (Fig. 16.1). The highest number of embryos (7.1) was generated using 4.54 μM TDZ. Higher concentration (13.62 μM) yielded fewer embryos (5.08), suggesting a saturation point. BA at a concentration of 1.33 μM also gave a statistically significantly higher number of embryo produced (5.7) compared to the control (2.8). A statistically significant reduction of embryo formation was observed with NAA and a combination of NAA+TDZ and 2,4-D+TDZ, similarly to what observed in *Oncidium flexuosum* (Sampaio et al. 2010).

It is well known that several factors controlling SE depend on species, explant type, and developmental stage. For the Orchidaceae family, many different inducing factors had been used: auxins, cytokinins, stress factors, and light. Among these, auxins like 2,4-D and NAA inhibit the formation of embryos in *Dendrobium* (Chung et al 2005), *Phalaenopsis* (Kuo et al. 2005), *Oncidium flexuosum* (Sampaio et al. 2010), and in protocorms of *Cattleya maxima* (Cueva et al. 2015). In addition, cytokinins stimulate SE in many orchid species. The most effective cytokinins is TDZ, as reported in *Oncidium* sp. (Chen et al. 1999; Chen and Chang 2001; Su et al. 2006; Sampaio et al 2010; Chen and Hong 2012). In a previous work, we found that the use of TDZ alone can induce somatic embryo formation in protocorms of *C. maxima* (Cueva et al 2013).

16.3.1 Auxin Polar Transport Inhibitors' Influence on Somatic Embryo Induction

With both TIBA and NPA, at all tested concentration, the number of embryo per leaf explant decreased significantly when compared to the control (Fig. 16.2). A treatment with 2.5 μM TIBA gave the lowest number of embryo per explant; furthermore, this treatment produced a high percentage of necrosis in leaf explants of *C. maxima*.

Previous studies with auxin polar transport inhibitor in somatic embryogenesis showed both promoting and inhibiting effects, depending on the dosage and tissue used. Inhibitory effects had been found in carrot (Nissen and Minocha 1993), *Geranium* (Hutchinson et al. 1996), *Panax ginseng* (Choi et al. 1997), and *Eleutherococcus senticosus* (Choi et al. 2002), while promoting effects had been reported for *Pisum sativum* (Tetu et al. 1990).

Specifically for orchids, Chen and Chang (2004) found that in *Oncidium* the presence of 0.1–0.5 μM TIBA does not affect the number of embryos compared to the control. However, higher concentrations of TIBA (1–10 μM) significantly inhibit somatic embryos production. This observation is consistent with the previous study that TIBA and NPA are able to reduce somatic embryo formation in leaf explants of *Eleutherococcus senticosus* (Choi et al. 2002). This latter finding suggests that an adequate internal auxin gradient, which is under control of auxin polar transport, is required for somatic embryo formation in orchids (Vanneste and Friml 2009; Sun et al. 2013). The synthetic polar auxin transport (PAT) inhibitors TIBA and NPA can change the internal auxin gradients by interfering with PIN protein activity (Muday and DeLong 2001; Dhonukshe et al. 2008).

16.3.2 Effect of Light/Darkness Regime during the First Month of Culture on Somatic Embryogenesis

In previous studies, it has been reported that the absence of light results in a higher percentage of embryogenesis in *Oncidium flexuosum* (Sampaio et al. 2010), *Phalaenopsis amabilis*, and *Phalaenopsis nebula* (Gow et al. 2009). Therefore, the SE test described above was repeated with leaf explants incubated in darkness and 16 h light photoperiod for the first month of culture. As shown in Fig. 16.3, in this condition, TDZ gave the highest number of embryo per explant (7.7). Conversely, in the light, TDZ produced not statistically significant higher number of embryos (4.1) than the control (3.12). Consistently with the first PGR assay, all the auxins produced a lower number of embryos in darkness but not as lower as in light conditions.

To identify which factor was the most important for embryo formation, we performed a combined GLM analysis. As shown in Table 16.1, the most significant factor for the production of embryos by leaf explant of *C. maxima* is darkness, as it produces more embryos compared to light. Among PGRs, TDZ affected the embryo formation, in both light and dark conditions. As we reported earlier, auxins decrease these responses in both illumination conditions assayed.

Sampaio et al. (2010) in a study with the tropical orchid, *Oncidium flexuosum*, found that darkness is a crucial factor for PLB regeneration from leaf explants. They reported that the use of 1.5 μM

Fig. 16.2 Effect of auxins polar transport inhibitors on embryo production by leaf explants of *Cattleya maxima*. Asterisks show the significances of each treatment after the analysis with general linearized models (GLMs) with negative binomial distribution. $p>0$ '***'; $p>0.001$ '**'; $p>0.01$ '*'

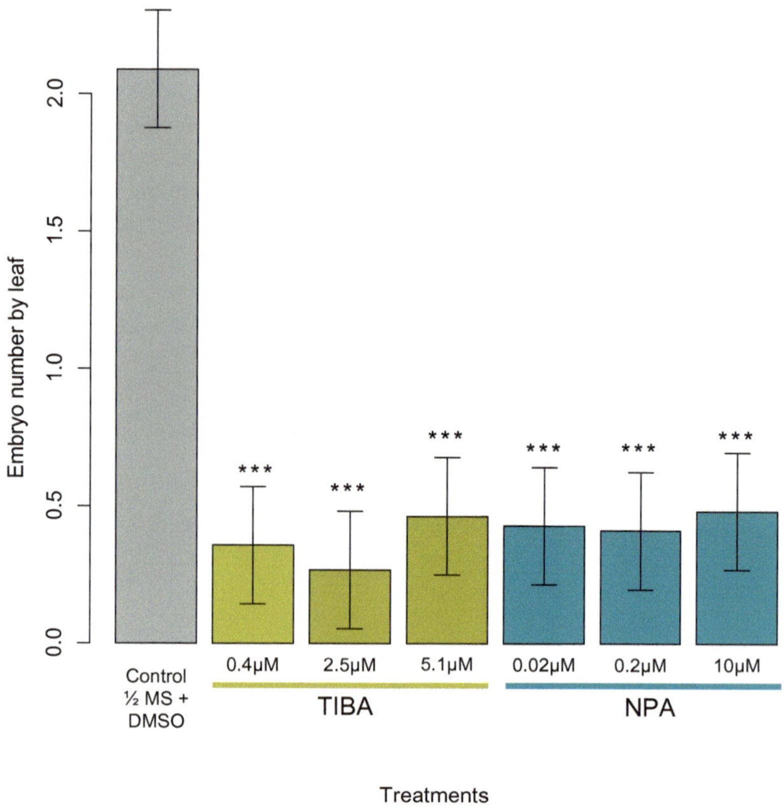

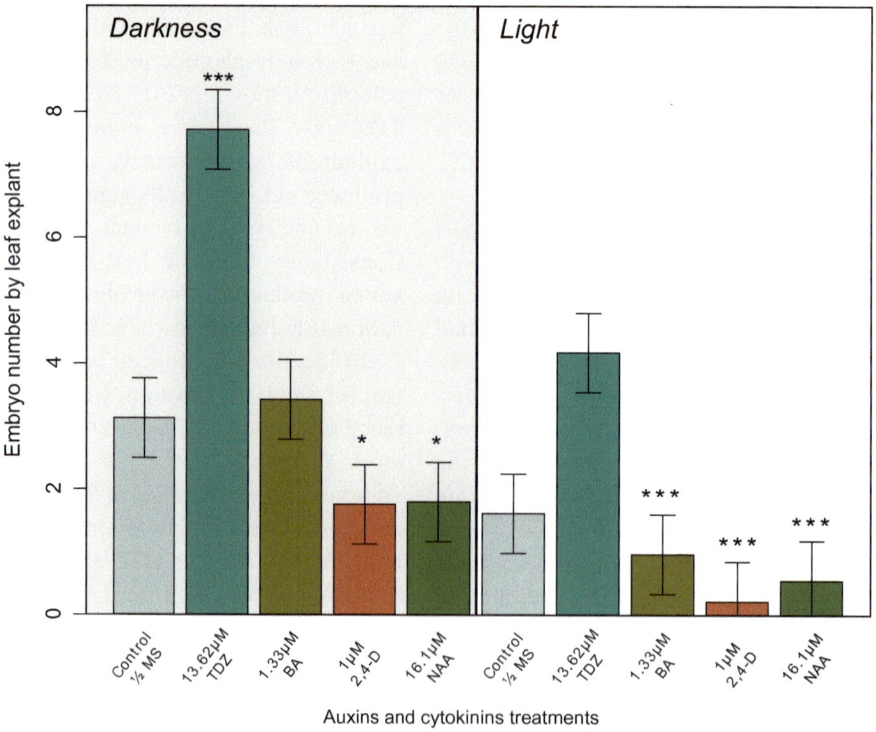

Fig. 16.3 Effect of light/dark regime during the first month of culture on the number of embryo by leaf explant in *C. maxima*. Asterisks show the significances of each treatment after the analysis with general linearized models (GLMs) with negative binomial distribution. $p>0$ '***'; $p>0.01$ '*'

Table 16.1 Summary table of the GLM, using negative binomial distribution, for number of embryos per explant in response to PGRs and light regime

Source of variation	Estimate	*Standard error*	*p*
Light regime			
Darkness	1.1.5	0.19	8.45e-10 ***
16 light photoperiod	−0.67	0.29	0.020614 *
Plant growth regulators			
TDZ	0.88	0.25	0.000386 ***
BA	0.07	0.26	0.773347
NAA	−0.57	0.28	0.044942 *
2,4-D	−0.59	0.28	0.037442 *

TDZ in darkness produced the highest embryo (PLB) number from leaf-tip explant. Also with *Phalaenopsis amabilis* (blue) and *Phalaenopsis nebula* light delayed embryogenesis (Gow et al. 2009). Our results show that darkness is a crucial factor for somatic embryo formation from leaf explants in *C. maxima*, while light reduces embryo formation with all PGRs tested (Table 16.1).

TIBA gave a lower number of embryos compared with the control (½MS + DMSO) in both light culture conditions (Fig. 16.4). However, the inhibitory effect of TIBA is higher in light condition than in darkness (Fig. 16.4). Having observed that auxin polar transport inhibitors reduce embryo formation, we analyzed how TIBA interacts with light. We observed that TIBA concentration is the most significant factor that influences the number of embryo formed by leaf explant (Table 16.2).

16.4 Site of Embryo Formation

In the case of orchids, leaves are one of the most used explant for somatic embryo formation. SE can occur in various parts of the leaf explant: after 30 days of in vitro culture in darkness and photoperiod, leaf explants of *C. maxima* formed embryos at the leaf tip (Fig. 16.5a), cut end of the leaf (Fig. 16.5b), and adaxial leaf surface (Fig. 16.5d), and also we could observe the formation of embryos at the site of an unintentional cut (Fig. 16.5c). In all cases, the cut end had the best embryogenic response (Fig. 16.6). Also when auxin polar transport inhibitors were used, the major number of embryos were produced at cut ends. Contrarily, in the case of *Oncidium*, the leaf tips had more embryogenic competence than the cut end and other parts of the leaf explant (Hou and Yang 2009).

The histological analysis showed visible somatic embryo formation at the cut end of the leaf (Fig. 16.7d). Embryogenic cells originated also from the epidermal cell layers of leaf explants (Fig. 16.7a). These cells are generally grouped together and can vary in size and number (Espinosa et al. 2010). The first morphological changes were visible at 30 days with the formation of proembryos resulting in globular embryos (Fig. 16.7b, c).

16.5 *C. maxima SERK* Gene (CmSERK) Is More Expressed at Cut End of the Leaf

Since its discovery in carrot (Schmidt et al. 1997), SERK gene of many other species has been characterized. The first SERK orchid gene was isolated from *Cyrtochilum loxense*, another Andean orchid species (Cueva et al. 2012), while a partial characterization of *SERK* gene of *C. maxima* was recently carried out (Cueva et al. in preparation). In this work, we analyzed the expression of *CmSERK* gene in two tissues of *C. maxima* leaf explants. It was observed that this gene is more expressed in the leaf cut end than in the leaf tips (Fig. 16.8), supporting the correlation between the higher competence of the mesophyll cells of leaf ends and *CmSERK* gene expression. Other studies proved that SERK gene expression is related to the ability of cells to achieve somatic embryo formation (Nolan et al. 2003, 2011; Ma et al. 2011; Huang et al 2010; Pérez-Nuñez et al. 2008; Cueva et al. 2012). Besides, *CmSERK* gene is more expressed in the presence of cytokinins (TDZ) rather than auxins (NAA). On the other hand, TIBA-treated samples showed a low level of *CmSERK* gene expression (Fig. 16.8).

Fig. 16.4 Effect of light regime during the first month of culture on the number of embryos formed by leaf explant of *C. maxima* using the auxin polar transport inhibitors TIBA. Asterisks show the significances of each treatment after the analysis with general linearized models (GLMs) with negative binomial distribution. $p>0$ '***'; $p>0.01$ '*'

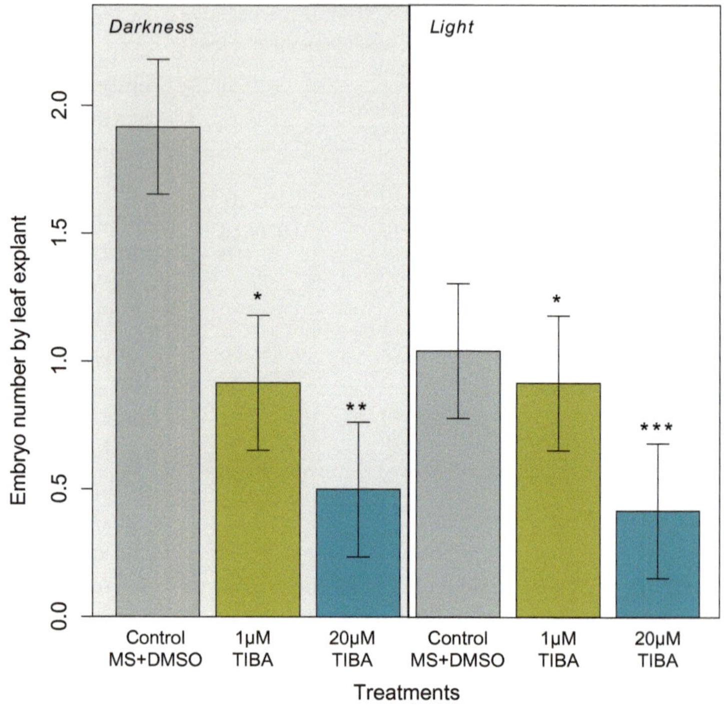

Table 16.2 Summary table of the GLM, using negative binomial distribution, for number of embryos by explant in response to TIBA and light regime

Source of variation	Estimate	*Standard error*	*p*
Light regime			
Darkness	0.65	0.24	0.00719**
16 h light photoperiod	−0.61	0.37	0.09751
TIBA concentration			
1 µM	−0.74	0.38	0.00148**
20 µM	−1.34	0.42	0.773347

**$p>0.001$

Taken together, these results confirmed that for *C. maxima*, the competence for somatic embryo formation is related to SERK gene expression also, consistent with what reported for other species.

16.6 Concluding Remarks

This study confirms that auxin levels are crucial for somatic embryogenesis in *C. maxima*, consistently with auxin sensitivity of embryo formation and development in other orchids (Novak et al. 2014). Both the concentration and the distribution of auxin affected somatic embryogenesis.

The establishment of the apical–basal axis during embryogenesis is dependent on the asymmetric distribution of auxin mediated by PIN proteins (Ying-Hua Su et al. 2011), which in turn is partially under the control of light through the induction of tissue-specific *PIN* expression (Sassi et al. 2013). Impairing the establishment of the polar auxin transport with TIBA and NPA, specific inhibitors of PIN carriers, reduced the formation of embryos, suggesting that a

Fig. 16.5 Place of somatic embryo formation in leaf explants of *Cattleya maxima* after 30 days of in vitro culture. (**a**) Somatic embryos formed on the leaf tip. (**b**) Somatic embryos formed on the leaf cut end. (**c**) Somatic embryos formed in unintentional cut leaf. (**d**) Somatic embryos formed on the adaxial surface leaf

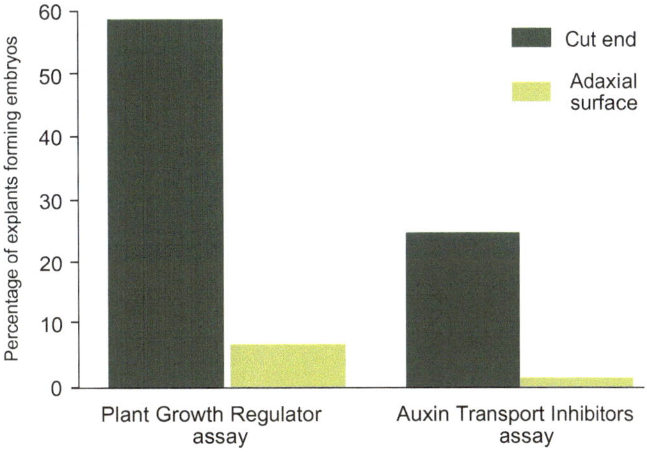

Fig. 16.6 Percentage of explant forming somatic embryos in different sites of leaf explants of *Cattleya maxima* after 30 days of in vitro culture

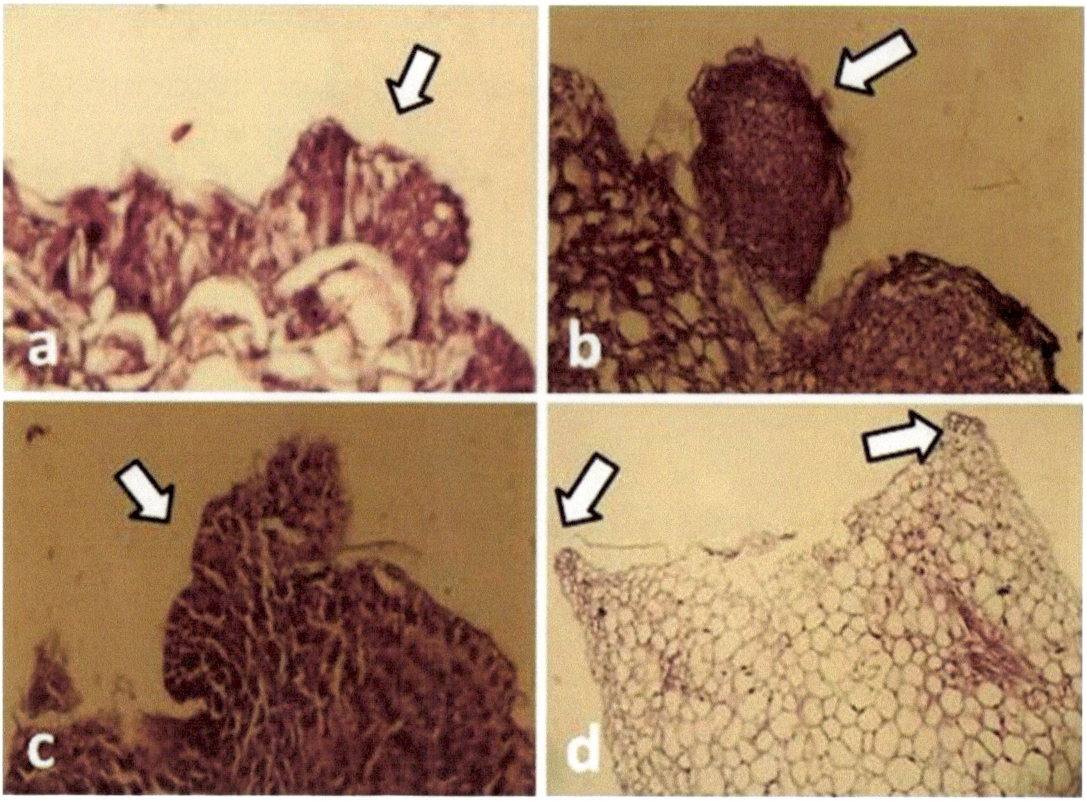

Fig. 16.7 Histology of direct somatic embryogenesis from leaf explants of *C. maxima*. (**a**) Embryogenic cells originate in the epidermal cell layer of the leaf. (**b**) Formation of a globular embryo. (**c**). Formation of somatic embryos on the leaf cut end. (**d**) Embryo formation at the leaf cut end

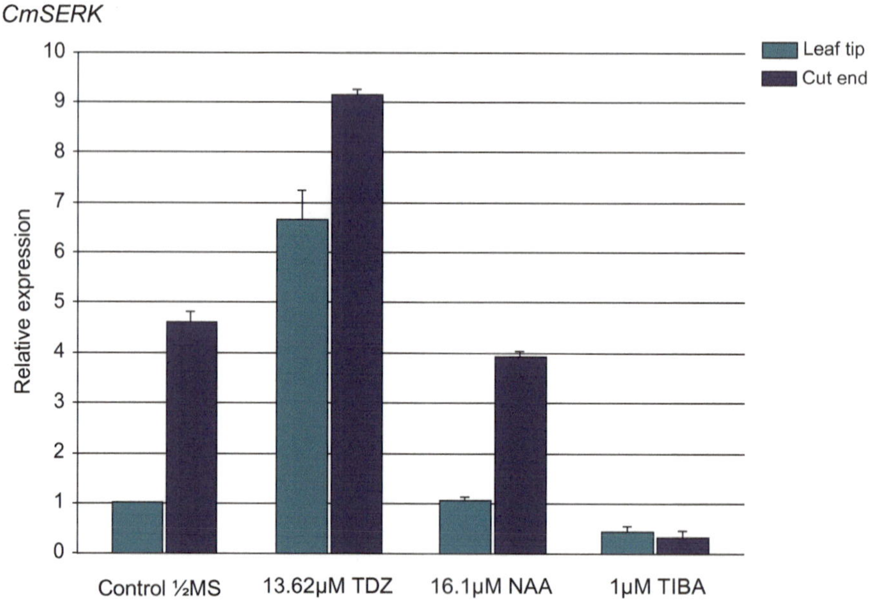

Fig. 16.8 Relative expression of *CmSERK* gene as determined by reverse transcription-qPCR analysis. Results shown are mean ± standard deviation of two independent samples calibrated to expression in leaf tip in control ½MS

physiological concentration of auxins, either endogenous or exogenous, is required for somatic embryogenesis.

Increasing concentration of synthetic auxins (2,4-D and NAA) did not have an effect on SE until a concentration of 5 uM of NAA when the explants were grown in the dark (Fig. 16.1). However, 1 uM of 2,4-D was sufficient to reduce SE in the presence of light (Fig. 16.3, right panel), consistent with the role of light in PAT and suggesting that an excess of auxin may disbalance the auxin–cytokinin ratio optimal for early embryogenesis (Müller and Sheen 2008).

The presence of exogenous cytokinin (TDZ) enhanced somatic embryogenesis, until a given threshold limit reached. The cytokinin effect was synergic with dark growing conditions.

Embryos were produced principally in cut-end leaf, where *CmSERK* gene is expressed at relatively higher levels. Coherently with the production of embryos, *CmSERK* gene is more expressed in the presence of TDZ and less expressed in the presence of PAT inhibitors.

The effect of light and PGRs on *SERK* expression, taken together with the overall effects on somatic embryogenesis, supports the idea of a hormone-dependent *SERK*-mediated pathway for somatic embryogenesis under the control of light presence and auxin–cytokinin balance in *C. maxima*.

Acknowledgments We thank Dr. Quintana-Ascencio (University of Central Florida) for advice on graphics and data presentation and Carlos Iñiguez-Armijos (Universidad Técnica Particular de Loja) for helpful training on the graphics editor Inkscape.

References

Abrahamsson M, Valladares S, Larsson E, Clapham D, von Arnold S (2012) Patterning during somatic embryogenesis in Scots pine in relation to polar auxin transport and programmed cell death. Plant Cell Tiss Org Cult 109:391–400

Arditti J (2008) Micropropagation of Orchids, 2nd edn. Blackwell Publishing, Malden

Chen JT, Chang WC (2001) Effects of auxins and cytokinins on direct somatic embryogenesis from leaf explants of Oncidium 'Gower Ramsey'. Plant Growth Regul 34:229–232

Chen JT, Chang WC (2004) TIBA affects the induction of direct somatic embryogenesis from leaf explants of Oncidium. Plant Cell Tiss Org Cult 79:315–320

Chen J, Chang W (2006) Efficient and repetitive production of leaf-derived somatic embryos of *Oncidium*. Biol Plant 50:107–110

Chen J-T, Hong P-I (2012) Cellular origin and development of secondary somatic embryos in *Oncidium* leaf cultures. Biol Plant 56(2):215–220

Chen JT, Chang C, Chang WC (1999) Direct somatic embryogenesis on leaf explants of *Oncidium* Gower Ramsey and subsequent plant regeneration. Plant Cell Rep 19:143–149

Choi YE, Kim HS, Soh WY, Yang DC (1997) Developmental and structural aspects of somatic embryos formed on medium containing 2,3,5-triidobenzoic acid. Plant Cell Rep 16:738–744

Choi YE, Ko SK, Lee KS, Yoon ES (2002) Production of plantlets of *Eleutherococcus sessiliflorus* via somatic embryogenesis and successful transfer to soil. Plant Cell Tiss Org Cult 69:35–40

Chugh A, Khurana P (2002) Gene expression during somatic embryogenesis-recent advances. Curr Sci 83:715–730

Chung HH, Chen JT, Chang WC (2005) Cytokinins induce direct somatic embryogenesis of *Dendrobium chiengmai* Pink and subsequent plant regeneration. In Vitro Cell Dev Biol Plant 41:765–769

Chung HH, Chen JT, Chang WC (2007) Plant regeneration through direct somatic embryogenesis from leaf explants of *Dendrobium*. Biol Plant 51:346–350

Crawley MJ (2007) The R Book. Wiley, Chichester

Cueva A, González Y (2009) In vitro germination and somatic embryogenesis induction in *Cyrtochilum loxense*, an endemic, vulnerable orchid from Ecuador. In: Pridgeon AM, Suarez JP (eds) Proceedings of the second scientific conference on Andean Orchids. Universidad Técnica Particular de Loja, Loja, Ecuador, pp 56–62

Cueva A, Concia L, Cella R (2012) Molecular characterization of a *Cyrtochilum loxense* Somatic Embryogenesis receptor-like Kinase (SERK) gene expressed during somatic embryogenesis. Plant Cell Rep 31:1129–1139

Cueva A, Guachizaca I, Cella R (2013) Combination of 2,4-D and stress improves indirect Somatic Embryogenesis in *Cattleya maxima* Lindl. Plant Biosyst 149:235–241

Cueva A, Guachizaca I, Cella R (2015) Combination of 2,4-D and stress improves indirect Somatic Embryogenesis in *Cattleya maxima* Lindl. Plant Biosyst 149:235–241

Dhonukshe P, Grigoriev I, Fischer R, Tominaga M, Robinson DG, Hasek J, Paciorek T, Petrasek J, Seifertova D, Tejos R, Meisel LA, Zazimalova E, Gadella TWJ, Stierhof YD, Ueda T, Oiwa K, Akhmanova A, Brock R, Spang A, Friml J (2008) Auxin transport inhibitors impair vesicle motility and actin cytoskeleton dynamics in diverse eukaryotes. Proc Natl Acad Sci 105:4489–4494

Espinosa E, Laguna A, Murguía J, Iglesias L, García B, Escobedo L, Martínez Y, Barredo SB (2010) Un protocolo de embriogénesis somática para la regeneración y Caracterización in vitro de Laelia anceps spp dawsonii. Rev Fitotec Mex 33:323–332

Fang-Yi J, Do Y, Liauh Y, Chung J, Huang P (2006) Enhancement of growth and regeneration efficiency from embryogenic callus cultures of *Oncidium* Gower Ramsey by adjusting carbohydrate sources. Plant Sci 170:1133–1140

Gaj M (2004) Factors influencing somatic embryogenesis induction and plant regeneration with particular reference to Arabidopsis thaliana (L.) Heynh. Plant Growth Regul 43:27–47

Gow W, Chen J, Chan W (2009) Effects of genotype, light regime, explants position and orientation on direct somatic embryogenesis from leaf explants of Phalaenopsis orchids. Acta Physiol Plant 31:363–369

Hakman I, Hallberg H, Palovaara J (2009) The polar auxin transport inhibitor NPA impairs embryo morphology and increases the expression of an auxin efflux facilitator protein PIN during *Picea abies* somatic embryo development. Tree Physiol 29:483–496

Hong PI, Chan JT, Chang WC (2008) Promotion of direct somatic embryogenesis of *Oncidium* by adjusting carbon sources. Biol Plant 52:597–600

Hou C, Yang C (2009) Functional analysis of FT and TFL1 orthologs from orchid (*Oncidium* Gower Ramsey) that regulate the vegetation to reproductive transition. Plant Cell Physiol 50:1544–1557

Huan LVT, Takamura T, Tanaka M (2004) Callus formation and plant regeneration from callus through somatic embryo structures in *Cymbidium* orchid. Plant Sci 166:1443–1449

Huang X, Lu XY, Zhao JT, Chen JK, Dai XM, Xiao W, Chen YP, Chen YF, Huang XL (2010) MaSERK1 gene expression associated with somatic embryogenic competence and disease resistance response in banana (Musa spp.). Plant Mol Biol Rep 28:309–316

Hutchinson MJ, Murch SJ, Saxena PK (1996) Morphoregulatory role of thidiazuron: evidence of the involvement of endogenous auxin in thidiazuron-induced somatic embryogenesis of geranium (*Pelargonium x hortorum* Bailey). J Plant Physiol 149:573–579

Ikeda M, Umehara M, Kamada H (2006) Embryogenesis-related genes; its expression and roles during somatic and zygotic embryogenesis in carrot and Arabidopsis. Plant Biotechnol 23:153–161

Ishii Y, Takamura T, Goi M, Tanaka M (1998) Callus induction and somatic embryogenesis of *Phalaenopsis*. Plant Cell Rep 17:446–450

Krapiec PV, Milaneze M, Pires M (2003) Effects of different combinations of growth regulators for bud induction from seedlings of *Cattleya walkeriana* Gardner (Orchidaceae). Acta Scientiarum: Biol Sci 25:179–182

Kuo HL, Chen JT, Chang WC (2005) Efficient plant regeneration through direct somatic embryogenesis from leaf explants of *Phalaenopsis* 'Little Steve'. In Vitro Cell Dev Biol Plant 41:453–456

Liao YK, Liao CK, Ho YL (2008) Maturation of somatic embryos in two embryogenic cultures of *Picea morrisonicola* Hayata as affected by alteration of endogenous IAA content. Plant Cell Tiss Org Cult 93:257–268

Ma J, He Y, Wu C, Liu H, Hu Z, Sun G (2011) Cloning and molecular characterization of a SERK gene transcriptionally induced during somatic embryogenesis in *Ananas comosus* cv. Shenwan. Plant Mol Biol Rep 6:1–9

Muday GK, DeLong A (2001) Polar auxin transport: controlling where and how much. Trends Plant Sci 6:535–542

Müller B, Sheen J (2008) Cytokinin and auxin interaction in root stem-cell specification during early embryogenesis. Nature 453:1094–1097

Nissen P, Minocha SC (1993) Inhibition by 2,4-D of somatic embryogenesis in carrot as explored by its reversal difluoromethylornithine. Physiol Plant 89:673–680

Nolan K, Irwanto R, Rose RJ (2003) Auxin up-regulates MtSERK1 expression in both *Medicago truncatula* root-forming and embryogenic cultures. Plant Physiol 133:218–230

Nolan K, Kurdyukov S, Rose R (2011) Characterisation of the legume SERK-NIK gene superfamily including splice variants: Implications for development and defence. Plant Biol 11:44–60

Novak SD, Luna LJ, Gamage RN (2014) Role of auxin in orchid development. Plant Signal Behav 9(10), e972277. doi:10.4161/psb.32169

Pérez-Nuñez MT, Souza R, Saenz L, Chan JL, Zuniga-Aguilar J, Oropeza C (2008) Detection of a SERK-like gene in coconut and analysis of its expression during the formation of embryogenic callus and somatic embryos. Plant Cell Rep 28:11–19

Pfaffl MW (2001) A new mathematical model for relative quantification in real-time RT-PCR. Nucleic Acids Res 29:2003–2007

R Development Core Team (2004) R: A language and environment for statistical computing. R Foundation for Statistical Computing, Vienna

Reinert J (1958) Untersuchungen über die Morphogenese an Gewebekulturen. Ber Dtsch Bot Ges 71:15–15

Sampaio J, Stancato G, Appezzato B (2010) Direct regeneration of protocorm-like bodies (PLBs) from leaf apices of *Oncidium flexuosum* Sims (Orchidaceae). Plant Cell Tiss Org Cult 103:411–416

Sassi M, Wang J, Ruberti I, Vernoux T, Xu J (2013) Plant Signal Behav 8: doi: 10.4161/psb.23355

Schiavone FM, Cooke TJ (1987) Unusual patterns of somatic embryogenesis in domesticated carrot: developmental effects of exogenous auxins and auxin transport inhibitors. Cell Diff 21:53–62

Schmidt ED, Guzzo F, Toonen MA, de Vries SC (1997) A leucine-rich repeat containing receptor-like kinase marks somatic plant cells competent to form embryos. Development 124:2049–2062

Steward FC, Mapes MO, Mears K (1958) Growth and organized development of culture cells.

II. Organization in cultures grown from freely suspended cells. Am J Bot 45:705–708

Su YJ, Chen J, Chang W (2006) Efficient and repetitive production of leaf-derived somatic embryos of *Oncidium*. Biol Plant 50:107–110

Sun J, Linlin Q, Chuanyou L (2013) Hormonal control of polar auxin transport. In: Rujin C, Baluska F (eds) Polar Auxin transport. New York Dordrecht London, Springer, 337 pp

Tetu T, Sangwan RS, Sangwan BSN (1990) Direct somatic embryogenesis and organogenesis in cultured immature zygotic embryos of *Pisum sativum* L. J Plant Physiol 137:102–109

Tokuhara K, Mii M (2001) Induction of embryogenic callus and cell suspension culture from shoot tips excised from flower stalk buds of *Phalaenopsis* (Orchidaceae). In Vitro Cell Devl Biol Plant 37:457–461

Vanneste S, Friml J (2009) Auxin: a trigger for change in plant development. Cell 136:1005–1016

Ying-Hua Su, Yu-Bo Liu, Xian-Sheng Zhang (2011) Auxin–Cytokinin interaction regulates meristem development. Mol Plant 4:616–625